GALILEO
AND HIS SOURCES

GALILEO
AND HIS SOURCES
The Heritage of the Collegio Romano in Galileo's Science

William A. Wallace

PRINCETON UNIVERSITY PRESS
PRINCETON, NEW JERSEY

*Published by Princeton University Press, 41 William Street,
Princeton, New Jersey 08540
In the United Kingdom:
Princeton University Press, Guildford, Surrey*

Library of Congress Cataloging in Publication Data will be found on the last printed page of this book

ISBN 0-691-08355-X

This book has been composed in Linotron Bembo

Clothbound editions of Princeton University Press Books are printed on acid-free paper, and binding materials are chosen for strength and durability. Paperbacks, although satisfactory for personal collections, are not usually suitable for library rebinding

*Printed in the United States of America by
Princeton University Press, Princeton, New Jersey*

CONTENTS

PART TWO
Science at the Collegio Romano

CHAPTER 3

CHAPTER 4

PART THREE
Galileo's Science in Transition

CHAPTER 5

CHAPTER 6

TABLES

PREFACE

STUDIES of the life and works of Galileo Galilei have flourished throughout the twentieth century because of the splendid National Edition of his writings, edited by Antonio Favaro, whose twenty-one volumes were brought to completion in 1909. The edition has stood the test of time, except for two shortcomings detected only in the past decade. The first is the insufficient attention it paid to the manuscript fragments associated with the *Two New Sciences*, Galileo's masterwork of 1638 in which he laid the foundations of modern mechanics. These fragments have been the subject of intensive recent research. Through their study it is now known that most of the experimental work on which the new science of local motion was based had been completed some thirty years before the publication of that work, while Galileo was still a professor at the University of Padua. The second shortcoming is the treatment accorded by Favaro to Galileo's early Latin notebooks. One of these he regarded as a scholastic exercise and deemed it not worthy of inclusion in the National Edition; another he did include but as one of Galileo's *Juvenilia*, dating it from the latter's student days at the University of Pisa. The sources on which these were based, as those behind a third notebook containing writings on motion, eluded Favaro, but this was for him a trivial problem whose solution would have little bearing on the genesis of the new science.

These Latin notebooks of Galileo, and their sources, are the subject of detailed study in this volume. Contrary to Favaro's estimate, all three were written around 1590 while Galileo was embarking on his teaching career at the University of Pisa and were of seminal importance for his later work. More surprising still, the first two notebooks were effectively copied from the lecture notes of young Jesuit professors, all about Galileo's age and then teaching at the Collegio Romano, while the third employs concepts that derive unmistakably from the other two. Galileo's early science, on this accounting, was in essential continuity with that being developed contemporaneously by Jesuit scholastics. Many of the terms and expressions he uses in these notebooks continue to recur in his later manuscripts and published writings, so much so that one may rightfully regard them as the heritage of the Collegio Romano whose elements still survive in the *nuova scienza* of 1638.

These results obviously have profound implications for anyone interested in the origins of the Scientific Revolution, and particularly in the continuity between late medieval or scholastic science and that of the early seventeenth century. They also may be expected to shed fuller light on Galileo's later relationships with the Jesuits, especially those surrounding his trial and condemnation by the Roman Inquisition in 1633. Neither of these topics, however, is addressed specifically in this volume. Its aim is rather to provide the groundwork on which such studies can be based by supplying hitherto unavailable evidence that serves to establish, beyond reasonable doubt, the sources and dating of Galileo's early Latin compositions.

To this end part 1 of the study is devoted exclusively to textual analysis, concerned mainly with the notebooks containing Galileo's logical and physical questions, and showing remarkable parallels that can be exhibited between these and Jesuit lecture notes. Part 2 expands the study to flesh out in fuller detail the logic and natural philosophy being taught at the Collegio Romano in the last decades of the sixteenth century; attention here is focused on methodological problems relating to Galileo's science and on the subject matter with which it was mainly concerned, local motion. Part 3 then applies these findings to a reexamination of all of Galileo's writings, divided into those before 1610 and those afterward, to show how his early concepts continue to reappear throughout, though undergoing gradual changes, until they emerge in revised form in the new science of 1638.

The literature of Galileo is vast, and a good part of it inevitably deals with his early period and his relationships to possible precursors and predecessors. Practically all of this was published without knowledge of the discoveries set forth in this volume. To attempt a revision or correction of alternative views on the subject would therefore be a thankless, if not hopeless, task. Because of this circumstance I have not attempted to juxtapose my thesis with divergent positions currently accepted or proposed by other historians and philosophers of science. My bibliographic references are sparse, limited for the most part to primary sources and to recently published materials that corroborate, or shed light on, my own findings.

The greatest difficulty in the research reported herein was that offered by the manuscript containing Galileo's logical questions, never transcribed by Favaro and thus not appearing in the National Edition. I had begun transcribing this myself when I heard that the work was already being done. Through the good services of Alistair C. Crombie I was put in contact with Adriano Carugo, who graciously made available his transcription of the manuscript. Later I found that the manuscript had been transcribed independently by William F. Edwards, who generously contributed the results of his work also. Readings of the manuscript given in chapter 1 to show parallels with the source from which they derive are based on Edwards' transcription, which supplies line as well as folio numbers and thus is more suitable for purposes of references. All texts that are given, however, have been checked against the manuscript and represent my own readings. The punctuation, spelling, and paragraphing are also my own, usually adapted somewhat to agree with the parallel text against which the passage is exhibited. With regard to the manuscript containing the physical questions, I have already made an extensive paleographical study of this in conjunction with my English translation of it, and have used this fully in chapter 2 of the volume.

The Jesuit sources on which Galileo's notes are based are widely scattered, some preserved in printed editions and others only in manuscript form. Early on in my investigations I was assisted immeasurably by Edmond Lamalle, archivist of the Roman Archives of the Society of Jesus, who supplied me with a list of all manuscripts preserved in the Fondo Curia, many of which contain copies of lectures given in the early years of the Collegio Romano. Vincenzo Monachino, librarian of the Gregorian University in Rome, made the more relevant of these codices available for study, and later for microfilming, so that detailed comparisons could be made with Galileo's notes. Other lists were supplied by Charles H. Lohr, director of the Raimundus-Lullus-Institut of the University of Freiburg, which served to identify Collegio manuscripts now preserved in other European collections. Charles B. Schmitt, of the Warburg Institute of the University of London, supplied additional invaluable bibliographical information. All manuscripts cited in the bibliography have been studied by me *in situ*; some have further been microfilmed and later used to set up the textual parallels exhibited in part 1 of this study.

The particulars of the materials covered in courses at the Collegio Romano, discussed in part 2, are summarized from printed sources or from the manuscripts treated in part 1. For those that have pagination or foliation, appropriate references are given in the standard manner. Many of the codices, unfortunately, lack foliation, and these are cited in terms of the major divisions of the subjects treated, so as to facilitate verification and further study by those who may be interested.

In part 3 the major project is to show how the terminology and conceptual structures of Galileo's Pisan manuscripts are preserved, and continue to be used and developed, in his later writings, both Latin and Italian. Modern English translations are not helpful for this task, and on that account I have paraphrased most of the passages to which I would direct attention, and then have given, in parentheses, Galileo's wording in its original language. Extensive use has been made of the National Edition for this purpose. In order to facilitate reference and at the same time reduce the number of notes, citations of that edition are given parenthetically in the text by volume, page, and line number.

My debts are many. Apart from the persons I already mentioned, I am deeply grateful to the librarians and curators of the collections with which I have worked, who have rendered invaluable assistance in making their materials available for study. To the many Galileo scholars here and abroad who are my coworkers in uncovering and interpreting manuscripts in the Galileiana font of the Biblioteca Nazionale in Florence, I offer sincere thanks and appreciation for being able to use the result of their labors. Helpful suggestions for the improvement of the volume have been given by Jean Dietz Moss, of the Department of English, and Bernard K. Harkins, of the Leonine Commission, of The Catholic University of America. Finally, the financial assistance of the National Science Foundation, through Research Grant No. SES 79-24825 and previous awards, has borne the major portion of the expenses incurred throughout the study; without its help the study could never have been undertaken.

Washington, D.C.
November 7, 1983

William A. Wallace

PART ONE
Textual Analysis

CHAPTER 1

Sources of Galileo's Logical Questions

MUCH is already known about the life of Galileo Galilei, particularly his discoveries and the polemics surrounding the publication of his principal works on astronomy and the science of mechanics. Considerably less is known about his early period, that from his birth at Pisa in 1564 to his perfection of the telescope at Padua in 1609, when the intellectual foundations were laid for most of his later work. Historians are aware that his preliminary studies were made at the Monastery of Vallombrosa near Florence, after which he entered the University of Pisa in 1581 with the intention of pursuing a career in medicine. It is known that he left the University of Pisa in 1585 without obtaining a degree, apparently to pursue a new-found interest in mathematics. By 1589 he had achieved sufficient status to obtain a lectureship in mathematics at the University of Pisa, where he taught until 1591. The next year he moved to the University of Padua as professor of mathematics, to succeed Giuseppi Moleti, and there he remained for eighteen years, doing research on the science of motion that has recently been fairly well documented.[1] What he accomplished before he moved to Padua is considerably more problematic. Several treatises ostensibly composed by him at Pisa are still extant: a mathematical treatise, in Latin, on the center of gravity of solids; a smaller work, in Italian, on "the little balance" (*La Bilancetta*); some measurements of specific gravities and a few notations on a text of Archimedes; and, finally, a fairly extensive series of Latin compositions dealing with questions on logic, on the universe and the elements, and on local motion.

These last-named Latin compositions have been the subject of recent

[1] The best account of the chronology and details of Galileo's scientific work is Stillman Drake, *Galileo at Work: His Scientific Biography* (Chicago-London: The University of Chicago Press, 1978), which should be consulted to flesh out the information provided throughout this volume. Drake did the pioneering work to remedy the first shortcoming in Favaro's National Edition mentioned in the preface, and reports most of his findings in *Galileo at Work*. At the time of its writing, however, the research that would detect the second shortcoming was not yet completed, and his book takes no account of the findings reported herein.

study, the fruits of which set the theme for this volume.[2] They are conserved in three manuscripts now in the Galileiana Collection of the Biblioteca Nazionale in Florence, all written in Galileo's hand. These are: MS 27, containing questions based on Aristotle's *Posterior Analytics*, henceforth referred to as the logical questions (LQ); MS 46, containing questions relating to Aristotle's *De caelo* and *De generatione*, henceforth referred to as the physical questions (PQ), and some memoranda on motion; and MS 71, containing drafts of a dialogue and some treatises on local motion, usually referred to as the *De motu antiquiora* so as to distinguish it from other tracts *De motu* composed by Galileo in later life. The first two manuscripts show signs of being derived and possibly copied from other sources, whereas the third contains more independent compositions, though these are now known to draw extensively on the memoranda on motion in the second manuscript.

The previously accepted evaluation of these three manuscripts—one requiring substantial revision in light of the latest research—derives from Antonio Favaro, the editor of the National Edition of *Le Opere* of Galileo, whose first volume appeared in 1890.[3] In that volume Favaro decided to publish only the second and third manuscripts, the second (MS 46) under the title *Juvenilia* and the third (MS 71) under the title *De motu*.[4] The first manuscript (MS 27), for reasons that historians of medieval and Renaissance science find hard to understand, he effectively excluded from the National Edition, providing only a brief description and some excerpts in volume 9 under the title *Saggio di alcune esercitazioni scolastiche di Galileo*.[5] Apparently Favaro was himself unacquainted with Aristotle's *Posterior Analytics* and we knew nothing of the role played by this work in the development of scientific methodology. Reading Viviani's reconstruction of Galileo's life, written in 1654, he therefore focused on a statement to

[2] The researchers who have contributed most substantially, apart from the author, are Raymond Fredette for his work on the *De motu antiquiora* manuscripts and Alistair Crombie and Adriano Carugo for their work on the manuscripts dealing with logic and the questions on the universe and the elements.

[3] Antonio Favaro, ed., *Le Opere di Galileo Galilei*, 20 vols. in 21 (Florence: G. Barbèra Editore, 1890-1909).

[4] Favaro, *Le Opere di Galileo Galilei*, vol. 1, pp. 7-177 and 243-419, henceforth cited as *Opere* 1:7-177 and 1:243-419, respectively.

[5] *Opere* 9:273-292. Whether Favaro was completely honest in this exclusion, or whether he was motivated by an anti-clerical and anti-medieval bias, has not yet been fully investigated. Some remarks bearing on his rejection of the works of Raffaello Caverni and Pierre Duhem that might suggest the latter are given in W. L. Wisan, "The New Science of Motion: A Study of Galileo's *De motu locali*," *Archive for History of Exact Sciences* 13 (1974): 109, 146, 153.

the effect that Galileo received instruction in logic from a priest-instructor of Vallombrosa (*un Padre Maestro Volambrosano*), but that his fine intellect found such matter tedious, fruitless, and unsatisfying.[6] This, plus a few errors in Latinity, persuaded Favaro that these were merely scholastic exercises and thus of no value in understanding Galileo's later work. The second manuscript (MS 46) he accorded slightly more value because he believed that it could be dated, on the basis of internal evidence, as written in 1584, while Galileo was a student at the University of Pisa. Thus he transcribed and published it as a "youthful work," which he believed was based on lectures of Francesco Buonamici and other professors who taught at Pisa while Galileo was studying there.[7] Only the third manuscript (MS 71) did Favaro regard as meriting serious attention, since it was clearly related to Galileo's later work on motion. This he dated, following a well-documented tradition, as written "around 1590," while Galileo was already teaching mathematics at the University of Pisa.[8]

The surprising discovery of the past two decades is that the foregoing chronology is quite wrong. Rather than have the three manuscripts date from various periods in Galileo's life at which he would have been 15, 20, and 26 years of age, respectively, there is now considerable evidence to show that they were all written in conjunction with his first teaching appointment at the University of Pisa and so date from the years 1589 to 1591. This revision leaves intact the time of composition of MS 71 as "around 1590," but it reinstates the notation of an early curator of the Galileiana manuscripts to the effect that the physical questions of MS 46 were composed "around 1590" also.[9] The logical questions of MS 27, finally, must have preceded the contents of the other two manuscripts, but not by much, since it can be shown that they could not have been written before late 1588 or early 1589. Such proximity in time of composition encourages one to investigate the many internal relationships between them, as well as their cumulative influence on Galileo's later writings, and it is such an investigation that constitutes the burden of this volume.

The evidence for the new datings of MSS 27 and 46 is quite complex, requiring considerable textual analysis and comparison to give it probative force. Since the character of this evidence changes for the two manuscripts, it is thought best to present it in two chapters: that for the logical questions in this first chapter and that for the physical questions in the next. This procedure has the advantage that material common to both analyses can be covered at the outset, and the evidence for dating

[6] *Opere* 9:279.

[7] *Opere* 1:9-13.

[8] *Opere* 1:245-249.

[9] *Opere* 1:9.

the logical questions can then be added to that for dating the physical questions, which offers slightly more difficulty and so can benefit from this help. The accepted dating of MS 71 is not at question here, as already noted, and thus discussion of that manuscript is reserved for part 3 of this study, where it is situated with respect to Galileo's other early writings.[10]

The technique employed in these two chapters of part 1 consists of showing that both sets of questions are derived or extracted from various writings of Jesuit professors at the Collegio Romano dating from the last decades of the sixteenth century. These Jesuit writings are numerous and fall into two categories: some are printed books, others are manuscripts containing lecture notes for the courses then being taught at the Collegio. For the first, the authors and the dates of publication are easy enough to identify, although there is one case of plagiarism pertaining to the logical questions that complicates the reasoning employed in its instance. For the second, the problems are more numerous: the possible manuscript sources are many; in a few instances their authors or their dates of composition are uncertain; and it is quite likely that the manuscripts thus far discovered represent but a small fraction of those that were at one time available. Such difficulties effectively rule out definitive proof that Galileo took his notes from any one Jesuit manuscript or book. The many textual parallels between Galileo's writings and these possible sources, however, leave little doubt, as will be seen, that the logical questions were written around 1589 and the physical questions around 1590. Although printed books have been helpful in establishing the provenance of these questions, the evidence also points to Galileo's having used handwritten rather than printed sources in the composition of both, as will be explained in what follows.[11]

1. Logic at the Collegio Romano

The philosophy curriculum at the Collegio Romano during the period of interest was taught in a three-year cycle, with logic occupying the first year of the cycle, natural philosophy the second, and metaphysics the third.[12] As can be seen from data presented in table 1, it was more

[10] See chap. 5, sec. 2, henceforth abbreviated as sec. 5.2.

[11] See chap. 1, sec 3, subsec a, henceforth abbreviated as sec. 1.3.a; also sec. 2.4.

[12] See R. G. Villoslada, *Storia del Collegio Romano dal suo inizio (1551) alla soppressione della Compagnia di Gesù (1773)*, Analecta Gregoriana, (Rome: Apud Aedes Universitatis Gregorianae, 1954), vol. 66, pp. 84-115, for details of the program in philosophy at the Collegio Romano.

TABLE I
PROFESSORS AND COURSES IN PHILOSOPHY OFFERED AT THE COLLEGIO ROMANO IN THE ACADEMIC YEARS 1559 THROUGH 1598

Years	Logic	Natural Philosophy	Metaphysics	Mathematics
1559-1560	F. Toletus	H. Torres	B. Pererius	B. Torres
1560-1561	P. Parra	F. Toletus	B. Pererius	B. Torres
1561-1562	B. Pererius	P. Parra	F. Toletus	
1562-1563	I. Acosta	B. Pererius	P. Parra	
1563-1564	A. Gagliardi	I. Acosta	B. Pererius	
1564-1565	B. Pererius	A. Gagliardi	I. Acosta	C. Clavius
1565-1566	I. Ricasoli	B. Pererius	A. Gagliardi	C. Clavius
1566-1567	H. Sorian	I. Ricasoli	B. Pererius	C. Clavius
1567-1568	B. Sardi	H. De Gregorio	I. Ricasoli	C. Clavius
1568-1569	F. Ribera	B. Sardi	H. De Gregorio	C. Clavius
1569-1570	I. Della Croce	L. Maselli	B. Sardi	C. Clavius
1570-1571	A. Lisi	I. Della Croce	L. Maselli	C. Clavius
1571-1572	L. Romano	A. Lisi	I. Della Croce	B. Ricci
. . .				
1577-1578		A. Menu		C. Clavius
1578-1579			A. Menu	C. Clavius
1579-1580	A. Menu			C. Clavius
1580-1581		A. Menu		C. Clavius
1581-1582			A. Menu	C. Clavius
1582-1583				C. Clavius
1583-1584	I. Lorinus		A. Parentucelli	C. Clavius
1584-1585		M. De Angelis	A. Parentucelli	R. Gibbone
1585-1586	I. Lorinus	M. De Angelis	P. Valla	F. Fuligati
1586-1587	I. Caribdi	M. De Angelis	P. Valla	F. Fuligati
1587-1588	P. Valla	I. Caribdi	M. De Angelis	C. Clavius
1588-1589	M. Vitelleschi	P. Valla	I. Caribdi	C. Clavius
1589-1590	L. Rugerius	M. Vitelleschi	P. Valla	C. Clavius
1590-1591	A. De Angelis	L. Rugerius	M. Vitelleschi	C. Clavius
1591-1592	R. Jones	A. De Angelis	L. Rugerius	C. Clavius
1592-1593	F. Raimondi	R. Jones	A. De Angelis	C. Clavius
1593-1594	A. De Angelis	F. Raimondi	R. Jones	C. Clavius
1594-1595	R. Ciamberlini	A. De Angelis	F. Raimondi	C. Clavius
1595-1596	S. Del Bufalo	R. Ciamberlini	A. De Angelis	C. Grienberger
1596-1597	A. Eudaemon	S. Del Bufalo	R. Ciamberlini	C. Grienberger
1597-1598	R. Ciamberlini	A. Eudaemon	S. Del Bufalo	C. Grienberger

or less customary for a professor to begin the cycle with a particular class and then continue along with them throughout the remaining two years. This assured greater continuity in the course, and even allowed a professor to make up, in a following year, material he had been unable to cover in the year assigned. Mathematics was usually studied in the second year, and this additional load, together with the large amount of natural philosophy that was assigned, resulted in a fair amount of that discipline being left over for the third year. Textbooks were probably available for all the courses, but most professors preferred their own teaching notes, and students were encouraged to make *reportationes* of them for their personal use. Time was in fact set aside for Jesuit scholastics to complete their class notes by referring to those of the professor and other sources.[13] It seems likely that most professors deposited a finished set of lectures in the Collegio library after their course was completed; if so, repeated copying of such codices would explain the large number of *reportationes* of lectures from the Collegio Romano that are now found in European libraries.

A summary of the best available information about the professors, the courses they taught, and the years in which they taught them at the Collegio is assembled in table 1.[14] A gap of five years in the *rotulus* of professors occurs between the years 1572 and 1577, and there are occasional other *lacunae* down to 1585, after which the list is complete.

For logic, the first professor on the list is Franciscus Toletus (Toledo), who taught philosophy at the University of Salamanca previous to his entry into the Jesuits in 1558, and had come to Rome while still a novice in the Society to teach logic in the academic year 1559–1560. Toletus is of importance in that his teaching resulted in a series of textbooks, including a complete logic course that became available in 1576 and was reprinted many times to the end of the sixteenth century. This undoubtedly served as the basic text for lecturers at the Collegio during the two decades that followed its initial printing.[15] It also was expanded or supplemented by Ludovicus Carbone of Costacciora, one-time professor at Perugia, who reprinted the text along with his putative emendations

[13] Ibid., p. 89; see also pp. 85 and 107.

[14] Most of the information in this list is taken from the Elencho dei Professori of the Collegio prepared by Ignazio Iparraguirre and appended to Villoslada's *Storia del Collegio Romano*, pp. 321–336; some additions have been made by the author on the basis of manuscripts of lecture notes he uncovered in his researches. The names are generally given in Latin, following the usage of the *rotulus* of professors at the Collegio and the manuscripts studied. For purposes of uniformity, considering the multiple ways some of the names are rendered in the vernacular, the versions of the names recorded in the table are used throughout this work. All vernacular variants, however, are listed in the index.

[15] Villoslada, *Storia del Collegio Romano*, pp. 51–52, 99, 102.

under the title *Additamenta ad commentaria doctoris Francisci Toleti in logicam Aristotelis* at Venice in 1597. The *Additamenta*, as will be seen, was plagiarized from a course given by Paulus Valla at the Collegio Romano in the academic year 1587-1588. It assumes great importance in this study from the fact that Galileo apparently used the very set of notes Carbone plagiarized when preparing the logical questions of MS 27. Indeed, establishing this connection between Galileo and Valla *via* Carbone is the primary goal of the textual analyses offered in this chapter.

Considering the large number of professors and the many times logic was taught at the Collegio between 1559 and the end of the century, a surprisingly small number of lecture notes or printed editions have survived to the present day. Those that are known are given in table 2, with an indication of the year in which they were taught, the professor who gave the course, and whether it survives in manuscript or printed form or both.[16] The course offered by Ioannes Lorinus is the only one now extant both in manuscript and in print, and the two versions are practically identical—rather remarkable considering their separation by an interval of some thirty-six years. Valla's printed course, on the other hand, is known to be different from the manuscript version (now lost) that antedated it by some thirty-four years, for reasons that will be explained presently. It is the most complete of the printed texts, filling two large folio volumes and treating in detail practically every subject that can be considered in logic. The remaining courses, those of Vitelleschi, Rugerius, Jones, and Eudaemon, are not as compendious as Valla's, although Rugerius comes closest to it in the quantity of material he

TABLE 2
EXTANT VERSIONS OF LOGIC COURSES TAUGHT AT THE COLLEGIO ROMANO, 1559 TO 1597

Years	Professor	Manuscript	Printed Edition
1559-1560	F. Toletus	—	Venice 1576
	[L. Carbone: Additamenta]	—	Venice 1597
1583-1584	I. Lorini	1584	Cologne 1620
1587-1588	P. Valla	—	Lyons 1622
1588-1589	M. Vitelleschi	1589	—
1589-1590	L. Rugerius	1590	—
1591-1592	R. Jones	1592	—
1596-1597	A. Eudaemon	1597	—

[16] Details pertaining to the manuscripts and printed books indicated in the table are given in the bibliography at the end of this volume.

covered. Vitelleschi and Jones approximate Lorinus's course in content, though with somewhat different emphases, and Eudaemon offers a less developed treatment, covering the subject in the most cursory fashion of all.

Since each of these courses sheds light on the way logic was taught at the Collegio, and on this account can aid in understanding the materials used by Galileo when writing his logical questions, their respective contents will be sketched before examining Galileo's manuscript in the following section.

a TOLETUS AND CARBONE

Toletus was one of the earliest philosophy professors at the Collegio Romano.[17] There he taught logic, as already noted, in 1559-1560, and then continued on in the philosophy cycle, teaching natural philosophy in 1560-1561 and metaphysics in 1561-1562, after which he passed to the teaching of theology. He achieved great eminence as a theologian and was made a cardinal in 1594, two years before his death. Some of his manuscripts are preserved in the archives of the Gregorian University in Rome, including a commentary on the *Posterior Analytics* of Aristotle.[18] This is not as complete as the similar commentary contained in the printed edition, and has little interest for its relation to Galileo's text on that account. The logic course that was printed, however, gives a good idea of the extent of the teaching of this subject at the Collegio, and apparently set the syllabus there for the last three decades of the sixteenth century.

Toletus's logic usually appeared in two volumes, the first a slim book of less than one hundred pages entitled *Introductio in dialecticam Aristotelis*, which contained a quick overview of all of logic, including the essentials of formal logic and topics similar to those contained in the *Summulae* tradition, and the second a book of some four or five hundred pages entitled *Commentaria una cum quaestionibus in universam Aristotelis logicam*, which detailed all topics in logic of philosophical interest.[19] As the course at the Collegio developed, the first volume formed the basis for a two- or three-week introduction to the science of logic in which the students were drilled in the uses of terms and the forms of reasoning, so that they

[17] For some particulars of Toletus's life and works, see Carlos Sommervogel et al., *Bibliothèque de la Compagnie de Jésus*, 11 vols. (Brussels-Paris: Alphonse Picard, 1890-1932), vol. 8, cols. 64-82; his role at the Collegio Romano is covered in Villoslada, *Storia del Collegio Romano*, passim.

[18] Archivum Pontificiae Universitatis Gregorianae, Fondo Curia (henceforth abbreviated as APUG-FC), Cod. 37, In libros Posteriorum.

[19] These were printed at Venice, Apud Iuntas, in 1576 and went through many printings thereafter. Details are given in Sommervogel, *Bibliothèque*, note 17 *supra*.

would have a general familiarity with the terminology and procedures employed by the logician. The second volume then enabled the professor to build on this foundation and at the same time cover most of the classical texts in logic in lectures extending over the remainder of the academic year. The structure of the second volume is of interest for what follows, and thus is indicated below in schematic form, with an indication of the approximate number of pages allotted to each tract:

A general division of the arts and the sciences, followed by five questions on the nature of logic as a science. (20 pages)

An introduction to the *Isagoge* of Porphyry, concerned with the "five universals," including a brief commentary on the text, interspersed with questions on the following topics:
Four questions on universals.
Two questions on first and second intentions.
Three questions on the subject of the book, viz., predicables, and how these are divided.
Five questions on genus, two on the individual, one on species, and discussions of disputed points on property and accident. (60 pages)

An exposition of the *Categories* of Aristotle, including a general discussion of the categories (more frequently called "predicaments") and their relation to metaphysics, and then detailed discussions of the antepredicaments, each of the predicaments in particular, and the postpredicaments. Among the questions relating to the predicaments are the following:
Six questions on substance.
Two questions on quantity.
Four questions on relation.
Five questions on quality.
Eight questions on the last six predicaments. (120 pages)

A very brief discussion of Gilbert de la Porrée's *De sex principiis*. (8 pages)

An exposition of Aristotle's *Perihermenias*, including a commentary and questions on various topics, with particular attention being devoted to questions of signification and truth, the noun, the verb, the expression (*oratio*), the enunciation or proposition and its various kinds, and modes and modal propositions. (70 pages)

A commentary with a large number of questions on the two books of Aristotle's *Posterior Analytics*; the questions are not numbered systematically as in the earlier expositions of the *Isagoge* and the *Categories*, but are interspersed at their proper places in the commentary. The more noteworthy topics that are discussed include:

First book: the structure of the treatise, the foreknowledge required for demonstration, the demonstrative regress, the requirements for demonstration, its kinds, and comparisons between them.

Second book: the kinds of scientific question, the relation between demonstration and definition, definitions and their relations to causes, and principles and how they are known. (200 pages)

It may be observed that Toletus covered the major classical texts in this way, with the possible exception of Aristotle's *Prior Analytics.* The omission is only apparent, however, since the main topic discussed in that work—the syllogism and its various figures and modes—is covered in the first volume already mentioned, where some twenty pages are allotted to this topic. Likewise, the discussion of probable argument and of fallacies, corresponding to Aristotle's *Topics* and *De sophisticis elenchis,* is covered there in very summary fashion, receiving a combined treatment of some twenty pages also.

Undoubtedly this presentation was improved upon by later authors who either worked up new sets of notes or else composed additions that would serve as supplements to Toletus's basic texts. One such author, as already mentioned, was Ludovicus Carbone, who published a number of emendations in his *Additamenta* of 1597.[20] The date of publication of this work is clearly not that of its composition, for Carbone notes in his preface that he is having these additions printed in order to make them available to others, lest they never see the light of day, although he had been holding them in the hopes of eventually producing a complete logic text himself.[21] At the same time he published an *Introductio in logicam*

[20] The full title reads: *Additamenta ad commentaria D. Francisci Toleti in Logicam Aristotelis. Praeludia in libros Priores Analyticos, Tractatio de Syllogismo; de Instrumentis sciendi; et de Praecognitionibus, atque Praecognitis.* Auctore Ludovico Carbone a Costacciaro, Academico Parthenio, et in Almo Gymnasio Perusino olim publico Magistro. Cum Privilegiis (Venetiis, Apud Georgium Angelerium, 1597), henceforth cited as *Additamenta.*

[21] Ibid., Auctor lectori, fol. a2r-v. The Latin reads as follows. "Additamenta haec, amice Lector, ut minus elaborata atque polita, nunquam mihi e manibus extorqueri permisissem, ni plus aliis, quam mihi soli viverem: quod qui faciunt, ea saepe faciunt, quae alias facturi non essent. Ut itaque nonnullorum votis satisfacerem, ea in lucem publicam venire permisi, quae una cum completo de re Logica opere, alio tempore, edere cogitabam. Habeo enim Commentaria in omnes libros Logicos non poenitenda, quibus, dum gravioribus operibus scribendis sum addictus, extremam manum imponere non queo: faciam tamen aliquando, favente Deo. Interim ab horum additamentorum editione, occasione sumpta, eodem tempore Introductionem in Logicam, seu totius Logicae Compendium, ni fallor, absolutum, foras dedi. Uno etenim opusculo, quod nunc sub praelo est, non solum quae in universo Aristotelis Organo continentur; sed etiam omnia quae ab ipso praetermissa, et ab aliis superaddita fuere, complexi sumus. Quare, qui nostra leget, disciplinam, alias Tironibus non facilem, facile sibi comparare poterit; si ea quae Aristoteles ad iuvenum ingenia minus

(Venice, 1597), apparently for the same reason, and released for publication a similar work, *Introductio in universam philosophiam*, which appeared in 1599—perhaps posthumously, as he seems to have died in 1597.[22] A prolific writer, Carbone issued a large number of works on theology and rhetoric between 1587 and 1597, including an *Introductio in sacram theologiam* in 1589 and an edition of Toletus's *Introductio in dialecticam* in 1588.[23] He was not a Jesuit, but he studied at the Collegio Germanico annexed to the Collegio Romano and probably had access to the materials taught there. Despite the late date of their printing, there is good evidence that the additions were composed a full decade earlier—surely after 1576, since this was the date of publication of the work for which they are seen as supplements, and most probably around 1588, when Carbone was reissuing the first part of Toletus's course, for reasons that will be explained presently.

Carbone's *Additamenta* is proposed as an aid to the study of the two

---

accommodate scripsit, praetermittere, et ad alias artes se conferre volet: cum quae ille fuse et obscure docuit, nos breviter, et perspicue, quantum per rem ipsam licuit, et quae ille omisit, nova quadam ratione, novis in re Logica ingeniis accommodata, tradiderimus. Etenim, logicis tradendis, eam tenere solemus viam, atque rationem, quae multorum annorum experientia edocti, iis qui primo ad Dialecticam cognoscendam accedunt, maximam lucem affert, et ad reconditas difficultates, quae in disserendi ratione tractantur intelligendas, maxime idoneos reddit. Quam, non modo discipuli, eius utilitates animadvertentes, saepe approbarunt, sed etiam viri eruditi, ut hoc dicam, et vere quidem, commendarunt. Et ne, dum haec nostra rudia Additamenta excusare deberem, aliud opusculum laudare velle videat, qui eo uti voluerint, nos in tanta de eadem re librorum multitudine, non omnino frustra tempus, operamque consumpsisse facile animadvertet. Ideo autem, de hac lucubratione in praesenti meminisse volui, ut multa quae in tractatio de Syllogismo hic omisi, ex ea promenda esse monerem. Ibi enim de Syllogismorum regulis communibus, de formis, et speciebus in particulari, deque aliis argumentationum speciebus scripsimus, de quibus hoc loco fere nihil Illud ad extremum Lectorem monuerim, in his Additamentis, interdum me dicere, de re aliqua, alias dixisse, aut esse dicturum, de qua in eis non egi, innuendo ea commentaria, de quibus supra dicebam: quae faxit Deus, ut in eius gloriam et studiosorum utilitatem aliquando cedant: in quos fines, quidquid elaboramus, cedere percupimus."

[22] Charles H. Lohr, in his "Renaissance Latin Aristotle Commentaries," *Renaissance Quarterly* 28 (1975): 698, indicates that he died at Venice after 1597; the card file at the Vatican Library gives his date of death as 1597. Bibliographic references are given by Lohr, ibid.

[23] Ibid., pp. 698-699. His other titles, all preserved in the Vatican Library, include: *Fons vitae et sapientiae, vel ad veram sapientiam acquirendam hortatio* (Venice, 1588); *De praeceptis Ecclesiae opusculum utilissimum* (Venice 1590); *De dispositione oratoria, disputationes xxx* (Venice, 1590); *De elocutione oratoria libri iiii* (Venice, 1592); *De octo partium orationis constructione libellus* (Venice, 1592); *De caussis eloquentiae libri iiii* (Venice, 1593); *Divinus orator, vel De rhetorica divina libri septem* (Venice, 1595); and *Introductio ad catechismum sive doctrinam Christianam* (Venice, 1596).

Analytics and deals with topics that are only summarily treated in Toletus's *Introductio* and *Commentaria*. The work contains an introduction in nine chapters, or questions, dealing with the nature of analysis and how it is covered in the *Prior* and the *Posterior Analytics*, after which come three lengthy treatises, one on the syllogism, another on instruments of knowing, and the final one on the foreknowledge required for demonstration. The *Tractatio de syllogismo* comprises fourteen chapters or questions, the *Tractatio de instrumentis sciendi*, sixteen, and the *Tractatio de praecognitionibus et praecognitis*, twenty-five. The latter treatise is particularly full, and contains much matter that is almost identical with that found in Galileo's logical questions, as will be shown in detail later. It is also noteworthy that Carbone's *Introductio in universam philosophiam* devotes over two hundred pages to a treatise, *De scientia*, that is very similar to matter covered in logic courses at the Collegio during the period of interest here, and that could have been worked up at the same time as the additions to Toletus's logic.

b. LORINUS

Ioannes Lorinus (Laurinis, de Lorini, 1559-1634) was born in Avignon, entered the Society of Jesus, and at the age of twenty-four or twenty-five taught the course in logic at the Collegio Romano during the academic year 1583-1584.[24] A codex containing his lecture notes, dated 1584, is preserved in the Vatican Library; the inscription on the first folio bears the legend "Ioannis Laurinis, S.I., Logica," but the "au" in his family name is crossed out (in another hand) and the letter "o" written over it.[25] This change in spelling is significant, for Villoslada lists a Ioannes Lacerino as teaching logic at the Collegio in 1585-1586, and his "Lacerino" is evidently a misreading of Laurino as originally inscribed in the rotulus of professors. In later documents at the Collegio Romano, Lorinus, who enjoyed a long and fruitful career there as a professor of Scripture, is usually referred to as Lorino, or Lorini, or de Lorini. His logic course, as already observed, was published subsequently under the title *In universam Aristotelis logicam* (Cologne, 1620).

The *reportatio* of the 1584 course is lengthy, containing 560 folios, the first half of which follows pretty much the outlines of Toletus's coverage of logic up to the *Posterior Analytics*. The exposition of the latter work begins on fol. 283v and continues until fol. 418r, where the notation is made: Posteriorum analyticorum quaestionum finis. Following this comes

[24] See Lohr, "Renaissance Latin Aristotle Commentaries," *Renaissance Quarterly* 31 (1978): 544, for biographical and bibliographical references.

[25] Biblioteca Apostolica Vaticana, Cod. Urb. Lat. 1471.

a treatise *De scientia* (fols. 420r–509r) and then a disputation *De definitione* (fols. 513r–560r), with which the course concludes. The hand in which the manuscript is written is poor, and thus it is fortunate that its entire contents are reproduced, almost verbatim, in the printed edition of 1620, itself a quarto volume of 691 pages. Since the published work is more readily available, and certainly more readable, than the manuscript, the former has been used to provide the following outline of the course:

An introduction to logic generally, based mainly on the three acts of the mind (pp. 1–55).

A disputation on the nature of logic (pp. 56–88).

An exposition of Porphyry's teachings, interspersed with commentary, including:
- An introduction (pp. 89–92).
- Ten questions on universals (pp. 92–115).
- Seven questions on genus, two on species, five each on difference, property, and accident (pp. 115–162).

A summary of the *Praedicamenta* of Aristotle, comprising:
- An introduction on the nature of the categories (pp. 163–181).
- A discourse on the antepredicaments, focusing mainly on analogy (pp. 182–203).
- Treatments of the individual predicaments, including twelve questions on relation (pp. 204–275).
- A discourse on the postpredicaments, centered on the concept of opposition (pp. 276–299).

A commentary on Aristotle's *Perihermenias*, detailing the contents of the chapters of each book (pp. 300–343).

A commentary on the *Prior Analytics* of Aristotle containing two disputations, one on the syllogism in general and the other on its various figures (pp. 344–378).

A commentary, with questions, on the *Posterior Analytics* of Aristotle, containing:
- An exposition of the text (pp. 379–410).
- A treatise on demonstration, made up of disputations or parts concerned with foreknowledge and its various kinds (pp. 411–457), with the nature of demonstration (pp. 458–523), with the species or kinds of demonstration (pp. 524–552), and with the properties of demonstration, focusing mainly on the demonstrative regress (pp. 553–557).

A treatise on science, made up of disputations or parts dealing with its existence (pp. 559-581), with science as an act (pp. 582-585), with science as a habit, including the various kinds of science, comparisons between them, and how they are subalternated one to the other (pp. 585-621), and with science as compared to other intellectual habits (pp. 622-639).

A treatise on definition, made up of sections dealing with its existence as an instrument of knowing (pp. 640-648), with its nature and kinds (pp. 649-659), with its relationships to the thing defined (pp. 660-676), with methods of defining (pp. 677-684), and with comparisons between definition and demonstration (pp. 685-691).

From the above list it can be seen that Lorinus structured his entire treatment of the *Posterior Analytics* as though it were a tract on demonstration, apparently not adverting to the fact that the second book of that work is also concerned with definition, as Toletus was clearly aware, and so reserving his own treatment of definition for the last part of his course on logic. He does have an extensive treatment of foreknowledge, moreover, and touches on many of the topics related to it that are taken up by Carbone in his additions to Toletus's text. And if one compares the amount of space he devotes to foreknowledge and to demonstration, respectively, approximately in the ratio of 1 to 2, one will find that this is precisely the ratio that governs the distribution of similar matter in Galileo's logical questions. Thus, as will become clear in what follows, these notes composed in 1584 and subsequently published in 1620 show similarities with Galileo's composition and probably represent, in their manuscript form, an intermediate stage of development for the notes on which he ultimately drew.

c. VALLA AND CARBONE

Paulus Valla (de Valle, Vallius, the latter being the spelling he employed in the printed version of his logic course), began to teach at the Collegio at the age of twenty-four, when he commenced the course in metaphysics in 1585.[26] The problems relating to his teaching of natural philosophy and to his composition of the *Tractatus de elementis* will be discussed at length in the following chapter.[27] Suffice it to mention here that his treatment of the elements and their qualities shows the closest agreement with Galileo's physical questions of any extant *reportatio*. In fact, although he does not have the scope of coverage of other authors,

[26] Sommervogel, *Bibliothèque*, vol. 8, col. 418.

[27] Secs. 2.1.b, 2.3.a, and 2.3.b.

in the portions that survive his composition is the best candidate for being the exemplar Galileo used in composing the portions of the physical questions dealing with the elements and their qualities. Unfortunately, none of his logic course, which he taught in 1587-1588, is now available in manuscript form. There are some very helpful remarks in the prefaces he wrote to the two volumes of his *Logica*, however, that shed light on this and other materials pertaining to his teaching at the Collegio Romano, and these deserve careful examination.[28]

The first volume, which was printed at Lyons along with the second volume in 1622, bears the inscription that it is concerned with the "old logic" whereas the second is concerned with the "new." Both are large folio volumes with relatively small print and contain seven hundred pages each. In his preface to volume 1, Valla first notes the procedure he will use in presenting each tract, explaining that he will start with various opinions and the reasons offered in their support, then present the position he regards as true (explaining Aristotle's text and the teaching of St. Thomas in the process), and finally show how all the divergent opinions can be explained in the light of his solution. He goes on to say that he plans to put out an entire course in philosophy along these lines, which will be made up of ten, or perhaps twelve, similar volumes. The first two are the logic volumes being introduced; following these will come two more expounding the *Physics*, then one each devoted to the *De caelo*, the *De generatione* (this containing everything pertaining to the elements, their qualities, and their motions), and the *Meteorology*; and finally three explaining the teaching of the *De anima* and the *Parva naturalia*, one on the soul in general and the vegetative soul, another on the sensitive soul, and a concluding one on the intellective soul. All of these are now ready to be printed, he writes, but he would like to add to them two more volumes expounding the *Metaphysics*, and then the entire course will be complete. Unfortunately Valla died in 1622, the year in which the logic volumes were printed. The latter, it may be noted, were approved for publication by two Jesuit censors, Ioannes Chamerota and Ioannes Lorinus, on June 24, 1612.[29] Some notations from a censorship report on his commentary on the *Physics*, dated September 18, 1621, and signed by Marcus Van Doorne, Ioannes Lorinus, and Ioannes Chamerota,

[28] The first volume is entitled *Logica Pauli Vallii Societatis Iesu duobus tomis distincta: Quorum primus artem veterem, secundus novam comprehendit. Tomus Primus. Quid in universo hoc opere praestetur sequens Epistola ad Lectorem docebit. Cum privilegio Regis* (Lugduni: Sumptibus Ludovici Prost, Haeredis Rouille, 1622), henceforth cited as *Logica*. The preface is on fols. 3r-4v. The title of the second volume is similar, except that it reads: *Tomus secundus. Quid hoc secundo tomo contineatur, sequens pagina docebit*. The preface follows the title page. Excerpts from both prefaces are translated *infra*, at notes 32-34.

[29] Archivum Romanum Societatis Iesu, Fondo Gesuitico, Cod. 654, fols. 275-281.

are preserved in the Roman Archives of the Society of Jesus.[30] An early bibliographical register lists this commentary as printed at Lyons in 1624, but its existence has not been definitively established.[31]

In addition to these remarks *Ad lectorem*, Valla outlines the contents of the two volumes and then explains that he will start by prefacing a brief introduction to the whole of logic. Of this he writes:

> We preface, I say, an *Introductio* that was explained by us thirty-four years ago [i.e., 1588] in the Collegio Romano and given to our hearers shortly thereafter. This work, with very little of the fruits of our labors changed in it, was published at Venice by some good author, who added some preliminary matter and made some inversions (or rather perversions) of its order that, in my judgment, achieve no better result. We wish to warn you the reader of this, so that, should you come across this book, you will recall that he took it from us. And since he stole this and similar matter from us and from the writings of our Fathers [i.e., other Jesuits], perhaps he should have added the author's name to these books, had he known it or thought it due us.[32]

Valla then announces that his second volume will contain his expositions of the *Prior* and *Posterior Analytics*, and to this he will add a *Disputatio de scientia*, of which he further remarks:

> The same thing happened to this *Disputatio* as I explained happened to the *Introductio*. But this we have now so enlarged and perfected that it would hardly be recognized by anyone except the author as the fetus of the same.[33]

This intimation of plagiarism is, of course, most interesting, but no less so than the additional charges Valla makes in the preface to the second volume of his *Logica*. By the time he came to its composition he had decided to append four complete tractates to his commentaries on the *Analytics* rather than the *De scientia* alone. These he enumerates as *De*

[30] Ibid., Cod. 644, fols. 464-467. The author wishes to thank Edmond Lamalle for furnishing him with copies of both these censorship reports.

[31] Lohr, "Renaissance Latin Aristotle Commentaries: Authors So-Z," *Renaissance Quarterly* 35 (1982): 209. In a private communication to the author, Dr. Lohr has expressed doubt that the printed work ever existed.

[32] *Logica*, vol. 1, fol. 4r. The Latin reads as follows. "Illam inquam praemittemus introductionem, quam ante annos quatuor, et triginta Romano in Collegio a nobis explicatam, Auditoribusque traditam non multo post, ab Auctore valde sane pio, nostrique laboris peramico paucissimis immutatis, Venetiis editam in lucem cognovi, Prooemiolis quibusdam additis, nonnullisque inversis, vel potius perversis, quod ad ordinem attinet (meo iudicio) non meliorem effectam, de quo te lectorem monitum voluimus, ne si forte in libellum illum incidas, veniat in animum, haec nos ab illo detraxisse, cum tamen ille vero haec, et alia huiusmodi a nostris, nostrorumque Patrum scriptis arripuerit, additurus forte, et Auctoris nomen illis in libellis, vel cognovisset, vel si id nobis gratum existimasset."

[33] Ibid. ". . .illud idem cum eodem accidit, quod in Introductione accidisse exposui. Verum ita nunc locupletavimus, melioremque effecimus, ut vis ab alio quam ab ipso Auctore, idem tamquam foetus dignosci valeat."

praecognitionibus, De demonstratione, De definitione, and *De scientia.* Concerning the order of these tractates, he now alerts the reader to the following:

> About twenty years ago [i.e., around 1602], a certain individual—possessing a doctorate, having published a number of small books, and being otherwise well known—had a book printed at Venice in which he took over and brought out under his own name a good part of what we had composed in our *De scientia* and had taught at one time, thirty-four years before this date [i.e., in 1588], in the Roman *gymnasio.* And having done this, this good man thought so much of other matters we had covered in our lectures that he took from them, and claimed under his own name, a large part of *De syllogismo, De reductione, De praecognitionibus,* and *De instrumentis sciendi,* and proposed these as kinds of *Additamenta* to the logic of Toletus, especially to the books of the *Prior Analytics.* He further saw fit to publish, again under his own name, our *Introductio* to the whole of logic, having changed only the ordering (disordering it, in my judgment), along with the introductions and conclusions. I wish you to know this, my reader, so that, should you see anything in either, you will know the author. I say, "should you see anything in either," for we have so expanded our entire composition that, if you except only the opinions (which once explained we have not changed), hardly anything similar can you see in either. So in those works you have what he took from me, in this what I have prepared more fully and at length.[34]

The import of these statements for fixing the date of Galileo's logical questions, needless to say, cannot be overestimated. Earlier it was remarked that Carbone's treatment of *De praecognitionibus* in his *Additamenta* of 1597 is very similar to Galileo's corresponding tract in his logical questions. Now Valla is attesting, in no uncertain terms, that most of that tract was plagiarized by Carbone from lecture notes that he himself had made available to students at the Collegio Romano in 1588 or shortly thereafter. For Carbone is the only person who could possibly fit the description of the unnamed "good man" in Valla's two prefaces. As has previously been indicated, Carbone's *Additamenta* to Toletus's logic contains treatises entitled *De syllogismo, De instrumentis sciendi,* and *De prae-*

[34] Vol. 2, fol. 1. "Annos fere ante viginti, Vir quidam, doctoris etiam laurea donatus, variisque in lucem editis libellis, satis notus, impressum Venetiis libellum, in lucem edidit, in quo multa, ex nostris de Scientia lucubrationibus, Romae olim, quatuor fere, et triginta ante hanc diem annos, in Romano gymnasio expositis, transtulit, suoque nomine evulgavit; quod etiam factu dignum existimavit, vir probus, aliis in nostris Romanis praelectionibus. Quare ex iis additiones quasdam in logicam Toleti, ad libros praesertim Priores, multa de syllogismo, resolutione, praecognitionibus et instrumentis sciendi, sibi, suoque nomini vindicavit, vel usurpavit: immo vero, et Introductionem nostram, in totam logicam, mutato, immo potius (meo iudicio) perturbato ordine, ac prooemiolis, et epilogis auctam, suo etiam nomine, imprimendam censuit. Hoc tibi notum volui, lector, ut si quid in utroque videris, Autorem noveris. Dixi si quid in utroque videris; ita enim locupletavimus omnia, ut sententias, (quas semel bene susceptas non immutavimus) si excipias; vix quid simile in utro videre possis. Habeat igitur ille, quae sibi usurpavit, habeas tu haec, quae longe meliora paravimus."

cognitionibus, precisely those alleged to be stolen, and his little book, fitting Valla's term *libellus* and entitled *Introductio in universam philosophiam* (printed in 1599, twenty-three years earlier than Valla's *Logica*), has the treatise *De scientia*, which, in Valla's recollection, was printed by this person under his own name "about twenty years ago." Valla admits that Carbone might have made a few changes in the notes, and also insinuates that he helped himself freely to the notes of other Jesuits, undoubtedly of the Collegio also, but otherwise he is clearly claiming that most of his materials are Valla's own. Not only this, but because of Carbone's plagiarism, Valla has felt compelled to change his own expression and to develop his treatises beyond the state in which they were when Carbone had access to them. It goes without saying that these are invaluable pieces of information to take into account when setting up textual comparisons between Galileo and Valla and Carbone, and thus for determining the date of Galileo's writing of the logical questions. And, so far as the content is concerned, even though the wording has been changed and the treatises expanded, one will henceforth expect a consistent uniformity of teaching in the logical writings of Valla, Carbone, and the young Galileo.

In view of the importance of Valla's textbook, a summary of the contents of the two volumes is given here in schematic form, similar to those already provided for the works of Toletus and Lorinus, after which follows a more detailed description of his treatise *De praecognitionibus*.

Volume 1. The Old Logic

An introduction, giving a brief description of logic, and then a summary treatment of the three operations of the mind, namely, that dealing with the term and definition (18 chapters), that dealing with the proposition and its various types (22 chapters), and that dealing with the syllogism, including the topical and sophistical (25 chapters).

Prolegomena to the study of logic as a science, divided into two parts containing, respectively:

Four questions on the nature of a "second intention" or of an *ens rationis*.

Four questions on the nature of logic as a science (the third question, on the object of logic, consists of 27 chapters and is an extremely full treatment of the subject). (43 pages)

An exposition of the *Predicables* of Porphyry, consisting of three parts containing, respectively:

Six questions on universals in general.

An extensive treatment of the various predicables, discussed under the rubric of "universals in particular," containing two questions on genus, four on species, two on difference, and one on property and accident.

A discussion of the points predicables have in common, and those in which they differ. (134 pages)

An exposition of the *Praedicamenta* of Aristotle, containing:

A prologue, with two questions, inquiring about the purpose of the book and its relation to logic.

A treatise on the antepredicaments, containing two questions on equivocation and analogy and six questions on what can be located in the predicaments.

A detailed examination of each of the predicaments or categories, with:

Nine questions on substance.
Nine questions on quantity.
Eleven questions on relation.
Three questions on quality.
One question on the remaining six predicaments.

A treatise on the postpredicaments, discussing mainly the various kinds of opposition and contrariety. (349 pages)

An exposition of the *De interpretatione* or *Perihermenias* of Aristotle, made up of three parts, as follows:

Three questions on vocal expression (*vox*) and truth.
Five questions on the noun and the verb.
Five questions on enunciations, their properties, and their opposition. (92 pages)

Volume 2. The New Logic

An exposition of the two books of the *Prior Analytics* of Aristotle, as follows:

An introduction, containing three questions, discussing the nature of, and the connection between, the books of the *Analytics* in general.

An explanation of the teachings of the first book, containing:

Three questions on the syllogism.
Three questions on the conversion of propositions.
Three questions on the figures of syllogisms and their contents.

An explanation of the second book, mainly in the form of an exposition, in 35 chapters, of the uses and defects of various figures of the syllogism. (119 pages)

An introduction to the first book of the *Posterior Analytics*, containing two questions on the object of the books and a brief exposition of the first chapter. (16 pages)

A disputation on foreknowledge (*De praecognitionibus*), made up of:

Three questions on foreknowledge and the foreknown, in general.
Five questions on the foreknowledge of principles.
Two questions on the foreknowledge of properties.
Two questions on the foreknowledge of subjects.
Two questions on the sources of assent to knowledge in general. (39 pages)

Exposition of the remainder of the first book. (12 pages)

A disputation on demonstration (*De demonstratione*), made up of:

Five questions on the principles of demonstration in general.
Twenty-five questions on the nature and conditions of demonstration.
Five questions on the kinds of demonstration.
Three questions on demonstrative regress and circularity in argument. (162 pages)

Exposition of the second book of the *Posterior Analytics*. (25 pages)

A disputation on definition (*De definitione*), made up of:

Seven questions on the nature of definition.
Three questions on the properties of definition as an instrument of knowing. (32 pages)

A disputation on science (*De scientia*), made up of:

Twelve questions on habits in general.
Seven questions on the essence or quiddity of science.
Five questions on the properties of science.
Nine questions on the divisions of sciences and their subalternation.
Seven questions on the comparison of science with other intellectual habits. (286 pages)

In the above listing, it can be seen that Valla devoted fourteen questions to his *Disputatio de praecognitionibus* and that these cover thirty-nine pages of text. Actually each question is made up of a number of chapters, with the result that there are sixty-nine chapters in the disputation as a whole. We will recall that Carbone's *Tractatio de praecognitionibus* contains only twenty-five chapters; each of these poses a single question, and the whole covers twenty pages of text in a smaller format. Thus Valla's claim to have expanded his treatment over what was available to Carbone is borne out. A comparison of the specific queries in their respective treatises *De praecognitionibus*, moreover, shows that Carbone's first six questions (1-6) treat essentially the same matters as Valla's questions on foreknowledge in general; his next five (7-11) correspond to Valla's questions on the foreknowledge of principles; his next two (12-13), to the foreknowledge of properties; his next six (14-19), to the foreknowledge of subjects; and his final six (20-25), to the sources of assent to knowledge in general. The material has been rearranged slightly, but there can be no doubt that the teaching is the same, and ultimately derives from the same source.[35]

d. VITELLESCHI

Mutius Vitelleschi (Vitelleschus) began his teaching career at the Collegio at about the same age as Lorinus and Valla, namely, at twenty-five, being entrusted with the logic course in 1588-1589, which launched him on his illustrious career in the Society of Jesus.[36] (It is perhaps noteworthy that Vitelleschi was born in 1563 and so was only one year older than Galileo; thus he would have been teaching this matter at approximately the same age as Galileo was appropriating it from whatever exemplar was available to him.) A manuscript of Vitelleschi's logic course is preserved in the Vatican Library, and this shows that he covered essentially the same content as Lorinus and Valla, though in slightly less detail than either.[37] The *reportatio* was made by a student of his, Torquatus Ricci, in a codex of 445 folios. The first 33 of these present the customary introduction to all of logic, including the syllogism with its various figures and modes, after which come 52 folios on the nature of logic as a science. Following this, Vitelleschi devotes 69 folios to Porphyry and the predicables, 86 folios to the *Categories* of Aristotle, and 45 folios to the *Perihermenias*. Unlike most of his predecessors, at this point he goes into the *Prior Analytics*, and explains its subject matter in 34 folios, 22 of which are devoted to the syllogism. Then he embarks on the *Posterior Analytics*, covering this in 87 folios, practically all of which are devoted to the first book; into his exposition he has inserted a brief treatise *De scientia* of some 13 folios. The concluding tract takes up the division of the sciences in 15 folios, obviously intended to complement the earlier insertion of *De scientia*. It seems doubtful that Vitelleschi covered all of this material in the year devoted to logic, for his course in natural phi-

[35] Sec. 1.1.a.

[36] Sommervogel, *Bibliothèque*, vol. 8, cols. 848-852.

[37] Biblioteca Apostolica Vaticana, Cod. Borgh. Lat. 197.

losophy, taught the following year at the Collegio, begins with the same *Tractatio de scientiarum divisione*—an indication that he probably resumed with matter he was unable to finish the previous year.

Vitelleschi's treatment of *De praecognitionibus* is very brief, and shows little similarity with the corresponding portions of Galileo's logical questions. His treatise *De demonstratione*, on the other hand, is quite full, and contains a number of parallels with Galileo's corresponding questions. Since similar parallels are also to be found in the writings of Lorinus and Valla on this subject, there seems little reason to regard Vitelleschi's notes as an independent source of Galileo's composition.

e. RUGERIUS

Ludovicus Rugerius (de Rugeriis, Ruggiero, Rugerio) taught the logic course at the Collegio in 1589-1590 and has left one of the most complete records of his teaching among all of the materials that have been preserved.[38] The notes of his *Logica* are bound in two codices, the folios of which are numbered consecutively, and each of which give indications of the numbers of lectures delivered and the dates on which the principal tracts were begun.[39] His course followed the tradition already sketched, but develops each portion of it in considerable depth. Valla's course could well have been the inspiration behind his work, for, of all the surviving notes, those of Rugerius come closest to the scope and detail of the *Logica* Valla eventually printed. The following summary, similar to that already given of Valla's work, conveys some idea of its contents:

Volume 1

Compendium of logic: begun November 3, 1589, containing 29 lectures summarized in 48 folios, treating of the three operations of the mind, methodology in general, and ways of finding middle terms.

First disputation. On the nature of logic: begun November 27, 1589, containing 30 lectures summarized in 50 folios, discussing the type of knowledge logic is, its object, and its end.

Second disputation. On universals: begun December 22, 1589, containing 32 lectures summarized in 57 folios, interpreting the *Isagoge* of Porphyry and explaining universals in general as well as the five individual types, namely, genus, species, difference, property, and accident.

[38] Lohr, "Renaissance Latin Aristotle Commentaries," *Renaissance Quarterly* 33 (1980): 705.

[39] Staatsbibliothek Bamberg, Msc. Class., Cod. 62-1 and 62-2. The remainder of his philosophy course follows in Cod. 62-3 through 62-7.

Third disputation. On the categories: begun January 25, 1590, containing 65 lectures summarized in 132 folios, explaining the *Categories* of Aristotle, with sections devoted to the antepredicaments, to the individual predicaments, namely, substance, quantity, relation, quality, and the six remaining ones, and to the postpredicaments.

Fourth disputation. On enunciations (or propositions): begun March 23, 1590, containing 23 lectures summarized in 42 folios, exposing the *Perihermenias* of Aristotle, and containing treatises on words and concepts, on the parts of propositions, on the essence and properties of propositions, and on future contingents.

Volume 2

Fifth disputation. On the syllogism: begun April 2, 1590, containing 31 lectures summarized in 61 folios, explaining the *Prior Analytics* of Aristotle, and containing tractates on the syllogism in general and its principles, on the parts of the syllogism, and on the figures of the syllogism.

Sixth disputation. On demonstration: begun May 21, 1590, containing 49 lectures summarized in 104 folios, and expounding the teaching contained in the first book of the *Posterior Analytics.* This is subdivided into three main treatises, one on foreknowledge, another on the conditions required for demonstration, and a third on the kinds of demonstration. Each treatise is divided into questions, in turn divided into *quaesita*, with the result that there are 20 questions and 81 *quaesita* (similar to Valla's chapters) in the exposition of the first book.

Seventh disputation. On definition: no date for the beginning of this, but it contains 12 lectures summarized in 26 folios and expounds the teaching of the second book of the *Posterior Analytics.* It is divided into two tractates, the first on definition as an instrument of knowing and the second on definition as the middle term of demonstration; each tractate contains three questions and cumulatively they comprise 17 *quaesita*.

Eighth disputation. On science: begun July 6, 1590, containing 33 lectures summarized in 70 folios, and made up of three tractates, 14 questions, and 41 *quaesita*; the first tractate discusses science in a broad sense, the second, how it is related to intellectual habits generally, and the third, science in the strict sense and its various kinds, including subalternated sciences.

Rugerius finished the 310th and last lecture of his logic course on August 24, 1590, which means that he had covered this number of lectures in less than ten months. If time is allowed for the Sabbath rest and brief recesses at Christmas and Easter, together with other feast days, he must have had double lectures on most days of the school year to cover this much matter.

As we will see, the late date of Rugerius's composition, which would hardly have been available in finished form until late 1590 or early 1591, effectively rules it out as an exemplar from which Galileo could have worked. Yet his extensive notes do give testimony to the continuing tradition in logic at the Collegio Romano, and particularly to its emphasis on the *Posterior Analytics* and the way in which this work influenced the understandings of science and its method then current at the Collegio.[40]

f. JONES

Robertus Jones was an English Jesuit of the same age as Galileo.[41] He entered the Society when nineteen and began teaching logic at the Collegio at twenty-seven, offering the course there in 1591-1592. His immediate predecessor was Alexander de Angelis, none of whose *reportationes* survive; before the latter, Rugerius, Vitelleschi, and Valla had all taught the same course in the years just preceding. Jones's complete course survives in a codex of 384 folios, now preserved in the Biblioteca Casanatense in Rome, and its content is somewhat similar to Rugerius's, though much abbreviated.[42] Like his predecessors, Jones begins with a general introduction to logic and the whole of philosophy, to which he devotes 56 folios. After this come expositions of Porphyry's *Introduction*, covering universals and predicables; of Aristotle's *Categories*, with sections on the antepredicaments, the individual predicaments, and the postpredicaments; and of Aristotle's *Perihermenias*, touching on the first and second acts of the mind, future contingents, and modal propositions; to these are devoted 29, 57, and 23 folios, respectively. The last two systematic treatments are those of the *Prior* and *Posterior Analytics*, accorded 22 and 124 folios each. The final 71 folios of the codex are then devoted to commentaries on selected texts in the various books of Porphyry's *Isagoge* and Aristotle's *Prior* and *Posterior Analytics*.

[40] Some particulars of Rugerius's teaching on *suppositio* and its relation to scientific reasoning are given by the author in his *Prelude to Galileo: Essays on Medieval and Sixteenth-Century Sources of Galileo's Thought*. Boston Studies in the Philosophy of Science (Dordrecht-Boston: D. Reidel Publishing Company, 1981), vol 62, pp. 158-159; consult also the index for further references to Rugerius

[41] Sommervogel, *Bibliothèque*, vol 4, Addenda, p. x; also Lohr, "Renaissance Latin Aristotle Commentaries," *Renaissance Quarterly* 30 (1977): 736

[42] Cod. 3611 (A.D. 1592).

Jones divides his systematic treatments into disputations, the disputations into parts, and the parts into questions. The materials relating to the two *Analytics*, as already noted, resemble those covered in the second volume of Rugerius's logic, with one exception. Jones's fifth disputation, like Rugerius's, is on the syllogism and the foreknown, in ten questions; and his seventh, on demonstration, in twelve questions. Thus Jones separates his exposition of foreknowledge from that of demonstration, reflecting an emphasis that is found in Lorinus, Valla, and Vitelleschi and that also shows up in Galileo's composition. Jones's eighth and ninth disputations then parallel Rugerius's seventh and eighth, considering definition and science in eight and fourteen questions, respectively.

Being even later than Rugerius's course, it is unlikely that Jones's logic notes had any direct influence on Galileo. Parallels between his teachings and those contained in MS 27, as in the cases also of Vitelleschi and Rugerius, are probably best explained in terms of the continuing tradition in logic at the Collegio Romano, similar to that in natural philosophy, which will be fully documented in the next chapter.

g. EUDAEMON-IOANNIS

Andreas Eudaemon (Eudemon), or, as he is usually referred to in the codices containing *reportationes* of his lectures at the Collegio, Andreas Eudaemon Ioannis Graecus (suggesting that he was Greek and used a double family name, Eudaemon-Ioannis), taught the logic course at the Collegio in 1596-1597, after which he continued on in the philosophy cycle, teaching natural philosophy in 1597-1598 and metaphysics in 1598-1599.[43] His logic course is preserved in the archives of the Gregorian University in Rome, and reveals that basically the same content as was covered by Toletus continued to be taught at the Collegio until the end of the sixteenth century.[44] The codex containing Eudaemon's logic has pagination, 912 pages in all, and is divided as follows:

Introduction to logic	pp. 1-93
On the object of logic	94-170
Porphyry's *Introduction*	171-331
Aristotle's *Categories*	332-536
Aristotle's *Perihermenias*	537-585
Aristotle's *Prior Analytics*	586-605
Aristotle's *Posterior Analytics*	606-912

[43] Lohr, "Renaissance Latin Aristotle Commentaries," *Renaissance Quarterly* 29 (1976): 725.

[44] Cod. APUG-FC 511.

Apart from this general similarity, however, the expositions of foreknowledge and of demonstration are not very detailed and bear little resemblance to the fuller treatments of the professors considered earlier in this section. (The same is true of Eudaemon's lectures in natural philosophy, which show little affinity with Galileo's physical questions, as will be noted later in chapter 5.[45]) He is mentioned here merely to suggest a date *post quem* for lecture notes that could almost certainly *not* have served as an exemplar for Galileo's logical questions.

2. Galileo's Questions on Logic (MS 27)

The codex containing Galileo's questions on logic, MS 27, is described by Favaro in the National Edition.[46] As already noted, some of its contents were transcribed by him, including the list of questions it contains, the text of the first question, and the text of a brief passage introducing the treatise on demonstration. The questions are obviously incomplete, for the first page of the manuscript already discusses foreknowledge and the foreknown *in particulari*, suggesting that Galileo had previously written about these topics *in generali* or *in communi*. Indeed, internal references to questions that had already been treated but are not extant in the codex give clear evidence that this is so, and that some folios of Galileo's treatises pertaining to the exposition of the *Posterior Analytics* have been lost. Similar references in Galileo's physical questions indicate that he had composed (or at least planned to compose) expositions of Aristotle's *Categories* or predicaments, and particularly those of substance, quality, and action and passion.[47] Thus it is possible that Galileo had written out, and at one time possessed, a fairly complete set of notes on logic. And although the last question in the manuscript ends half-way down the verso side of a folio, there is clear indication that Galileo intended to continue beyond the treatise on demonstration and to have another treatise on science; he mentions this explicitly in his introductory remarks to the former treatise, saying that he will have much to say about this (i.e., science) "in the following treatise."[48]

Of the materials that have survived, however, there is the major portion of a *Tractatio de precognitionibus et precognitis*, together with a *Tractatio*

[45] Sec 5.4.c.

[46] *Opere* 9:279–282

[47] See the author's English translation, *Galileo's Early Notebooks: The Physical Questions. A Translation from the Latin, with Historical and Paleographical Commentary* (Notre Dame: University of Notre Dame Press, 1977), pars A19, K157, N4, N7, S12, and Y6, with their respective commentaries

[48] The English translation of this passage is given *infra*, sec. 1.3.b.

de demonstratione in its entirety. The titles of the disputations and questions making up these treatises are listed in table 3 in Latin and in English translation; on the right are given the folios of MS 27 where the questions are found, and on the left a conventional designation that will henceforth be used to refer to the various questions.

Note that there are twenty-seven questions in all, eleven in the treatise on foreknowledge and sixteen in the treatise on demonstration. There are also some peculiarities in the composition of the manuscript that should be pointed out. The first is that this codex, like MS 46, shows numerous signs of copying, with words being crossed out and appearing elsewhere, occasional repetitions that have to be deleted, and similar indications along the lines of the paleographical commentary in *Galileo's Early Notebooks: The Physical Questions*. Another is that the size of Galileo's handwriting changes markedly throughout the composition of the manuscript, with his hand being large and flowing from LQ2.1 through 6.4, but decreasing in size with each succeeding question; then getting pronouncedly smaller from LQ6.5 through 6.11; and finally becoming extremely small and cramped for the last disputation, consisting of LQ7.1 through 7.3. There are also a number of omissions, three in the titles of the disputations and four in the numbering of the questions. These are possible clues suggesting that Galileo was not making a slavish copy of his exemplar, but rather was attempting to rearrange materials under alternative headings. Apparently he lost track of his enumeration of questions between LQ3.2 and 3.3, because no third question is indicated for the third disputation of the first treatise. Again, questions LQ6.5 through 6.8 have not been numbered, although LQ6.4 is indicated as the fourth question and LQ6.9 as the tenth, whereas in the expected numeration LQ6.9 should have been ninth. Galileo's confusion here might have been caused by the fact that two intervening questions, LQ6.6 and 6.8, actually are double questions, and either could have been counted twice by him when attempting to arrive at the numbering of LQ6.9.

If one searches diligently through all of the corresponding questions in the authors surveyed in the previous section, one can see that there are many possibilities of derivation of Galileo's teaching in the logical questions from the various redactions of the logic course at the Collegio Romano. So great are the similarities, however, between Galileo's notes and the materials plagiarized by Carbone from Valla's lectures of 1587-1588, that it does not seem necessary to canvass all these possibilities. In what follows attention will be focused on the more obvious similarities, for these, when taken in conjunction with the information already presented, should serve to establish the provenance and dating of the logical questions with a high degree of accuracy.

TABLE 3
GALILEO'S LOGICAL QUESTIONS (MS 27)

No.	Question	MS 27 folios
	[*Tractatio de praecognitionibus et praecognitis*. (Treatise on foreknowledge and the foreknown.)]*	—
	[*Disputatio prima. De praecognitionibus et praecognitis in communi*. (On foreknowledge and the foreknown in general.)]	—
	De praecognitionibus et praecognitis in particulari. (On foreknowledge and the foreknown in detail.)	4r
	Disputatio secunda. De praecognitionibus principiorum. (On the foreknowledge of principles.)	4r
LQ2.1	Quaestio prima. An in omnibus principiis praecognoscendum sit quaestio an sit? (Whether the question *an sit* must be foreknown for all principles, i.e., whether such principles must be foreknown to be true?)	4r
LQ2.2	Quaestio secunda. An de primis principiis praecognoscendum sit quid nominis? (Whether the meanings of the terms in first principles must be foreknown?)	4v
LQ2.3	Quaestio tertia. An principia sint actualiter vel habitualiter praecognoscenda? (Should principles be foreknown actually or habitually?)	5v
LQ2.4	Quarta quaestio. An principia in scientiis sint ita nota ut nulla ratione probari possunt? (Should the principles of sciences be so evident that they cannot be proved by any reasoning?)	6r
	Tertia disputatio. De praecognitionibus [*subiecti*]. (On foreknowledge [of the subject].)	6v
LQ3.1	Quaestio prima. Quid intelligat Aristoteles nomine esse quando dicit de subiecto debere praecognosci an sit? (What does Aristotle mean by the term existence when he says that the existence of the subject must be foreknown?)	6v
LQ3.2	Quaestio secunda. An scientia possit demonstrare de suo obiecto adaequato esse existentiae? (Can a science demonstrate the *esse existentiae* of its adequate subject?)	8r

* Expressions enclosed in brackets do not appear in the manuscript but have been reconstructed and inserted by the author on the basis of internal evidence (references or indications elsewhere in the manuscript) or of external evidence (similar expressions in sources on which the manuscript may have been based).

Expressions enclosed in parentheses are the author's translations of Galileo's Latin, made for the convenience of the reader.

TABLE 3 (*cont.*)

LQ3.3	Quaestio quarta [sic]. An scientia possit demonstrare an sit subiecti scientiae partialis? (Can a science demonstrate the existence of the partial subject of a science?)	9r
LQ3.4	Quaestio quinta. An scientia possit ostendere quid rei sui subiecti et reddere propter quid illius? (Can a science manifest the quiddity of its subject and give the *propter quid* of it?)	10r
LQ3.5	Quaestio ultima. Quid intelligat Aristoteles per praecognitionem quid est, quando de subiecto dicit praecognoscendum esse quid est quod dicitur? (What does Aristotle mean by the foreknowledge of the quiddity when he says that the quiddity of the subject must be foreknown?)	10v
	Disputatio ultima. De praecognitionibus passionis et conclusionis. (On the foreknowledge of the property and the conclusion.)	11r
LQ4.1	Quaestio prima. An de passione praecognoscendum sit quia est? (Must the existence of the property be foreknown?)	11r
LQ4.2	Quaestio secunda. An conclusio cognoscatur simul tempore et natura cum cognitione praemissarum? (Is the conclusion known at the same time and in the same order as the knowledge of the premises?)	11v
	Tractatio de demonstratione. (Treatise on demonstration.)	13r
	Disputatio prima. [*De natura et praestantia demonstrationis.* (On the nature and importance of demonstration.)]	13r
LQ5.1	Quaestio prima. De definitione demonstrationis. (On the definition of demonstration.)	13r
LQ5.2	Quaestio secunda. An demonstratio sit nobilissimum omnium instrumentorum, vel definitio? (Is demonstration the most noble of all instruments [of knowing], or is definition?)	14r
	Disputatio secunda. [*De proprietatibus demonstrationis.* (On the properties of demonstration.)]	17v
LQ6.1	Quaestio prima. An demonstratio constet ex veris? (Must demonstration consist of true premises?)	17v
LQ6.2	Quaestio secunda. An demonstratio debeat constare ex primis et prioribus? (Must demonstration be made from premises that are first and prior?)	18v
LQ6.3	Quaestio tertia. Quid intelligat Aristoteles per propositiones immediatas quando docet demonstrationem debere constare ex illis? (What does Aristotle mean by immediate propositions when he says that demonstration must be made from them?)	19v
LQ6.4	Quaestio quarta. An omnis demonstratio constet ex immediatis, et quomodo? (Is every demonstration made from immediate premises, and how?)	20v

TABLE 3 (*cont.*)

No.	Question	MS 27 folios
LQ6.5	[Quaestio quinta.] An omnia principia immediata per se nota ingrediantur quamcumque demonstrationem? (Do all immediate principles that are self-evident enter into every demonstration?)	21r
LQ6.6	[Quaestio sexta.] An demonstratio constet ex notioribus, et an cognitio praemissarum sit maior et perfectior cognitione conclusionis? (Is demonstration made from premises that are more known, and is knowledge of the premises better and more perfect than that of the conclusion?)	22r
LQ6.7	Quaestio [septima.] An demonstratio debeat constare ex propositionibus necessariis et de omni et quomodo? (Must demonstration be made from propositions that are necessary and universal, and how?)	23v
LQ6.8	[Quaestio octava.] Quot sint regulae cognoscendarum propositionum quae in primo et secundo modo continentur, et an sint plures quam duo praedicandi modi? (What are the rules for knowing propositions that are contained in the first and second modes, and are there more than two modes of predicating?)	26v
LQ6.9	Quaestio decima. Qui sint modi demonstrationi inservientes? (What are the modes that function in demonstration?)	27r
LQ6.10	Quaestio undecima. Quid sit praedicatum universale, et quae propositiones sub illo contineantur? (What is a universal predicate, and what propositions are contained under it?)	27v
LQ6.11	Quaestio ultima. An perfecta demonstratio debeat constare ex propositionibus per se universalibus et propriis? (Must perfect demonstration be made from propositions that are per se, universal, and proper?)	28r
	Tertia disputatio. De speciebus demonstrationis. (On the kinds of demonstration.)	29r
LQ7.1	Quaestio prima. Quot sint species demonstrationis? (How many kinds of demonstration are there?)	29r
LQ7.2	Quaestio secunda. In quo conveniant et differant demonstratio propter quid et quia, et de huius divisione. (How are demonstrations *propter quid* and *quia* the same and how do they differ, and concerning this division?)	30v
LQ7.3	Quaestio tertia. An detur regressus demonstrativus? (Is there such a thing as a demonstrative regress?)	31r
	[Finis]	31v

3. Sources and Dating

The extant disputations of Galileo's treatise *De precognitionibus et precognitis* are three in number, being concerned, respectively, with foreknowledge of principles (LQ2.1-2.4), with foreknowledge of the subject (LQ3.1-3.5), and with foreknowledge of the property and the conclusion (LQ4.1-4.2). All three sets of questions show close agreement with corresponding portions of Carbone's treatise on foreknowledge in the *Additamenta*, and thus, on the basis of Valla's charge, with *reportationes* of the latter's logic course taught at the Collegio in 1588. In what follows, the identification of Carbone with Valla will be signalled by designating the text as that of Valla-Carbone and by assigning it to the year 1588 rather than 1597, the date of publication of Carbone's *Additamenta*.

a. TREATISE ON FOREKNOWLEDGE

The Latin text of Galileo's first question (LQ2.1) is transcribed in the National Edition, and this corresponds to chapter 8 of Carbone's treatise.[49] Galileo raises the question whether the *an sit*, that is, the existence, of all principles that enter into a demonstration must be foreknown, which is the scholastic way of inquiring whether the truth of such principles must be manifest, in the sense that a principle exists only if it is known to be true. (Carbone states the question more directly by inquiring whether all principles that enter into a demonstration must be true.) It seems not so for Galileo, since one can know a conclusion from principles that are immediate or proper for the subject matter without knowing first principles, and thus one need not know *all* principles before demonstrating; again, particular sciences have perfect knowledge of their conclusions without claiming knowledge of first principles, and thus *all* principles need not be foreknown. Despite these arguments, Galileo proposes two conclusions that effectively reply to the question in the affirmative: (1) first principles must be foreknown *in some way*, for otherwise knowledge of the conclusion would be unattainable; and (2) proximate and immediate principles must be foreknown also. He then explains how first principles can be known, and says that this is in a variety of ways: (a) some are grasped immediately from their terms; (b) others are gotten by sense alone; (c) yet others are known by induction, division, and hypothetical syllogism; (d) still others by experience; and (e) some only from practice or habituation. The examples Galileo gives for each are the following: (a) the whole is greater than the part; (b) fire is hot; (c)

[49] *Opere* 9:291-292; *Additamenta*, fol. 42ra-va.

all teaching is based on pre-existent knowledge; (d) the medical adage that contraries are cured by contraries; and (e) the principles of moral science, which cannot be understood unless they are practiced. He replies, finally, to the objections by saying that his answer to the first difficulty is apparent from the first conclusion, while that to the second is twofold. Particular sciences do not usually claim knowledge of first principles, not that such principles are not necessary, but rather because they are presupposed as obvious and the one learning the sciences assents to them immediately; and first principles considered in a general way pertain to metaphysics, but when applied to this or that subject matter, they pertain to the particular sciences.

All of this information is contained in Carbone, and also in Valla's *Logica* of 1622, although arranged somewhat differently and explained at greater length.[50] Galileo's conclusions, for example, are given in the reverse order of Carbone's, but in terms that are very similar, as can be seen from the following parallel texts:

VALLA-CARBONE (1588)	GALILEO
{Secunda *positio, de primis principiis oportet praecognoscere aliquo modo* quod vera sint. *Quia non potest perfecte cognosci conclusio, nisi aliquo pacto cognoscantur principia regulantia illius*, cuiusmodi sunt *prima* scientiae *principia. Quare, ut exacta habeatur cognitio, aliqua ratione* talia principia cognoscenda erunt.}	Prima *conclusio. Prima principia sunt aliquo modo praecognoscenda ad hoc ut perfecte cognoscatur* ipsa conclusio, *quia non potest perfecte cognosci conclusio nisi cognoscantur* omnia *principia regulativa illius* a quibus *aliquo modo* pendet; sed pendet aliquo modo a *primis principiis*; ergo. . . .
Prima *positio, de principiis proximis* demonstrationis et de praemissis *praecognoscendum est* quod vera sint. [M: *primo Posteriorum capite secundo*] Primo probatur ex communi sententia et *ex Aristotele, qui* hoc expresse docet, immo *ait* esse magis cognoscenda quod vera sint quam conclusio.	Secunda *conclusio. Principia proxima* et immediate necessario *praecognoscenda sunt* ad hoc ut sciatur conclusio. *Quia Aristoteles, primo Posteriorum capite secundo, ait*
Deinde, conclusio cognoscitur ob principia proxima: *ergo* multo *magis ipsa debet esse cognita*, iuxta illud axioma, *Propter quod unumquodque tale, et illud magis.*[51]	*propter unumquodque tale, et illud magis; sed conclusio scitur propter principia; ergo ipsa* principia *magis scienda sunt.*[52]

Words that are common or more or less synonymous in the two passages have been italicized. The inversion of the order of the conclusions in

[50] These and other textual parallels are indicated in table 4 *infra*.

[51] *Additamenta*, fol. 42rb.

[52] MS 27, fol. 4r9–18. *Secunda conclusio* is a marginal notation between lines 13 and 14 of the folio.

Carbone's text is signalled by enclosing the rearranged matter in braces {}—a convention used throughout these textual comparisons; another convention, seen at lines 11-12, is that of enclosing marginal notations in square brackets and preceding them by a capital M, viz. [M: . . .]. Galileo's ordering probably reflects that of Valla's original, but his text is likely abbreviated from Valla's, which one would expect to be preserved more fully in Carbone's version. All of Galileo's examples, moreover, are found in Carbone, though these are given in the latter's chapters 7 and 10, where fuller explanations are provided of the different kinds of first principles and of how knowledge of each can be attained.

A similar example of textual parallels may be seen in Galileo's reply to the second question of this disputation, where he is inquiring whether the meanings of the terms that occur in first principles must be foreknown. Here he is extremely brief, explaining only the sense of the question and providing a clarificatory note, and then giving an affirmative answer with a short proof. Carbone, on the other hand, has a lengthy discussion of this problem in chapter 7 of his treatise, providing a number of divisions and definitions, six different opinions on the matter, and eight conclusions with their respective proofs. On inspection, however, one finds that Carbone's sixth conclusion is practically the equivalent of Galileo's position, as can be seen in the following parallel:

Valla-Carbone (1588)	Galileo
Sexta *positio*, ante quamlibet demonstratio *praecognoscere oportet quid nominis* terminorum qui sunt *in praemissis*. Primum *quia impossibile est praecognoscere eas esse veras, nisi cognoscatur quid significent termini* ex quibus componuntur. *Secundo ex* communi doctrina *Themistii, Philoponi, Averrois*, divi Thomae, et ipsiusmet *Aristotelis, qui hoc asserit.* [M: Cap. 4 & 10, Com. II.33, *primo Posteriorum tex. 14*][53]	Hoc posito, *dico: de principiis* non solum *debere praecognosci* an sit, sed etiam *quid nominis. Probatur ex Aristotele, primo Posteriorum tex.* 5 et *14, ubi hoc* expresse *docet. Secundo* ratione, *quia cognitio an sit supponit praecognitionem quid nominis*; ergo, necessario ad hoc *ut cognoscamus an* prima principia *sint vera* debemus praecognoscere *quid significant*. Hoc etiam in illum tex. 14 docet et *Averroes* et *Philoponus* et *Themistius*.[54]

Here, as previously, Carbone has reversed the order of the two arguments, first giving the proof from reason and then citing authorities in its support, whereas Galileo has done the opposite and abbreviated the references. Otherwise both excerpts register close agreement and suggest derivation from a common source.

Galileo's next disputation, numbered by him as his third, is concerned with how much one can know about the subject whose properties are

[53] *Additamenta*, fol. 41vb.

[54] MS 27, fol. 5r2-8.

being proved before one can truly demonstrate them (LQ3.1–3.5). This, like the foregoing disputation, is quite relevant to his later attempts to construct a science whose subject would be local motions of particular types, and will be analyzed more fully in subsequent chapters.[55] For purposes of future reference, and in order to provide an extended passage to supplement the excerpts already given, the entire text of Galileo's first question (LQ3.1) is given below. Accompanying it in parallel column are selections taken from the more extensive treatment in Carbone, which is spread over three consecutive chapters. Galileo's question inquires about the meaning of the term *esse* when one says that the *an sit* (i.e., the existence, literally, whether it is) of the subject must be foreknown. Carbone's chapters address the same query, but in successive stages: he first raises a number of difficulties concerning the foreknowledge of the subject (Cap. 14. De quibusdam communibus dubiis circa praecognitiones subiecti); then asks Galileo's question about the foreknowledge of its existence (Cap. 15. Quid intelligat Aristoteles nomine esse cum ait illud de subiecto praecognoscendum; vel an de subiecto praecognoscendum sit esse essentiae vel existentiae), thus specifying two meanings for *esse*, viz., *esse essentiae* or *esse existentiae*; and finally offering his solutions to the difficulties raised (Cap. 16. De quibusdam dubitationibus circa ea quae proximo capite dicta sunt). The respective texts read as follows:

VALLA-CARBONE	GALILEO
Prima dubitatio, quid nomine *subiecti* intelligatur. Respondeo, hoc nomen *variis modis sumi*. Primo . . . Nono et ultimo, *pro ea re de qua demonstrantur passiones* in aliqua scientia: et hoc modo sumitur in praesenti tractatione.	Notandum est, *primo: subiectum multipliciter posse accipi* a nobis, autem in presenti sumi *pro illo de quo demonstrantur* aliquae *passiones*;
Secunda dubitatio, ad quam scientiam pertineat agere de subiecto . . . Primum dictum, agere de subiecto reali *universe sumpto pertinet ad metaphysicum*. . . . Secundum dictum, agere de subiecto reali *in particulari pertinet ad scientias* reales *particulares* . . .	de quo quidam *in communi sumpto tractare ad metaphysicum*, *in particulari* vero de hoc vel illo *ad scientias particulares spectat*.
. . . *animadvertendum est* philosophos *duo assignare esse: alterum* quod vocant *essentiae* sive naturae, *alterum existentiae*. . . . *Esse existentiae . . . duplex est, alterum actuale . . . alterum potentiale* . . .	*Notandum est* secundo: *esse duplex esse, alter essentiae, alter existentiae*; praeterea *esse existentiae duplex esse, vel actuale, vel potentiale*.
Sunt autem *duo in quibus omnes conveniunt: primum, in acquisitione scientiae opus fuisse subiectum esse in rerum natura; quia* nostra	Notandum est tertio: *authores* in hac quaestione *in duobus convenire. Primo: in principio acquisitionis scientiae fuisse necessariam*

[55] Secs. 3.1 b, 3.4.a, 5.2.a–c, 5.3 b, 5.4.d–e, 6 3.d–f, and 6.4 a.

scientia habet ortum a sensu, qui solum versatur circa ea quae actu existunt; secundum, in ipso etiam *scientiae progressu praecognoscendum esse* aliquo modo *ipsum subiectum esse. Sed dubium* est: *quid nomine esse intelligendum sit, et quid Aristoteles* intelligat cum *ait*, de subiecto praecognoscendum esse, quia est . . .

actualem existentiam rei; cuius ratio est, *quia*, cum *omnis* nova *cognitio ortum habeat ex sensu, qui versat tantum circa existentiam*, sequitur etc. *Secundo: in progressu scientiae esse praecognoscendum de subiecto esse. Differre autem quid nomine* huius *esse* secundi *intelligendum sit, de quo loquitur Aristoteles.*

De qua re hae sunt scriptorum sententiae. *Prima eorum qui dicunt esse praecognoscendum* esse: *esse* existentiae tantum. *Quoniam non possunt demonstrari passiones de subiecto nisi* supponamus *illud habere* certam et *determinatam* existentiam: igitur tale esse supponendum erit. Confirmatur, *quia ut res se habet ad esse, ita ad cognitionem: sed prius est rem* existere in se, quam habere ordinem ad passiones quae de ea cognoscuntur; ergo.

Prima opinio est *quorundam, qui dicunt esse praecognoscendum* de subiecto *esse* essentiae: {tum *quia non possunt demonstrari* aliquae *passiones de subiecto* aliquo, *nisi illud habeat determinatam* quidditatem}; tum *quia, sicut res se habet ad esse, ita ad cognosci*, ex Aristotele; sed *prius est rem* habere suam essentiam quam existentiam; ergo.

Secunda opinio aliorum, qui dicunt solum *praecognoscendum esse, esse potentiale*, hoc est, rem esse in suis causis. *Quoniam scientia* solum *probat passiones posse esse in subiecto*; quare *satis* erit *praecognoscere* subiectum *posse existere.*

Secunda opinio est *aliorum, qui volunt* de subiecto *esse praecognoscendum esse* existentiae *potentiale*, non actuale. Fundamentum est: *quia scientiae demonstrant passiones posse* tantummodo *inesse sui subiecti*; ergo esse tantummodo *existentiae potentiale* erit *praecognoscendum*, cum illud *sufficiat.*

Tertia . . . Quarta . . . Quinta . . .

Sexta *Averrois* et *Porphyrii* [M: *Libr. 1 Poster. com. 2*], referente Alberto, *Caietani et aliorum, qui dicunt de subiecto praecognoscendam esse actualem existentiam* in scientiis realibus, non autem in scientiis rationalibus. Antequam huic dubitationi satisfacio, unum et alterum subjicio fundamentum.

Tertia opinio est *Averrois* in *primo Posteriorum commento secundo, Porfiri, Caietani* in primum caput primi libri Posteriorum, quaes. 2, *et* multorum *recentiorum, dicentium esse praecognoscendum de subiecto* esse *existentiae actuale.*

Primum fundamentum: *tria sunt quae hic difficultatem faciunt: primum, an de subiecto semper praecognoscendum sit esse existentiae,* et quomodo; *cum* videamus *de multis esse scientiam quae non semper existunt. Secundum, cur* de aliqua re *non sufficiat tantum praecognoscere esse essentiae. Tertium, an aliqua possit esse demonstratio de subiecto*, cuius *nullum esse praesupponatur*. . . . His ita notatis, ad positiones accedo.

Animadvertendum est, *tria esse quae* quaestionem *hanc perdifficilem reddunt: primo, an de subiecto semper praecognoscendum sit esse existentiae* actuale, *quia multa sciuntur* a nobis semper, *quae* tamen *non semper existant; secundo, quare non sufficiat praecognoscere esse essentiae tantummodo* de subiecto; *tertio, quare in aliquibus demonstrationibus non* sit necessarium *praecognoscere an sit subiecti*. Quae omnia sequentibus questionibus declarabo.

Prima positio, de subiecto praecognoscendum est esse essentiae antequam passio de illo

Prima conclusio: esse essentiae praecognoscendum est de subiecto, ita ut, probetur. . . . Confirmatur . . . Secundo, *quia* hoc demonstratur in conclusione, sed *non potest demonstrari nisi* de subiecto *praecognoscamus esse essentiae*, ergo. . . .

nisi praecognoscatur, nulla potest haberi demonstratio. Probatur ex argumentis primae opinionis.

Secunda positio, in scientia reali debet praecognosci esse existentiae actuale, suis saltem locis et temporibus, remotis impedimentis. Dicimus 'suis locis et temporibus' quia multae sunt naturae quae non habent individua semper actu existentia, et ideo de his satis est praecognoscere quod suis temporibus existant. Primo, *quaestio an sit supponitur ab omnibus aliis* quaestionibus . . . ; *ergo hoc esse praecognoscendum est antequam* aliquid de subiecto probatur. Secundo, *illud esse de subiecto est praecognoscendum ratione cuius substat passionibus: at* subiectum non *substat suis passionibus* nisi per esse *existentiae actuale; ergo*. Tertio . . . Quarto . . . secundum [de subiecto praecognoscendum est esse potentiale] dici non potest, *quia* pari ratione de passione praecognosceremus esse, quod est *contra Aristotelem: nemo* enim *est qui aliquam passionem demonstret de subiecto* aliquo *qui non* praecognoscat et *supponat talem passionem esse possibilem*. Quinto, *omnis nostra cognitio est de* rebus *existentibus*, quamvis non sit de eis prout existunt; *ergo* ante scientiam *de subiecto noscendum erit hoc esse*. . . . Denique . . .

Secunda conclusio: in scientiis realibus praecognoscendum est esse existentiae actuale [M: *saltem suis locis et temporibus, remotis* suis *impedimentis*] de subiecto demonstrationis in quo vel passio ostenditur de illo vel aliquid aliud praedictatum

Probatur haec conclusio: quia *scientiae humanae sunt de existentibus; ergo de obiecto* illarum *debet praecognosci an sit*, ne fictitiae sint. Secundo, ex Porphyrio, quia *illud esse de subiecto praecognoscendum est ratione cuius substat* suis propriis *passionibus; sed substat suis* propriis *passionibus actualis existentiae; ergo*.

Tertio: quia *quaestio an sit est prima inter omnes*, requirit existentiam actualem rei; *ergo, ante* quaestionem qualis sit, necessario *praecognoscenda est an sit*. Quarto et ultimo: *quia* alias sequeretur *contra Aristotelem* nihil plus debere praecognosci de subiecto quam de passione, quia *nemo est qui, volens probare passionem de subiecto, non supponat illam posse inesse illi*.

Tertia positio, non est opus ante omnem demonstrationem praecognoscere de subiecto esse actuale existentiae. . . . Primo . . . Secundo, *in aliquibus demonstrationibus probatur subiecti existentia, ut* cum probatur existentia causae per effectum, *ut Deum et materiam esse*: igitur non supponitur esse.

Tertia conclusio: in scientiis non semper de subiecto praecognoscendam esse existentiam actualem. Patet *in illis demonstrationibus in quibus ostenditur existentia inesse subiecto: ut* videre est in illis in quibus aut *materia* prima *existere* per transmutationem *aut primum motorem dari* ex motus eternitate probatur.

Sed *diceres: cur ergo Aristoteles dixit de subiecto praecognoscendum esse quia est. Respondeo: eius dictum intelligendum esse de subiecto demonstrationis potissimae* qua probatur propter quid passio insit subiecto; aut *indefinite, hoc sensu, de aliquo subiecto praecognoscendum est, an sit*.

Dices: quare igitur Aristoteles dixit de subiecto debere praecognosci an sit. Respondeo: Aristotelem locutum esse de subiecto propriae demonstrationis. Secundo: Aristotelem locutum esse *indefinite; in hoc sensu de aliquo subiecto praecognoscendum est in sit*, non tamen de omni.

Quarta positio . . . Quinto positio, *in scientiis rationalibus, ut in logica, non est necesse praecognoscere de subiecto esse existentiae ac-*

Sit *quarta* et ultima *conclusio: in scientiis rationalibus*, si quae dantur, *qualis est logica* secundum multos, *de subiecto non est prae-*

tualis, sed solum esse obiectivum in intellectu. Ratio est, *quia hoc ens non habet aliud esse.*

cognoscendum esse existentiae actualis, sed tantummodo esse obiectivum, quia illud esse de subiecto praecognoscendum est *quod habet ipsum subiectum*; sed esse subiecti disciplinarium rationalium est esse *in intellectu*; ergo.

Ut adhuc ea quae docuimus magis clara fiant, *nonnullae dubitationes* nobis enucleandae sunt. Prima dubitatio . . .

Secunda dubitatio, *si Deus corrumperet omnes species*, solo homine relicto; *adhuc homo haberet earum scientiam, et tamen non* haberent *existentiam* actualem vel possibilem . . . *Respondeo*, ideo *homo haberet* rerum destructarum *scientiam* . . .Adde etiam, quia *si tolleretur impedimentum voluntatis divinae* possent actualiter existere. unde non ab: strahunt ab omni existentia, et *ita* de illis *praecognosci posset esse*, sublatis impedimentis, *et hoc satis est*. Tertia dubitatio . . .

Obiicit . . . secunda opinio:
{Obiicitur quarto. *si Deus destrueret omnes species* corporeas, *daretur adhuc scientia de illis, et tamen non* posset praesupponi *existentia* illarum; ergo. *Respondeo*: hoc dato, *dari scientiam*;

tamen necesse *esse praecognosci* tunc temporis an sit subiectorum *remoto* tamen *impedimento voluntatis divinae*;

quod satis est.}

Quarta dubitatio, *existentia est propria individuorum* et singularium; *atqui singularium*, cum sint varietati subiecta, *non habetur scientia* . . . *Respondeo, hanc vel illam existentiam convenire particularibus* . . . Cum ergo species conserventur in individuis, necessario dabitur aliquod individuum in quo existant. Praecognoscitur itaque *natura universalis, non in abstracto sed in aliquo individuo* singulari, *nullo* tamen *determinate assignato*, quod quidem scientiis minime repugnat. Quo *loco animadverte, si finem naturae spectemus*, potius *existentia* est propter *speciem quam* propter *individuum; quia natura intendit conservare individua causa specierum*. Quinta dubitatio

Obiicit tertio: *existentia est propria individuorum; at scientia non considerat individua*, ergo, neque existentiam. *Respondeo: hanc vel illam existentiam* in particulari *esse propriam individuorum*, existentiam tamen, quae sequitur *naturam universalem non* quidem *in abstracto sed in aliquo individuo indefinite*, esse propriam speciei.

Hic autem nota: existentiam speciei magis intendi a natura quam individuorum, quia natura magis perficiatur speciebus quam individuis

Sexta dubitatio, scientia non *demonstrat passiones* actu esse in *subiecto*, sed *tantum posse inesse* . . . *ergo* . . . *Confirmatur*, quod de aliquo subiecto *demonstratur* est passio in natura *necessaria: at huiusmodi propositio abstrahit a* quacunque *existentia: igitur. Respondeo, esto scientam demonstrare tantum passiones posse inesse subiecto; tamen quia illas probat de subiecto reali* et vere *existentiali, ideo de subiecto praecognoscendum est esse*. Adde etiam, quod *scientiae frequenter demonstrant passiones vere* et realiter *inesse subiectis*, saltem suo determinato tempore. *Ad confirmationem*, tales *propositiones dupliciter considerari possunt: primo ut dicunt* solam

Obiicitur, sexto: scientiae ostendunt tantummodo proprietates posse inesse subiectis; ergo, etc. *Confirmatur*: quia scientiae tantum *probant* aliquas propositiones *necessarias* esse; *sed huiusmodi propositiones abstrahunt* ab esse *existentiae; ergo*, et ipsae scientiae. *Respondeo: esto scientiae ostendant tantummodo proprietates posse inesse subiectis; tamen, quia ostendunt posse inesse subiectis realibus, ideo de illis praecognoscendum est esse existentiae.* Respondeo, secundo: *scientias ut plurimum probare passiones actu inesse* suis *subiectis. Ad confirmationem* respondeo: *propositionem dupliciter posse sumi: vel secundum connexionem terminorum, et sic abstrahit ab* omni *existen-*

connexionem terminorum, et secundum hanc considerationem abstrahunt ab existentia; secundo ut in illis denotatur praedicatum vere et realiter *inesse subiecto, et sub hac consideratione negarem abstrahere ab existentia* saltem communi.

tia; vel secundum quod dicit praedicatum inesse

subiecto, et sic nego abstrahere existentiam.

Septima *dubitatio, mathematica* est scientia realis quae *cum abstrahat ab esse non potest illud supponere . . . Respondeo, mathematica abstrahit ab esse in demonstrandis passionibus* de subiecto; *non tamen* abstrahit *a praecognitione existentiae subiecti. . . cum* omnes *scientiae* reales *versentur circa* aliquod obiectum *actu existens . .*

Obiicitur, ultimo: *mathematica, cum abstrahat ab ente* et bono, *non supponit esse existentiae* de suo obiecto. *Respondeo: mathematicam in demonstrandis passionibus abstrahere ab* omni *existentia, praecognoscere tamen esse* sui *subiecti*, quia *cum* sit *scientia* humana *versatur circa existentia.*

Octava *dubitatio*, omnes *scientiae abstrahunt ab existentia: igitur non praecognoscunt illam de suis subiectis. Respondeo, scientias abstrahere ab esse existentiae si spectemus rationem formalem ipsarum:* quia *cum versentur circa universalia* formaliter *non possunt considerare* subiectum, *ut* formaliter *existit: sed si consideremus conditionem sine qua non* ipsius subiecti, *nego abstrahere ab existentia.*

{*Obiicit* primo . . . : *scientiae abstrahunt ab existentia; ergo non* poterunt *praecognoscere existentiam suorum subiectorum. Respondeo: si spectemus rationem formalem scientiarum, illas* quidem *abstrahere ab existentia* subiectorum: *cum enim considerent universalia, non possunt illa ut existentia cognoscere; si autem attendamus conditionem sine qua non, nego* illas *abstrahere ab existentia.*

Nona *dubitatio, si Deus nullam* omnino *speciem creasset, nec facere decrevisset*, adhuc *saltem Angeli possent habere scientiam proprietatum* illarum specierum, et tamen nulla ratione de illis posset praecognosci esse subiecti; . . *Respondeo primo, hic esse sermonem de scientiis humanis*, quae versantur circa res existentes. *Secundo, tali posito casu, Angeli non cognoscerent proprietates esse re in subiectis, . . . sed* tantum cognoscerent *posse inesse, sicuti* ipsa species *possibiles sunt*

Obiicitur quinto: *etiamsi Deus nihil procreasset neque procreare decrevisset* quidquam, *Angeli tamen possent cognoscere proprietates* aliquas corporeae naturae convenientes; ergo.
Respondeo, primo: hic nos loqui de scientiis humanis.
Secundo, respondeo: *quod tunc Angeli cognoscerent* illas *proprietates posse inesse subiectis, non tamen inesse actu, sicuti* talis etiam natura *possibilis est.*

Decima *dubitatio, res esse, est contingens, . . at scientiae non versantur circa contingentia . . . Respondeo, hanc vel illam existentiam particularem esse contingentem*, non tamen *existentiam speciei* universe consideratam. Nam *supposito universo, necesse est* species *saltem suo tempore existere.*[56]

Obiicit secundo. *rem existere est contingens; sed contingentia scientiae reiciunt*; ergo. Respondeo: duplicem esse existentiam; *alteram particularem huius vel illius individui*, et hanc *esse contingentem; alteram speciei*, et haec consideratur a scientia, quae *supposito universo necessaria est saltem suis temporibus.*}[57]

[56] *Additamenta*, fols. 45vb-48ra.

[57] MS 27, fols. 6v25-8r18. The marginal addition in Galileo's manuscript shown on the right at lines 72-74 above is marked by a caret to be inserted after "de subiecto demonstrationis" on line 74; it has been moved by the author to show the agreement of the

A comparison of lines 1-27 shows that Galileo begins with three notations, the first of which (lines 1-12) correspond to the first two difficulties contained in Carbone's chapter 14, and the second and third of which (lines 13-27) abbreviate two animadversions with which Carbone begins chapter 15. Following this comes a listing of *sententiae* or opinions by both authors (lines 29-53), with Carbone enumerating six in all and Galileo reporting only three. Noteworthy here is a difference in the respective expositions of the first opinion (lines 30-38), for Galileo explains this as favoring *esse essentiae* for the meaning of *esse*, whereas Carbone, in a sentence whose Latin leaves something to be desired (lines 30-31), favors *esse existentiae*. It is difficult to judge in this case which text more faithfully reflects Valla's original, since neither identifies those who hold the opinion: for Carbone it is merely *eorum qui dicunt* . . . and for Galileo, the equivalent *quorundam qui dicunt*. . . . Following the opinions, Carbone has two *fundamenta*, but again Galileo employs only one of these, the first, which he proposes as an *animadversio* (lines 54-64), but whose wording is strikingly like that of Carbone's first *fundamentum*.

Lines 65-119 give the statements, five for Carbone and four for Galileo, in which both writers couch their reply to the question asked; Carbone identifies these as *positiones* and Galileo as *conclusiones*, but otherwise their contents are almost identical. The first statement is that the *esse essentiae* of the subject, that is, its nature or definition, must be foreknown for there to be demonstration (lines 65-70). Carbone's proof is much longer than Galileo's (much of it is omitted above) and makes no reference to arguments in support of the first opinion, which Galileo simply invokes. The second conclusion for both builds on this first, and maintains that in the case of the real sciences, as opposed to the rational or logical sciences, the actual existence of the subject must be known in addition. Galileo's manuscript contains a marginal addition at this place, signalled by [M: . . .], and his insertion is found almost exactly in Carbone's text (lines 72-73)—suggesting that the words were found in Galileo's source, which he thought to omit in his abbreviation and then found necessary for the passage's sense. Carbone even explains the reason for adding the qualification—and this could have alerted Galileo to its importance—since it is possible with natural things that they not be *always* existent, but only at certain places and times, and when impediments that might prevent their existence from occuring have been removed. Galileo's recognition of the significance of *impedimenta* in this context is

passage with the text on the left. Similarly, Galileo's treatment of objections, beginning at line 121 of the parallel passages, has been rearranged to agree with the ordering as found in the *Additamenta*; sections that have been rearranged are enclosed in braces The content of Galileo's teaching is summarized in sec. 3.1.b *infra*.

extremely important, as will be seen later, because of his repeated use of this term when attempting to justify his "new science" of motion.[58] Carbone further offers six arguments in support of his assertion, whereas Galileo reduces these to four. The correspondence can be verified, however, for Galileo's first proof (lines 79-81) abbreviates Carbone's fifth (lines 92-96), his second parallels Carbone's second (lines 81-85), his third (lines 88-91) is equivalent to Carbone's first (lines 78-81), and his fourth (lines 91-96) again parallels Carbone's fourth (lines 85-92). This is probably another case where Galileo followed the original ordering for Valla's arguments, whereas Carbone rearranged them to suit his own purposes.

The third conclusion is then very similar in the two authors, since both point out that the actual existence of a subject is not necessary for every demonstration, for sometimes the subject's existence can itself be demonstrated; again, both respond to a tacit objection drawn from the text of Aristotle (Carbone has *diceres* and Galileo the simpler *dices*) and give the same two replies to it (lines 97-110). Finally, Galileo's fourth and last conclusion, which is Carbone's fifth, makes explicit an exception for the actual existence of logical entities when these are the subject of demonstration (lines 111-119)—an exception both had hinted at in restricting their second conclusions to the "real sciences."

The resolution of difficulties that follows these conclusions disposes of ten in Carbone's version, which he puts in chapter 16, and of seven in Galileo's version. These can be paired up as follows: Carbone's first difficulty has no counterpart, but his second corresponds to Galileo's fourth (lines 120-130), wherein both refer to the "impediment of the divine will" (*impedimentum voluntatis divinae*); Carbone's third again has no parallel, but his fourth is the same as Galileo's third (lines 131-146); yet again Carbone's fifth has no equivalent, but his sixth is identical with Galileo's sixth (lines 147-167), both exhibiting the same structure of objection, confirmation, response, additional response, and reply to the confirmation; Carbone's seventh difficulty corresponds to Galileo's seventh (lines 168-175), and again has the same structure; Carbone's ninth is equivalent to Galileo's fifth (lines 185-195), yet again with similar structure; and finally Carbone's tenth corresponds to Galileo's second (lines 196-202), both making reference to "the supposition of the universe" (*supposito universo*), a teaching related to the "impediment of the divine will" that will be discussed in chapter 3 when treating the role of *suppositiones* in scientific proof.[59]

[58] Texts wherein references are made to *impedimenta* are given in chaps 5 and 6 *infra*. See especially secs. 5.1.c, 5.2.a-c, 5 3.a-b, 5.4.a, 5.4.e, 6.1.b, 6 1.e, 6.2 b, 6 3.b, 6.3 d-f, and 6.4.a.

[59] Secs 3 1.a, 3 2 a, and 3.4 c.

The final disputation in Galileo's treatise on foreknowledge is composed of only two questions, the first considering how much one must know about the property being demonstrated before one can actually demonstrate it (LQ4.1), and the second inquiring how knowledge of the conclusion is preceded by knowledge of the premises (LQ4.2). The beginning of the first question is of interest because of the close agreement it shows between Galileo's composition and chapter 13 of Carbone's treatise on foreknowledge, the first several paragraphs of which are shown below in parallel column:

VALLA-CARBONE (1588)	GALILEO
Cap. 13. *An de passione praecognoscere oporteat, an sit.*	Q. 1. *An de passione praecognoscendum sit quia est.*
Circa hanc praecognitionem illud dubitationem facit, quod Aristoteles nihil de ea dixerit; et tamen *videtur esse necessarium praecognoscere passionem existere, cum* ut plurimum *eius praesupponatur existentia de qua propter quid sit probatur.* Confirmatur: *quia antequam probemus passionem inesse subiecto, opus est ut illam* in praemissis *alicui attribuamus; ergo, praecognoscendum est* eam *esse.* . . .	*Videtur quod sic, quia de passione supponimus semper esse, et deinde quaerimus propter quid insit* alicui. Secundo, *quia antequam demonstramus passionem inesse* alicui *subiecto* et propter quid insit illi, *debemus illam inscribere alicui, ergo* et *supponere esse.*
Primum notandum: haec dubitatio non solum intelligenda est de passione proprie sumpta . . . sed de quocunque praedicato quod potest in conclusione *de quovis subiecto concludi* . . .	*Notandum* est, *primo: hanc quaestionem esse intelligendam non solum de propria passione, sed* etiam *de illis omnibus quae demonstrantur de aliquo subiecto.*
{Tertium *notandum*: ex Averroe, *tria dantur demonstrationum genera.* Primum, quod vocatur demonstratio *quia*, cum *per effectum probatur causa.* Secundum, quod nominatur *propter quid*, cum *a priori* per causam *probatur effectus.* Tertium, quod ipse vocat demonstrationem *potissimam*, sive simpliciter, *qua simul probatur quid rei et esse ipsius* . . .}	*Notandum* est, secundo: *triplicem esse demonstrationem, quia, propter quid,* et *potissimam. Quia,* illa est quae *ab effectu causam demonstrat; propter quid,* est *quae ostendit propter quid passio* insit subiecto; *potissima,* est *quae et propter quid passio insit subiecto et existentiam illius probat*
Secundum *notandum*: proprietates, sive *passiones, sunt in duplici differentia: aliae quae uni tantum subiecto* conveniunt *quocum convertuntur,* ut latrare cum cane . . . ; *aliae quae* insunt pluribus subiectis et ita *non convertuntur* cum eo subiecto de quo probari possunt, ut rotunditas respectu terrae . . [60]	*Notandum* est, tertio. *duplicem esse passionem: alteram reciprocam cum suo subiecto,* qualis est risibilitas respectu hominis, *alteram non convertibilem,* qualis est albedo respectu hominis. [61]

[60] *Additamenta*, fol. 45ra-rb.

[61] MS 27, fol. 11r14-26.

In the title (lines 1-2) Carbone speaks of the *an sit* (whether it is) of the property whereas Galileo has the equivalent *quia est* (that it is). As heretofore, Carbone has the longer treatment and orders the matter differently, with the result that his second (lines 26-32) and third (lines 17-25) notations correspond to Galileo's third and second respectively. He acknowledges the source of his third notation and Galileo does not, and offers different examples in his second notation—possibly his own improvement on Valla's original—but otherwise practically every word in Galileo's version can be accounted for in Carbone's treatise.

b. TREATISE ON DEMONSTRATION

The foregoing textual comparisons offer striking evidence that the first extant treatise of Galileo's logical questions derives from a source identical with that used by Carbone for his *Additamenta* of 1597, namely, the notes for Valla's course in logic given at the Collegio in 1587-1588. The provenance of Galileo's second treatise, *De demonstratione*, on the other hand, is more difficult to establish. The reason for this is that Carbone did not reprint this particular treatise in any of his works, nor does Valla mention it as being plagiarized in the prefaces of his *Logica* of 1622. Nonetheless, there are two clues that serve to tie Galileo's second treatise to the same source as his first, and these must now be examined. Such clues aside, one could still show evidences of similar content and other agreement between the treatise and corresponding tracts in Lorini, Vitelleschi, Rugerius, Jones, and the Valla text of 1622, but this recourse seems unnecessary if a direct connection to the 1588 lectures can be established.

Galileo begins the treatise on demonstration with a brief prologue that has been transcribed in the National Edition and may be translated as follows:

> Omitting the definition of science, which Aristotle treats quite sensibly and from which he begins his treatise—so that once one knows the end of demonstration, which is science, he will be better able to understand its nature and properties, and about this we will have much to say in the following treatise—I proceed to the treatise on demonstration itself. This contains within it three disputations: the first, on the nature and importance of demonstration; the second, on its properties; the third, on its kinds. When these are finished and done well, nothing that can be known about demonstration will remain to be desired.[62]

Here one has a precise identification of the three disputations that make up the treatise, the first two of which are not titled in Galileo's composition, although the third is.[63] Here also is the reference to the third

[62] The Latin text is given in *Opere* 9:279, transcribing MS 27, fol. 13r9-16.

[63] For details, consult table 3 *supra*.

treatise that Galileo plans for his logical questions, the *De scientia*, which follows immediately after *De demonstratione* and presumably has as its exemplar Valla's *De scientia*, as this was reprinted by Carbone in his *Introductio in universam philosophiam* of 1599.

This particular paragraph has no counterparts in the *reportationes* of logic notes that have survived from the Collegio Romano. Valla, however, does have a brief preamble to his extensive treatment of demonstration in volume 2 of his logic text, and in this makes a point that is very similar to that mentioned by Galileo in his first sentence. For purposes of comparison the two sentences are given below in Latin, in parallel column:

GALILEO

Tractatio de demonstratione, *omissa definitione scientiae, quam* sapientissime *tradidit Aristoteles, exordiens tractationem suam ab illa,* ut cognito nimirum *demonstrationis fine, qui est scientia,* melius et perfectius natura et proprietates illius elucescant; de qua multa tractatione sequenti disseremus, *aggredior tractatum ipsum demonstrationis.*[64]

VALLA (1622)

Quamvis enim *prius agendum videretur de scientia, de qua agit Aristoteles in principio primi libri, quia tamen scientia est effectus demonstrationis* et definitionis, et haec sunt duo instrumenta quibus scientia acquiritur, ideo *prius de illis agendum erit.*[65]

Directing attention to the italicized words, one may see that both passages express some reservation about beginning the tract on demonstration without first having had a treatise on science, since science is the end or effect of demonstration and thus can serve to clarify the latter's treatment. But this is only a reservation in both instances, for Galileo (or better, his source) decides to treat demonstration first and then science, whereas Valla, in his more mature development of the subject, prefaces tracts on both demonstration and definition to his treatise *De scientia.* The obvious inference is that Valla had formulated both prologues, one thirty-eight years previous to the other, and had changed his wording and the ordering of the subject matter in the intervening period.

A much better clue, however, is offered by the fact that Carbone, though he did not provide a copy of Valla's *De demonstratione*, did manage to reprint the latter's *De instrumentis sciendi* as part of his *Additamenta* to Toletus's logic of 1597. Now, if one considers the list of logical questions treated by Galileo (see table 3), one finds that the second question of the first disputation in the treatise on demonstration inquires whether demonstration, or definition, is the more noble instrument of knowing (LQ5.2). Should this question have a counterpart in Carbone's *De instrumentis*

[64] MS 27, fol. 13r9-16.

[65] *Logica*, vol. 2, p. 188.

sciendi, and should it agree with Galileo's composition, this would provide *prima facie* evidence that Carbone had done precisely what Valla had accused him of, namely, he had rearranged the materials—disordering them, in Valla's opinion—in addition to appropriating them as his own. Such, indeed, proves to be the case. And not only does this confirm the accuracy of Valla's claims in the prefaces of his *Logica*, but it supplies substantial proof that Galileo worked with the same materials as Carbone, not only in writing his *Tractatio de praecognitionibus et praecognitis* but in composing his *Tractatio de demonstratione* as well.

Galileo's LQ5.2 is quite long, about the same length as LQ3.1, which has been transcribed above in its entirety. Its counterparts in Carbone's *Additamenta* are found in two different places in the *Tractatio de instrumentis sciendi*, chapters 2 and 9, respectively. Rather than duplicate the analysis of LQ3.1, in what follows the similarity of structure in the two treatments will first be shown, and then the four conclusions to which Carbone and Galileo come will be placed in parallel column as a representative sample of the word coincidences that can be identified throughout the entire question. Carbone's structuring of the materials is given below on the left, identified by the first few words that introduce the passages and preceded by the folio and column numbers in the *Additamenta*, while Galileo's is given similarly on the right, preceded by the folio and line numbers as found in MS 27.

VALLA-CARBONE	GALILEO
Tractatio de instrumentis sciendi . . .	Tractatio de demonstratione
23vb Cap. 2 Quot sint sciendi instrumenta?	14r15 Q. 2. An demonstratio sit nobilissimum omnium instrumentorum, vel definitio?
24ra Primum notandum . . .	14r17 Notandum est primo . .
Secundum notandum . . .	Notandum est secundo . . .
24rb Prima positio . . .	24 Notandum est tertio: si spectemus instrumenta quae aliquo modo . . . inserviunt . . .
Secunda positio . . .	14b1 Si autem spectemus instrumenta quae immediate inserviunt . . .
24va Sed diceres . . .	8 Neque obijcias . . .
Tertia positio . . .	16 Si autem considerentur instrumenta quae perfecte et immediate deserviunt . . .

24vb Sed rursus obijceres . . .	20 Neque dicas . . .
28va Cap. 9 Utrum instrumentum sit praestantius, definitio an demonstratio?	
Gravis atque . . .	25 Notandum est quarto . . .
28vb De secundo . . .	15r 6 Prima opinio . . .
29ra {Quinto . . .	9 Primo . . .
28vb Quarto . . .	11 Secundo . . .
Tertio . . .	17 Tertio . . .
Primo . . .	20 Quarto . . .
29ra Sexto . . .	23 Quinto . . .
Quod si dicas . . . }	27 Dices . . .
Secunda sententia . . .	15v 2 Secunda opinio . . .
29rb Tertia sententia . . .	7 Tertia opinio . . .
29va {Secundum fundamentum . . .	12 Pro solutione . . . sciendum est, primo . .
28va Quod igitur ad primum attinet . . .	16 Sciendum est, secundo . . .
29va Primum fundamentum . . .	22 Sciendum est, tertio . . .
Tertium fundamentum . . . }	16r 6 Sciendum est, quarto . . .
29vb Prima positio . . .	10 [M: Prima conclusio] Dico, primo . . .
Primo . . .	Probatur, primo . . .
Quod si dicas . . . Secundo . . .	12 Dices . . . secundo . . .
Confirmatur . . .	15 Confirmatur . . .
Tertio . . .	19 Tertio . . .
Quarto . . .	25 Quarto . . .
30ra Quod si dicas . . .	16v 5 Obijcies . . .

30rb Secunda positio . . .	12 [M: Secunda conclusio] Dico, secundo . . .
Quod si dicas . . .	19 Obijcies . . .
Quod si quis diceret . . .	24 Obijcies, secundo . . .
30va Tertia positio . . .	17r 7 [M: Conclusio] Ex his infero, primo . . .
30vb Quarta positio . .	13 [M: Conclusio] Colligitur, secundo . . .
31ra {Ad quintum . .	17 Ad primum argumentum . . .
Ad quartum . . .	Ad secundum . . .
Ad ultimum . .	20 Ad tertium . . .
Ad primum . . .	24 Ad quartum . . .
Ad sextum . . . }	29 Ad quintum . . .
. . . Quia definitio posita . . .	30 Ad sextum . . .
31rb Sed hic restat . . .	17v 5 Dices . . .
Et ratio disparitatis . . .	10 Respondeo . . .

What is remarkable about the foregoing schema is that it identifies the entire articulation of Galileo's LQ5.2 with paragraphs in the two chapters of Carbone. The material is organized somewhat differently, for Galileo begins with four preliminary notations, three of which are found in Carbone's chapter 2 and the fourth at the beginning of his chapter 9. After this, both list three opinions or schools of thought on the respective merits of definition and demonstration. Arguments in support of the first opinion are given by both: Galileo offers five whereas Carbone has six, though in an order different from that found in Galileo (and thus enclosed in braces). After that Galileo has four animadversions (*sciendum est* . . .); these are included in the three *fundamenta* provided by Carbone, again arranged and sectioned differently. Then come the four conclusions, which, as will be seen below, are the same in each; both offer four proofs for, and answer one objection to, the first conclusion, and correspondingly answer two objections to the second conclusion. Finally, both conclude their presentations with responses to the arguments in support of the first opinion. Here Galileo makes a slip that gives evidence of his copying: he replies to a "sixth argument" (17r30), whereas he has offered only five in his text (15r9-27). The fact that Carbone has such a sixth argument argues for its being present in the original from which Galileo worked, and which he apparently was attempting to summarize—possibly on the same scale of reduction as his questions now stand in relation to the fuller treatment in Carbone.

The four conclusions subscribed to by Carbone and Galileo are given below in the same format as earlier.

VALLA-CARBONE (1588)	GALILEO
Prima positio: definitio secundum se considerata est demonstratione nobilior. . . . Primo, definitio facit scire rei *substantiam . . . demonstratio* vero *accidens . . .* [M: libro *secundo Posteriorum, capite secundo*] *at* multo *nobilior est substantia quam accidens; ergo,* si verum est . . . instrumenta sumere suam praestantiam a re quam faciunt *cognoscere,* definitio erit demonstratione praestantior. . . .	[M: *Prima conclusio*] Dico primo: *definitionem secundum se esse* longe *nobiliorem demonstratione.* Probatur, *primo:* quia *definitio substantiam, demonstratio accidens facit scire,* ex *secundo Posteriorum, capite secundo; sed nobilius est substantiam cognosci quam accidens; ergo. . . .*
Secunda positio: definitio etiam considerata *ut est in nobis, est demonstratione praestantior. Primum, quia licet definitio ut est in nobis aliqua ratione pendeat a syllogismo* divisivo quo eius partes inveniuntur, *tamen quia definitio est finis talis syllogismi et ab eo non pendet nisi instrumentaliter, ideo* dicendum est definitionem etiam ut in nobis est esse nobiliorem demonstratione. *Deinde quia* etiam respectu nostri *definitio versatur circa nobilius obiectum et modo quodam nobiliore,* quare suam retinebit nobilitatem, *etiam ut est in nobis. . . .*	[M: *Secunda conclusio*] Dico secundo: *definitionem, ut in nobis est, esse praestantiorem* ipsa *demonstratione.* Probatur conclusio: *tum quia, etiam si definitio ut est in nobis pendeat ab aliquo syllogismo, tamen quia est finis illius et pendet ab illo tantummodo instrumentaliter, ideo* etc.; *tum quia definitio, etiam ut est in nobis, versatur circa nobilius obiectum et nobiliori modo* quam demonstratio, hoc est sine mora. . .
Tertia *positio: instrumentum sciendi analogice dicitur de demonstratione et definitione. . . . Sunt autem analoga analogia proportionalitatis. Ut enim se habet definitio* ut est instrumentum *ad* ipsam *quidditatem quam facit cognoscere, ita demonstratio ad accidens cuius scientiam gignit.* Verum, cum *definitio sit primum analogatum,* ideo dictum est eam esse nobiliorem, et instrumenti rationem de ipsa primario et de demonstratione secundario dici. Reperitur etiam in his instrumentis analogia *attributionis, quia saltem demonstrationes potissimae* in ratione instru-	[M: *Conclusio*] ex his infero, primo: *definitionem et demonstrationem convenire analogice in instrumento sciendi,* ita tamen ut *definitio sit primum analogatum. Sunt autem analoga* tum *analogia proportionis* tum attributionis: proportionis, *quia sicut se habet definitio ad cognitionem quidditatis, ita demonstratio ad cognitionem passionum;* *attributionis, quia* cum definitio sit finis de-

menti *habent aliquam dependentiam* intrinsecam *a definitione*, ut est instrumentum sciendi.

monstrationis et *demonstratio, praesertim potissima*, in sui certitudine *pendeat a definitione*, ergo.

Quarta *positio: definitio est finis demonstrationis. Primum, quia omne imperfectum revocatur ad perfectum* tanquam ad finem; *at demonstratio*, ut est demonstratum, *est definitione imperfectior; igitur. Secundo*, definitio facit scire *substantiam* et demonstratio *accidens*; sed *hoc ordinatur ad illam* ut ad finem; *igitur et* instrumentum *quo scitur accidens ad* illud *quo scitur substantia.*[66]

[M: *Conclusio*] Colligitur, secundo: *definitionem esse finem demonstrationnis Tum quia omne imperfectum reducitur ad* aliquid *perfectum; sed demonstratio est quid imperfectum* si conferatur cum *definitione; ergo. Tum* quia, sicut *accidentia sunt propter substantiam*, ex Aristotele ubi supra, *ita etiam cognitio accidentium est propter cognitionem* ipsius *substantiae.*[67]

Carbone's exposition is lengthier than Galileo's, as usual, but the order is the same for both. The common elements in their reply are that definition in itself is more noble than demonstration; that definition, even as it is in us, is more important than demonstration; that definition is the end of demonstration; and that definition and demonstration are related analogously as instruments of knowing, but in such a way that definition is the primary analogate, by an analogy of proportionality as well as by one of attribution. The only difference is that Carbone presents all four *positiones* on their own merits, whereas Galileo sees his third and fourth as inferences or corollaries deriving from his first two conclusions.

With regard to the remainder of Galileo's questions in the treatise *De demonstratione*, it is a simple matter to show that they summarize the material on demonstration in volume 2 of Valla's *Logica* of 1622. All of Galileo's logical questions can therefore be tied to Valla's work, the eleven relating to foreknowledge and one relating to demonstration via the plagiarized *Additamenta* of Carbone, and the remaining fifteen on demonstration by direct comparison with the *Logica*. A complete correlation of all these materials is provided in table 4, with Galileo's MS 27 being referenced in terms of folio and beginning line number, Valla's lectures of 1588 as appropriated by Carbone in the *Additamenta* of 1597 by folio and column number, and Valla's own published text of 1622 by volume and page number. What is striking about the tabular arrangement is that it reveals that Valla followed much the same order of treatment in 1622 as he did in his original of 1588, whereas the version preserved in Carbone's *Additamenta* clearly shows the different ordering against which Valla protested in both his prefaces.

[66] *Additamenta*, fols. 29vb-30vb.

[67] MS 27, fols. 16r10-12, 16v12-15, 17r7-16

TABLE 4
TEXTUAL CORRELATIONS FOR THE LOGICAL QUESTIONS:
GALILEO, CARBONE, AND VALLA

No. of Question	Galileo MS 27	Valla-Carbone (1588) *Additamenta* (1597)		Valla *Logica* (1622)
LQ2.1	4r2	42ra-42va	c. 8	2:149
LQ2.2	4v14	40vb-42ra	c. 7	2:147
LQ2.3	5v1	42va-43ra	c. 9	2:150
LQ2.4	6r4	43vb-44rb	c. 11	2:150
LQ3.1	6v23	45vb-48ra	cc. 14-16	2:159, 163-165
LQ3.2	8r19	48ra-49ra	c. 17	2:160
LQ3.3	9r24	49ra-50ra	c. 18	2:161
LQ3.4	10r13	50ra-50va	c. 19	2:164
LQ3.5	10v13	38rb-39va	c. 4	2:164
LQ4.1	11r11	45ra-45rb	c. 13	2:156
LQ4.2	11v25	55rb-56vb	c. 25	2:153
LQ5.1	13r17			2:220
LQ5.2	14r15	28va-31va	cc. 2, 9	2:123, 406-409
LQ6.1	17v17			2:221
LQ6.2	18v2			2:224
LQ6.3	19v10			2:229
LQ6.4	20v6			2:235
LQ6.5	21r16			2:248
LQ6.6	22r7			2:250
LQ6.7	23v5			2:253
LQ6.8	26v1			2:257
LQ6.9	27r18			2:266
LQ6.10	27v23			2:273
LQ6.11	28r28			2:281
LQ7.1	29r14			2:299
LQ7.2	30v20			2:313
LQ7.3	31r6			2:343

c. DATING OF THE LOGICAL QUESTIONS

The foregoing evidence fixes quite well the date *post quem* of Galileo's logical questions. From what is known about the teaching schedule at the Collegio Romano—about which more details will be given in the following chapter—Valla did not complete his logic course until near the end of August 1588, and his notes, by his own admission, were not available until shortly after this.[68] Thus, allowing some time for transmission, Galileo could not have composed the logical questions until late in 1588 or some time in 1589, possibly while already teaching at the University of Pisa or while actively preparing himself for that post. The various ways in which Galileo could have obtained a copy of Valla's lectures on logic, and the relationship of his logical questions to the physical questions of MS 46, will be discussed at the end of the next chapter.

It goes without saying that the textual comparisons given above do not, in themselves, fix the date of Galileo's MS 27 with absolute certitude. Yet the cumulative effect of all the correlations that have been offered, plus those that could be given on the basis of table 4 if this were not tedious for author and reader alike, leaves little room for doubt. When esoteric questions of the kind given in table 3 are being asked, and near identical answers given, through page after page of MS 27 and the *Additamenta* of 1597, and again in reworked form in the *Logica* of 1622, the conclusion is inescapable that they derive from a common source. If plagiarism were not involved, to be sure, the case would be much simpler: one could say that MS 27 was not written until after 1597, when Galileo copied them from Carbone's printed work, or, much less likely—since MS 27 itself shows signs of copying—that Carbone's *Additamenta* were plagiarized from Galileo's notes. With the discovery of Valla's allegation in the prefaces to the two volumes of his *Logica* the story becomes more complicated. The additional alternative, now that the *Additamenta* are known to be Carbone's rearranged version of Valla's lecture notes of 1588, is that Valla himself might have copied these notes practically word-for-word from a yet earlier source. That such a source would be the young Galileo lacks all credibility. Further, a detailed survey of all the passages correlated above (plus many others transcribed by the author) does not show a single instance of close agreement with corresponding passages in the logic courses of Toletus, Lorinus, Vitelleschi, Rugerius, or Jones outlined earlier in this chapter. The same matter is touched on in each, and similar conclusions are reached, but never expressed in precisely the same terms. In this respect Valla's logic notes are much more idiosyncratic than his *Tractatus de elementis*, to be discussed in the next chapter, which exhibits many textual parallels with the notes of earlier and later professors at the Collegio Romano. Again, it seems hardly believable that Valla would plagiarize another's work (for such word-for-word agreement is indeed that) and then later complain of

[68] See the portion of the preface to vol. 1 of his *Logica* cited in sec. 1.1.c.

having been plagiarized himself. The fact that his Jesuit censors, one of whom was Lorinus, would have approved the *Logica* for publication in that circumstance effectively rules it out as a possibility.[69] Thus one is left with the only remaining alternative: Galileo somehow obtained a copy of Valla's lecture notes of 1588, similar in every respect to those used by Carbone, and from them wrote out the very interesting material contained in his MS 27.

This surprising result makes the logical question practically contemporaneous with the notes written in MSS 46 and 71, both of which date from "around 1590." It further suggests that these questions, which themselves contain a treatise on methodology, could have played a seminal role in the development of Galileo's science of motion up to 1610 and even beyond. To explore that possibility, the initial chapter of part 2 of this study (i.e., chapter 3) will lay out in fuller detail the teachings expounded in MS 27, locating these in the context from which they were extracted and supplementing them, where desirable, with additional materials drawn from Valla and his contemporaries at the Collegio Romano. Before setting out on that project, however, it is desirable to canvass the bearing of the findings now described on related research into the sources and dating of MS 46, the topic to be investigated in the following chapter.

[69] In approving the *Logica* for publication, Chamerota and Lorinus both indicate that the work should be regarded as that of a private author, and not as a course suitable for either the Society of Jesus or the Collegio Romano. The conclusion of their approval reads: ". . . existimamus operae precium fore si a Censoribus recognoscatur, tanquam privati Autoris opus, non autem ut cursus aut Societatis aut Collegii Romani. Datum in Collegio Romano, 24 Junii 1612."—Archivum Romanum S.I., FG Cod. 654, fol. 275. This would seem to confirm the idiosyncratic character of Valla's course in logic, and thus argue against its being derived from notes given previously at the Collegio Romano.

CHAPTER 2

Sources of Galileo's Physical Questions

PREVIOUS STUDIES of the materials in Galileo's MS 46, with its questions on the universe and the elements—called the physical questions mainly to distinguish them from the logical questions of MS 27—show that these are based on, and probably were extracted from, writings of Jesuit professors at the Collegio Romano.[1] These studies likewise reveal that the physical questions, unlike the logical questions, can be correlated with a wide range of passages in manuscripts and printed books dating from as early as 1566 and as late as 1597.[2] Indeed, the large number of textual parallels turned up unduly complicates the task of dating the source Galileo used as the exemplar for MS 46. Most of the difficulty is traceable to the fact that the studies were undertaken in an attempt to fix the time of writing of MS 46 alone, with no awareness that the date of Galileo's composing MS 27 could be established independently and with greater accuracy. Now that the evidence provided in the preceding chapter has been uncovered, the work of dating MS 46 is considerably simplified. Internal evidence shows that MS 46 was written after MS 27, and so the physical questions could not have been written by Galileo before 1589 or 1590.[3] The connection of MS 27 with Valla further suggests that this same Jesuit is the source behind MS 46 as well. As it turns out, the correlations with Valla's course in natural philosophy, though preserved in fragmentary form and only for its later parts, are better than those

[1] See the author's *Galileo's Early Notebooks: The Physical Questions* for an English translation of the contents of this manuscript, together with an identification of the main sources from which it derives. An earlier study identifying some of these sources is A. C. Crombie, "Sources of Galileo's Early Natural Philosophy," in *Reason, Experiment, and Mysticism in the Scientific Revolution*, M. L. Righini Bonelli and W. R. Shea, eds. (New York: Science History Publications, 1975), pp. 157-175 and 303-305. Additional details are to be found in Christopher Lewis, *The Merton Tradition and Kinematics in Late Sixteenth and Early Seventeenth Century Italy* (Padua: Editrice Antenore, 1980), and the author's *Prelude to Galileo.*

[2] The passages with which Galileo's questions can be correlated are described generally in Wallace, *Galileo's Early Notebooks*, pp. 12-21, and then specifically in the author's commentary in the same, pp. 253-303. Some parallel Latin texts are given in idem, *Prelude to Galileo*, part 3. Additional parallels are exhibited *infra*.

[3] See sec. 2.4 *infra*; also chap. 1, note 47, *supra*.

with any other author. Thus the initial project of this chapter becomes that of showing the correspondence between Galileo's questions and the portions of Valla's course that are extant. Once this is done, if it can be shown that Valla's teaching correlates well with that of other professors who taught just before and after him, the way will be opened to tying Galileo's remaining questions to the materials lacking in Valla that survive in the notes of his contemporaries. As a consequence, the lecture notes preserved in MS 46 will be seen to reflect the natural philosophy being taught at the Collegio "around 1590," and will prove of value for recreating the context in which Galileo developed his early science of motion.

It is this program that dictates the structure of this chapter. First Galileo's MS 46 will be described, and then the more likely sources on which it is based will be surveyed. Following this a series of textual comparisons will be provided, along with incidental clues that favor some sources over others. In the concluding section all of this evidence will be assessed to formulate some conclusions about the provenance of MS 46, the purpose Galileo had in mind in writing it together with MS 27, and the most likely time at which he did so.

1. Galileo's Questions on Natural Philosophy (MS 46)

The manuscript containing the physical questions, written in Galileo's own hand, is now in the Biblioteca Nazionale at Florence and bears the signature MS 46. The questions of interest are on folios 4r to 100v, and they are divisible into two general categories: those dealing with matters pertaining to the *De caelo* of Aristotle and those dealing with matters pertaining to the *De generatione et corruptione* of the same author. The manuscript has been transcribed and is printed in the National Edition of Galileo's works; the Latin text has been translated into English by the author, with a commentary correcting some of the readings given in the National Edition and providing paleographical details, such as evidences of copying, many of which connect Galileo's composition with the Jesuit lecture notes already mentioned.[4] For purposes of reference the titles of Galileo's questions are given in table 5 in Latin and in English translation, along with the folios of the manuscript on the right and the letter designation already used in the English translation, plus a new one similar to that used for the logical questions, on the left.

A few of the expressions in the table's listing of the contents of the

[4] *Opere* 1:15-177; Wallace, *Galileo's Early Notebooks*, English translation on pp. 25-251, commentary on pp. 253-303.

TABLE 5
GALILEO'S PHYSICAL QUESTIONS (MS 46)

No.	Question	MS 46 folios
	[*Introductio* (Introduction)]	
A PQ1.1	Quaestio prima. Quid sit id de quo disputat Aristoteles in his libris De caelo? (What is Aristotle's subject matter in his books *De caelo*?)	4r
B PQ1.2	Quaestio secunda. De ordine, connexione, et inscriptione horum librorum. (On the order, connection, and title of these books.)	6v
	Tractatio prima. De mundo. (On the universe.)	7v
C PQ2.1	Quaestio prima. De opinionibus veterum philosophorum de mundo. (On the opinions of ancient philosophers concerning the universe.)	7v
D PQ2.2	Quaestio secunda. Quid sentiendum sit de origine mundi secundum veritatem? (The truth concerning the origin of the universe.)	8v
E PQ2.3	Quaestio tertia. De unitate mundi et perfectione. (On the unity and perfection of the universe.)	10v
F PQ2.4	Quaestio quarta. An mundus potuerit esse ab aeterno? (Whether the universe could have existed from eternity?)	13r
	Tractatio [secunda]. De caelo. (On the heavens.)	16v
G PQ3.1	Quaestio prima. An unum tantum sit caelum? (Is there only one heaven?)	16v
H PQ3.2	Quaestio secunda. De ordine orbium caelestium. (On the order of the celestial orbs.)	22r
I PQ3.3	Quaestio tertia. An caeli sint unum ex corporibus simplicibus, vel ex simplicibus compositi? (Are the heavens one of the simple bodies or composed of them?)	26r
J PQ3.4	Quaestio quarta. An caelum sit incorruptibile? (Are the heavens incorruptible?)	30v
K PQ3.5	Quaestio quinta. An caelum sit compositum ex materia et forma? (Are the heavens composed of matter and form?)	34r
L PQ3.6	Quaestio sexta. An caelum sit animatum? (Are the heavens animated?)	51r
	[*Tractatus de alteratione.* (On alteration.)]	
M PQ4.1	[Quaestio prima. De alteratione. (On alteration.)]	57r
N PQ4.2	Quaestio secunda. De intensione et remissione. (On intension and remission.)	57r

TABLE 5 (*cont.*)

O PQ4.3	Quaestio ultima. De partibus sive gradibus qualitatis. (On the parts or degrees of qualities.)	63r
	Tractatus de elementis. (On the elements.)	65v
P PQ5.1	[*Prima disputatio. De elementis in universum.* (On the elements in general.)]	65v
Q PQ5.2	*Prima pars. De quidditate et substantia elementorum.* (On the quiddity and substance of the elements.) Prima quaestio. De definitionibus elementi. (On the definitions of an element.)	67v
R PQ5.3	Quaestio secunda. De causa materiali, efficiente, et finali elementorum. (On the material, efficient, and final cause of the elements.)	68v
S PQ5.4	Quaestio tertia. Quae sint formae elementorum? (What are the forms of the elements?)	69v
T PQ5.5	Quaestio quarta. An formae elementorum intendantur et remittantur? (Do the forms of the elements undergo intension and remission?)	71v
	[*Secunda pars. De distinctione elementorum.* (On the differentiation of the elements.)]	
U PQ5.6	[Quaestio unica. De numero et quantitate elementorum. (On the number and quantity of the elements.)]	75v
	Secunda disputatio. De primis qualitatibus. (On primary qualities.)	88v
V PQ6.1	Quaestio prima. De numero primarum qualitatum. (On the number of primary qualities.)	88v
W PQ6.2	Quaestio secunda. An omnes hae quatuor qualitates sint positivae, an potius aliquae sint privativae? (Are all four of these qualities positive, or are some privative?)	90r
X PQ6.3	Quaestio tertia. An omnes quatuor qualitates sint activae? (Are all four qualities active?)	92v
Y PQ6.4	Quaestio quarta. Quomodo se habeant primae qualitates in activitate et resistentia? (How are primary qualities involved in activity and resistance?)	96v
	[Finis]	100v

Note: See note in table 3.

manuscript are enclosed in square brackets to show that the titles are either missing from the manuscript or else are incomplete. A similar situation obtains with the manuscript containing Galileo's logical questions, already discussed, and with some of the manuscripts containing notes of the lectures given at the Collegio Romano, for example, Antonius Menu's, to be discussed presently. It is possible that Galileo worked from an incomplete set of notes, such as Menu's, or, more likely, and as seems to be the case with the logical questions, that he was working from a complete set but was reordering it in an attempt to abbreviate its more extensive coverage. In the latter event, he probably left the notes unfinished and did not return to number and title the questions properly.

A perusal of the table will show that questions PQ1.1 through 3.6 deal with problems posed by Aristotle's *De caelo et mundo*, and that questions PQ4.1 through 6.4 do the same for Aristotle's *De generatione et corruptione*. There are some slight differences in the two sets of questions that may be noted at this point. The first set contains an introductory section that is untitled and is made up of two questions (PQ1.1 and 1.2) preliminary to the first *Tractatio*, whereas the second set lacks this feature. (Such introductory sections sometimes occur in the Jesuit notes, usually under the title *Proemium* or *Exordium*.) Again, the basic unit of division in the first set is the *Tractatio*, which is subdivided into a number of *Quaestiones*. The basic unit of division in the second set, on the other hand, is the *Tractatus*, and this is subdivided into *Disputationes*, the first of which is again subdivided into a *Prima pars* and an unidentified *Secunda pars*, then into *Quaestiones*, and the second directly into *Quaestiones*. (This way of dividing the second set is identical with that of Valla in his corresponding *Tractatus de elementis*.) Otherwise the handwriting and the indications of copying are much the same, the only notable difference being that the first set is on paper without watermarks, whereas the paper of the second set is watermarked.[5] These differences might suggest a different procedure for dating the two sets of questions, about which more later.

2. Possible Sources

As already indicated, much is already known about the possible sources of the physical questions, with all evidence pointing to their derivation from teachings developed at the Collegio Romano in the latter part of

[5] The paper used by Galileo in his early autograph writings has been studied for watermarks by Adriano Carugo and A. C. Crombie; their results are presented in Crombie, "Sources of Galileo's Early Natural Philosophy," pp. 304-305. It is difficult to establish a chronology on the basis of watermarks alone, but the author has compared their findings

the sixteenth century. Four printed books have thus far been identified as containing materials similar to Galileo's: Franciscus Toletus's *Commentaria una cum quaestionibus in octo libros Aristotelis de physica auscultatione* (Venice, 1573); the same author's *Commentaria in librum de generatione et corruptione Aristotelis* (Venice, 1575); Benedictus Pererius's *De communibus omnium rerum naturalium principiis et affectionibus* (Rome, 1576); and Christopher Clavius's *In sphaeram Ioannis de Sacrobosco* (Rome, 1581).[6] There were later editions of all these books, but the dates given are the earliest editions that show similarities with Galileo's composition. With the possible exception of Clavius's work, however, there are much better correspondences between Galileo's questions and similar questions contained in *reportationes* of lectures given at the Collegio, making it more likely that Galileo's work is based on manuscript sources that, in turn, made use of these printed works, which surely were available to their authors.[7]

The lectures on natural philosophy that contain parallels to Galileo's text are those of eleven Jesuit professors who taught courses at the Collegio for a period of over thirty years, from 1565 to 1598. During that time, as was customary for the course in natural philosophy, these professors commented on Aristotle's *Physics, De caelo, De generatione,* and *Meteorology*. Not all of their lectures have survived, but those that are known at this writing are listed in table 6. Identification of relevant manuscripts listed in the table, together with the description of their contents as related to the materials contained in Galileo's physical questions, is to be found in *Galileo's Early Notebooks*.[8] Of the eleven professors, however, the first three and the last three, plus Parentucelli, seem only peripherally related to Galileo's text, and thus need not be dwelt on at length. The lectures of Menu, Valla, Vitelleschi, and Rugerius, on the other hand, show numerous parallels and coincidences with the words actually used by Galileo, and thus bear fuller investigation.

As described in the preceding chapter, the philosophy curriculum at the Collegio Romano was taught in a three-year cycle, with logic occupying the first year of the cycle, natural philosophy the second, and

---

with his own indications of dating for the materials contained in MSS 27, 45, 46, 70, and 71, and has discovered no inconsistencies between the two.

[6] Crombie, "Sources of Galileo's Early Natural Philosophy," pp. 160-171; Wallace, *Galileo's Early Notebooks*, pp. 12-15.

[7] Arguments supporting the derivation of Galileo's notes from manuscripts rather than from printed works are given in Wallace, *Prelude to Galileo*, pp. 194-217; with regard to Clavius's *Sphaera*, see also the author's commentary in *Galileo's Early Notebooks*, pp. 263-266, where copying errors are pointed out that would be hard to explain if the copyist were using a printed text rather than a handwritten source. See also note 82 *infra*.

[8] Wallace, *Galileo's Early Notebooks*, p. 13; a few additional manuscripts have been discovered since the appearance of that work, and these are included in the bibliography at the end of this volume.

TABLE 6
EXTANT VERSIONS OF PHYSICS COURSES TAUGHT AT THE COLLEGIO ROMANO, 1560 TO 1598

Years	Professor	Manuscript	Printed Edition
1560–1561	F. Toletus	—	Venice 1573, 1575
1565–1566	B. Pererius	Physics De caelo De generatione	Rome 1576
1567–1568	H. De Gregorio	Physics De caelo De generatione	
1577–1579	A. Menu	Physics De caelo De generatione Meteorology (portions)	
1583–1584	A. Parentucelli	Meteorology (portions)	
1585–1590	P. Valla	— De elementis Meteorology	Lyons 1624?
1589–1590	M. Vitelleschi	Physics De caelo (2 copies) De generatione Meteorology (2 copies)	
1590–1591	L. Rugerius	Physics De caelo De generatione Meteorology	
1592–1593	R. Jones	Physics De caelo Meteorology	
1596–1597	S. Del Bufalo	De caelo De generatione Meteorology	
1597–1598	A. Eudaemon-Ioannis	Physics De caelo (2 copies) De generatione (2 copies)	

metaphysics the third.[9] Because of the large amount of material to be covered in natural philosophy, and the additional courses in mathematics that were assigned to the second year, it was customary to postpone the treatment of the second book of the *De generatione* to the third year.[10] This procedure creates a difficulty when trying to determine the precise year in which a professor taught the materials covered in Galileo's physical questions, since the latter's PQ6.1 through 6.4 pertain to the matter of the second book of *De generatione* and thus were probably not taught during the year devoted to natural philosophy proper. Again, depending on how a professor organized his treatment of the elements (Galileo's PQ5.1 through 5.6), he might prefer to consider them in the context of the *De caelo*, and thus cover this tract during the year devoted to natural philosophy, or he might decide to postpone their treatment until taking up related matters in the *De generatione* during the year devoted to metaphysics. Surviving lecture notes reveal a difference of practice in this regard, which will have to be taken into account when dating the sources that are being analyzed.[11]

Of the four professors whose lecture notes show close similarities with Galileo's physical questions, namely, Menu, Valla, Vitelleschi, and Rugerius, the first two saw repeated service in the second and third years of the philosophy cycle and thus could have developed more than one set of notes for their courses. Vitelleschi and Rugerius, on the other hand, taught the cycle only once and left fairly complete records of their teaching, and thus pose no problems with regard to dating their materials.

a. MENU

Antonius Menu (Menutius) is listed as first teaching natural philosophy in 1577-1578 and metaphysics in 1578-1579.[12] His lecture notes for these two years survive in two manuscripts, one quite complete and containing most of his treatise on natural philosophy and metaphysics, the other containing the first three books of his exposition of the *Meteorology*.[13] In

[9] Sec 1.1.

[10] Villoslada, *Storia del Collegio Romano*, p. 102.

[11] Menu and Valla, for example, took up the elements in their commentaries on the *De generatione*, whereas Vitelleschi and Rugerius did so in their commentaries on the *De caelo*.

[12] Lohr, "Renaissance Latin Aristotle Commentaries," *Renaissance Quarterly* 31 (1978): 583; Villoslada, *Storia del Collegio Romano*, pp 327, 329.

[13] Ueberlingen, Leopold-Sophien-Bibliothek, Cod. 138 (A.D. 1577-1579); Pistoia, Biblioteca Forteguerriana, Cod Chiapelli 235. The Pistoia manuscript attributes the exposition of the first three books of the *Meteorology* to Antonius Maria Mendi, and the name on the cover is given as Menctii or Menutii. In view of the dating and the contents there can be no doubt that the author is Menu.

the first of these he makes reference to his notes on logic, which is a good indication that he taught the first year of the philosophy cycle also, probably in 1576-1577. Some dates are interspersed throughout the main manuscript, and these reveal that he began his treatment of the *Physics* on October 20, 1577, and that of the *De caelo* on May 4, 1578. There is a break in the lectures at June 10, 1578, after which topics resume that relate to the *De generatione*, the last portions of which were being treated on November 28, 1578. By January 5, 1579, he had begun a brief treatment of *De elementis in particulari*, after which he started the course in metaphysics, which was concluded on October 6, 1579. The manuscript containing the treatise on meteorology has a notation to the effect that these began on June 25, 1578, which would seem to indicate that Menu covered this material contemporaneously with the course on the *De generatione*.[14] His notes occupy the first half of the codex, and there is no colophon indicating when they were completed. Quite interesting, however, is the fact that the remaining folios of the codex record a continuation of his exposition, starting at Book 4, that was taught by Antonius Parentucelli beginning on July 11, 1583.[15] This agrees with the dates given on the *rotulus* of professors at the Collegio (see table 1, *supra*) and suggests that Parentucelli, who followed Menu in the third year of the philosophy cycle, used the latter's notes for at least part of his own teaching.

Menu's treatment of topics relating to the physical questions is much more complete than Galileo's, but there is a similarity between the manuscripts of the two in that both omit titles of sections or questions and occasionally fail to number them. In Menu's case this is understandable because the handwriting on the folios (possibly his own) is cramped, and he thanks students for lettering titles in the spaces left in the manuscript; it is clear that they never completed the task.[16]

A comparison of Menu's questions with Galileo's shows that Menu has at least some counterparts for twenty-one of the twenty-five questions, and that these follow roughly the same order in the two expositions.[17] Menu's method of division is somewhat similar to that adopted

[14] Cod. Chiapelli 235, fol. 1r: In primum librum Meteororum Aristotelis per R. D. Antonium Mariam Mendi Societatis Iesu xxv Junii anno 1578 Romae.

[15] Ibid., fols. 58r-116v.

[16] Unfortunately the ink that was used ran through the paper, making it difficult to read passages on the verso of the folios containing titles. The Ueberlingen manuscript lacks foliation, and thus cannot easily be referenced. There are many personal remarks scattered throughout the manuscript, including names and dates, that would amply repay study by an historian.

[17] For details see Wallace, *Galileo's Early Notebooks*, pp. 16-17, together with the commentaries on the various questions.

by Galileo in questions PQ3.6 to 6.4, namely, with *Tractatus* being the main division, which is divided directly into *capita* that are the equivalent of *quaestiones*, or else is subdivided into *sectiones* and *disputationes*, which in turn are subdivided into *capita*. The main differences are that Menu has no introduction to the *De caelo* (Galileo's PQ1.1 and 1.2), nor does he discuss the number and ordering of the heavenly orbs (Galileo's PQ3.1 and 3.2)—though earlier professors such as Pererius and De Gregorio had sections devoted to these topics. Galileo, on the other hand, has nothing to match Menu's treatment of the properties and activities of the heavens, nor does he discuss generation, action and passion, the transmutation of the elements, and the motive qualities of the elements, all of which are covered at length in Menu's lectures.

A detailed study of the contents of Menu's course reveals particularly strong correspondences with Galileo's treatment in the first part of the exposition of Aristotle's *De caelo*, particularly in the questions relating to the origin and to the unity and perfection of the universe (PQ2.1 through 2.3). There are also significant similarities in the second part of the same exposition, most notably in the questions dealing with the composition of the heavens (PQ3.3 and 3.5) and their possible animation (PQ3.6). In the matters relating to Aristotle's *De generatione* the correspondences are not so striking, although there are some notable similarities in the question dealing with the causes of the elements (PQ5.3) and with the positive character of the active qualities (PQ6.2). Overall, the agreement is remarkably good considering the broad range of questions treated. The divergences in teaching, moreover, are relatively minor: on particular ways of viewing the impossibility of the world's existing from eternity (PQ2.4.23-24); on how intelligences are to be understood as forms of the heavens (PQ3.6.9); and on the ability of elemental forms to undergo intension and remission (PQ5.5.5&9).[18]

b. VALLA

Valla was born in Rome in 1561 and died in 1622.[19] As can be seen in table 1, he is first listed as teaching metaphysics at the Collegio in 1585-1587, after which he began the philosophy cycle from the beginning,

[18] Galileo's teachings are given in English translation in ibid., pars. F23-24, L9, and T5 and T9. Reference to table 5 *supra* will enable one to locate these and other paragraphs readily, since question F in *Galileo's Early Notebooks* is PQ2.4, question L is PQ3.6, and question T is PQ5.5. Henceforth the number given after the second decimal point in the PQ convention designates the paragraph of the physical question as given in *Galileo's Early Notebooks*; thus PQ2.4.23 is simply an alternative way of writing F23.

[19] See chap. 1, note 26.

covering logic in 1587-1588, natural philosophy in 1588-1589, and metaphysics again in 1589-1590. Apart from the materials discussed in the previous chapter, there are only two evidences of Valla's teachings on natural philosophy at the Collegio Romano. The first is a work listed by Sommervogel as the following: "Tractatus de mixtis, inanimatis, imperfectis et perfectis Pauli de Valle Societatis Iesu in Collegio Romano anno 1585, studente et scribente D. Marco Casali, Placentino Cler. Reg. ante religionis ingressum."[20] This work has apparently disappeared, but its being noted by Sommervogel gives evidence that Valla taught the tract *De mixtis* at the Collegio in 1585, probably as part of the third-year material pertaining to Aristotle's *De generatione*, as already explained. If this is the case, Valla would then have been filling out the natural philosophy taught in the previous year by his predecessor, listed by Villoslada as Mutius de Angelis.

A more extensive set of teaching notes pertaining to natural philosophy is a commentary by Valla on Aristotle's *Meteorology*, to which he appended a lengthy treatise on the elements.[21] This is similar in content to the latter part of Menu's course on the *De generatione*, although for Valla it is his *Tractatus quintus* whereas for Menu it is the *Tractatus quartus*. The manuscript containing this material is undated, and there are several possibilities for their identification: (1) it may be a tractate preparatory to the *Tractatus de mixtis* mentioned by Sommervogel and so taught in 1585-1586; (2) it may be the latter part of Valla's course in natural philosophy taught for the second year of the philosophy cycle in 1588-1589; or (3) it may be a further development of the commentary on *De generatione* taught in 1586-1587 and again in 1589-1590 for the third year of the philosophy cycle. The first possibility is unlikely, since in the extant manuscript Valla makes reference to a lengthy treatise he had written earlier, *De mixtis imperfectis*, which is probably the 1585 *Tractatus de mixtis imperfectis* noted by Sommervogel.[22] The second possibility is also unlikely, since the material relating to the elements was taught by Valla in 1585-1586 as part of his commentary on *De generatione* in the third year of the philosophy cycle, not in the second. Moreover, in the projected

[20] Sommervogel, *Bibliothèque*, vol. 8, col. 418.

[21] Cod. APUG-FC 1710; the manuscript lacks foliation.

[22] The reference occurs at the beginning of Valla's *Tractatus septimus. De mixtis perfectis*, whose introductory paragraph reads as follows: "Tractatio de elementis tam in genere quam in particulari sequitur tractatio de mixtis perfectis et inanimatis. Ut Aristoteles dixit, elementa enim sunt elementa in ordine ad mixta, quia elementum est id ex quo aliquid fit inexistente indivisibili specie in aliam speciem, et ideo elementum essentialiter ordinatur ad mixtum. Et quamvis prius agendum esset de mixtis imperfectis, de quibus Aristoteles agit in tribus libris Meteororum, quoniam tamen multa a nobis de his mixtis alibi dicta sunt, ideo hic agendum erit de mixtis inanimatis perfectis. . . ."

series of philosophy textbooks he describes in the preface to the first volume of his *Logica* of 1622, he assigns the tract on the elements to his exposition of *De generatione* and not to that of *De caelo*, where it was occasionally treated by his colleagues.[23] By exclusion, therefore, one is left with the third possibility as the preferred option. This leaves open the likelihood that at one time there were two versions of Valla's *De elementis* apart from the work described by Sommervogel, one taught in 1586-1587 and the other in 1589-1590. (For reasons to be explained presently, it appears to the author that Galileo had access to a version slightly different from the one still extant, and that from it he prepared the second set of questions in MS 46.) With respect to the extant version, however, it might be noted that it is written in a hand remarkably similar to Galileo's in his MSS 27 and 46—apparently a coincidence, but nonetheless an indication that the two hands were contemporary.

A comparison of the questions contained in Valla's *Tractatus de elementis* with Galileo's *Tractatus de elementis* shows that they employ basically the same divisions: the complete unit is the *tractatus* for both, and this is divided by Valla into two *disputationes*; the second *disputatio*, for Valla, is subdivided into *quaestiones*, just as it is for Galileo. Valla's first *disputatio*, on the other hand, is subdivided into five *partes*, and each *pars* is then distributed into questions. Recall that Galileo's first division of his *tractatus* is his *prima pars*, and that this is also distributed directly into questions, following the same pattern here as Valla's. Already remarked is Galileo's failure to identify the divisions properly in his *Tractatus*;[24] since these are all found in Valla, it is not unlikely that Galileo's practice was based on the latter source.

With regard to the content of the two tractates, it is noteworthy that Valla's treatment of the elements is much more complete than Galileo's and is even more extensive than Menu's. Topics Valla treats that have no counterpart in Galileo's questions are the transmutation of the elements, their motive qualities, the individual elements in particular, and their relation to perfect compounds. His ordering of the subject matter is similar to Galileo's but departs from it occasionally: a notable instance is the quantity of the elements, which he considers after their transmutation and before their motive qualities. But in the questions he treats in common with Galileo there are many points of agreement.[25] Correlations are particularly strong throughout the entire treatment of the elements (PQ5.1-5.6), and only slightly less so for the questions dealing with the

[23] *Logica*, vol. 1, fol. 4v; see sec. 1.1.c *supra*.

[24] See sec. 2.1 *supra*.

[25] For details see Wallace, *Galileo's Early Notebooks*, pp. 17-18, together with the commentaries on the various questions.

primary qualities (PQ6.1-6.4). Excellent correspondences also characterize the questions covering the definitions of element (PQ5.2), the causes of the elements (PQ5.3), their forms (PQ5.4), and the positive nature of the primary qualities (PQ6.2). With the exception of a few passages on the causes of the elements (PQ5.3), all of Valla's correlations are better than Menu's for the corresponding questions. The divergences in teaching are again relatively minor: Valla seems partial to Flaminio Nobili's understanding of the forms of the elements, which Galileo rejects (PQ5.4.2); and he sides with Menu in holding that such forms can undergo intension and remission, again contrary to the position taken by Galileo (PQ5.5.5&9).[26]

These few disparities between Valla's *De elementis* and Galileo's corresponding tractate on the elements seem to be a good indication that the lecture notes that survive from Valla's courses on natural philosophy were not those used by Galileo when composing the physical questions. (This assumes that the physical questions in their entirety are as dependent on some exemplar as the logical questions are dependent on Valla-Carbone and questions PQ3.1 and 3.2 are on the *Sphaera* of Christopher Clavius, which seems a reasonable assumption at this point.) Since Valla taught the questions relating to the *De generatione* three times and, as already explained, in the third year of the philosophy cycle rather than in the second, this means that at one time he could have had three series of notes relating to the materials of *De generatione*, namely, those of 1585-1586, 1586-1587, and 1589-1590. For the questions relating to the *De caelo*, on the other hand, he would have had only one version, namely, that offered in 1588-1589, since this is the only time he taught such matter, and, of course, these lectures are not known to be extant. Which of Valla's lecture notes Galileo might have used, if indeed he did so, will be discussed later in this chapter.[27]

c. VITELLESCHI

Vitelleschi taught the philosophy cycle at the Collegio from 1588 to 1591, covering the material pertaining to natural philosophy in 1589-1590.[28] Four manuscript copies of his lectures survive, the first giving his questions on the *Physics* and the *De caelo*, the second his questions on the *De caelo* and the *De generatione*, and the third and fourth his questions on the *Meteorology*.[29] The first manuscript fortunately provides

[26] Ibid., pars. S2, T5, and T9.

[27] See sec. 2.3 *infra*.

[28] See table 1 *supra*; also chap. 1, note 36.

[29] These are listed in Wallace, *Galileo's Early Notebooks*, p. 308, note 16.

indications of the dates on which Vitelleschi completed the principal parts of the course, and from these we know that he had finished a preliminary tract on the division of the sciences by November 10, 1589, at which time he began the *Physics*, and that he had covered the third book of the *Physics* by February 23, 1590.[30] The remaining books of the *Physics* occupied him until May 2, 1590, when he started with the questions on the *De caelo*.[31] No further dates are given, but if one assumes that he continued to cover the material of the *De caelo* at the same rate as that of the last three books of the *Physics*, he would have completed the tract on the heavens by the first week in August 1590. Presumably he then began the questions on the *De generatione*, or else postponed their treatment to the first part of the third-year course (1590–1591), depending on when the *quies* occurred at the Collegio in the summer of 1590. One of the manuscripts containing his exposition of the first three books of the *Meteorology* records 1590 as the year of its composition, but no month is given, and thus it is of no help in deciding this issue.[32]

Vitelleschi's treatment of the *De caelo* and the *De generatione*, when compared with Galileo's reveals that the basic division for both is the *Tractatio*, and that, following the introductory section, there are two such *Tractationes* for both, dealing respectively with *De mundo* and *De caelo*. Whereas Galileo subdivides the *Tractatio* into *Quaestiones*, however, Vitelleschi divides it into *Disputationes*. Vitelleschi, moreover, has many more *Disputationes* than Galileo has *Quaestiones*, with the result that the former's treatment is much more complete than Galileo's, and even more complete than Menu's for the corresponding sections. In addition, Vitelleschi provides a third *Tractatio* entitled *De elementis* as part of his lectures on Aristotle's *De caelo*; this takes up problems relating to *gravitas* and *levitas* and to the motion of the elements that are not covered in Galileo's physical questions, though they are the subject of Galileo's consideration in his *De motu antiquiora*.[33] Finally, Vitelleschi continues the same method of division throughout his lectures on Aristotle's *De generatione*, beginning with a *Tractatio* on alteration and generation that is not titled or identified as such, and then providing very full *Tractationes* entitled *De augmentatione, De actione et passione, De mixtione*, and *De elementis ut sunt materia mixtionis*. Vitelleschi concludes the last *Tractatio* with the remark that this should suffice for his treatment of the elements, for matters that pertain to their shape and quantity have partly been

[30] Cod. SB Bamberg, Msc. Class. 70, fols. 27r and 197v.

[31] Ibid., fol. 285r.

[32] Biblioteca Nazionale Roma, Fondo Gesuitico, Cod. 747.

[33] Secs. 4.2.a and 5.2.c.

explained by him elsewhere (probably a reference to the third *Tractatio* of his lectures on the *De caelo*) and are partly presupposed from the course on the *Sphaera*, which was taught that year by Christopher Clavius.[34]

Vitelleschi's questions show extensive agreement with Galileo's, having counterparts for twenty-one of the latter's twenty-five questions, and in this being similar in coverage to Menu's treatment.[35] Both Menu and Vitelleschi lack Galileo's questions PQ3.1 and 3.2, possibly because these were amply covered by Clavius in his parallel course on the *Sphere* of Sacrobosco; otherwise Menu lacks Galileo's introductory questions PQ1.1 and 1.2, which have counterparts in Vitelleschi's notes, whereas Vitelleschi lacks Galileo's questions PQ5.3 and 5.5, dealing, respectively, with the causes of the elements and with whether or not their forms undergo intension and remission, which have counterparts in both Menu's and Valla's notes.

Detailed textual comparisons indicate good agreement between Galileo and Vitelleschi throughout the entire range of the physical questions, with particularly strong agreement in the treatises devoted to alteration (PQ4.1 through 4.3) and to the primary qualities of the elements (PQ6.1 through 6.4). The introductory section, absent in both Menu and Valla, shows fair agreement with Galileo's corresponding questions. The treatise on the universe and the heavens (PQ 2.1 through 3.6) agree less than Menu's with Galileo's similar treatises overall, but it is noteworthy that on particular questions, viz., whether the universe could have existed from eternity (PQ2.4) and whether the heavens are incorruptible (PQ3.4) and animated (PQ3.6), Vitelleschi's treatment is closer to Galileo's than is Menu's. Moreover, in the treatise relating to the *De generatione*, Vitelleschi shows much better agreement for the tracts on alteration and on primary qualities than does Menu, and is even slightly closer than Valla for the second of the two. Again the differences in teaching are quite insignificant: Vitelleschi disagrees with Galileo's views that no change has ever been observed in the heavens (PQ3.4.24) and that the heavens are composed of matter and form (PQ3.5.57 and related texts); he offers a different definition of resistance (PQ6.4.4); and he does not maintain, as does Galileo, that all qualities that are more active are also more resistive (PQ6.4.19).[36]

[34] Cod APUG-FC 392, last folio, which concludes with the sentence. "Et hec de elementis, nam quae spectant ad eorum figuram et quantitatem partim explicata sunt a nobis alibi, partim supponimus ex *Sfera*."

[35] Wallace, *Galileo's Early Notebooks*, pp. 18-19, provides a general conspectus; more details are given in the commentaries on the various questions.

[36] Ibid., pars. J24, K57, Y4, and Y19.

d. RUGERIUS

Rugerius is the final Jesuit professor whose writings are considered in this section.[37] He followed Vitelleschi in the philosophy cycle, teaching logic in 1589-1590, natural philosophy in 1590-1591, and metaphysics in 1591-1592. His complete philosophy course survives in a set of manuscripts that provide the most detail for any professor around this time about the content of the course and the dates on which various tracts were covered.[38] From these manuscript indications it is known that Rugerius began the *Physics* of Aristotle on November 3, 1590, and completed it on May 21, 1591; on the same date he began to treat questions relating to the *De caelo*, which he continued to teach until August 7, 1591, at which time he began the *De generatione*.[39] Coincident with his completing the second book of the *De caelo* in early July, and concurrently with his coverage of its remaining two books, he also taught a parallel course on the *Meteorology*.[40] There is no clear indication as to when the summer *quies* took place in 1591, but this was probably during September and October of that year. In any event the *De generatione* was resumed in November and brought to completion on February 12, 1592.[41] On that very day Rugerius began to comment on the fourteen books of Aristotle's *Metaphysics*, a task that kept him occupied until the end of August 1592; three days later, on February 15, 1592, he began a parallel commentary on the three books of the *De anima*, which ended on September 18, 1592.[42] From his marginal indications about the number of lectures devoted to particular topics, one can see that he gave 207 lectures on the *Physics*, 74 on the *De caelo*, 125 on the *De generatione*, and 99 on the *Metaphysics*. There are no such marginal notes for the courses on the *Meteorology* and the *De anima*, but the numbers of pages of notes devoted to their subject matters would correspond to 75 and 198 lectures, respectively, in the pagination of similar materials.

Rugerius's basic unit for dividing his course on natural philosophy is the *Disputatio*, and in the more extended tracts he further subdivides this into the *Tractatus*, which he divides in turn into the *Quaestio*. Usually Rugerius's *Quaestiones* are quite long, and when this is the case he subdivides them into smaller queries that he entitles *Quaesita*. Even a cursory examination of the many *Quaesita* relating to the subject matters of the

[37] See chap. 1, note. 38.
[38] Cod. SB Bamberg, Msc. Class. 62-1 through 62-7.
[39] Ibid., 62-3, fols. 2r, 355v; 62-4, fols. 3r, 114v, 120r.
[40] Ibid., 62-5, fol. 1r.
[41] Ibid., 62-4, fols. 138v, 281r.
[42] Ibid., 62-7, fols. 3r, 191r; 62-6, fols. 1r, 399r.

De caelo and the *De generatione* will show that these are far more numerous and detailed than those in Galileo's physical questions. Yet there are correspondences for twenty-one of Galileo's twenty-five questions in Rugerius's lectures, following much the same pattern as has already been seen with Vitelleschi.[43]

Textual comparisons manifest considerable agreement between Galileo and Rugerius over the entire range of questions, though not quite as good as that between Galileo and Vitelleschi. The one treatise for which Rugerius is superior to all others is the introductory treatise (PQ1.1 and 1.2). Agreement is also strong in the treatise on alteration (PQ4.1 through 4.3) and on primary qualities (PQ6.1 through 6.4), although in neither of these is it as good as with Vitelleschi. On individual questions within particular treatises, however, Rugerius sometimes shows better agreement than any other author. This is true of the question about whether the universe could have existed from eternity (PQ2.4) and of the question on the parts and degrees of quality (PQ4.3). Yet overall, in the questions relating to the *De caelo*, Rugerius's coincidences with Galileo's text are the lowest in number of all the professors whose notes are being compared. In matters relating to the *De generatione*, on the other hand, Rugerius's coincidences are comparable to Vitelleschi's and superior to Menu's, but considerably fewer than Valla's. Once again there are no extensive disagreements in doctrine: Rugerius departs from Galileo on a particular way in which the world could have existed from eternity (PQ2.4.24), on the fact that no alteration has ever been observed in the heavens (PQ3.4.24), on how intension is related to the essence of a quality (PQ5.6.34), and on the validity of Flaminio Nobili's views regarding the nature of resistance (PQ6.4.8).[44]

3. Textual Comparisons

Although the textual coincidences between Valla's extant *Tractatus de elementis* and Galileo's corresponding treatise are not as striking as those between his plagiarized *Additamenta* and Galileo's *De praecognitionibus*, there are nonetheless a significant number of passages in the *Tractatus* that register close agreement with Galileo's notes. A selection of these will now be exhibited to show how much they resemble the correspondences shown in the previous chapter, and thus argue for a dependence of the physical questions on Valla analogous to that established in chapter

[43] Details are given in Wallace, *Galileo's Early Notebooks*, pp. 19-20, and in the commentaries on the various questions.

[44] Ibid., pars. F24, J24, U34, and Y8.

1. As previously noted, the questions dealing with the elements show particularly good agreement, and thus the main illustrations for this section will be drawn from that treatise.

a. TREATISE ON THE ELEMENTS

Galileo's question on the definition of the elements (PQ5.2) begins with an enumeration of a number of ways of defining the elements, the opening paragraphs of which are set out below on the right, arranged in parallel column with the corresponding passages in Valla's lecture notes on the left:

Valla	Galileo
Elementum *potest dupliciter definiri* sicut quacumque alia res: *primo metaphysice, per genus et differentiam; secundo, physice, per* omnes *causas*: omnis enim *cognitio*, ex *primo Physicae in initio* et ex *octavo Metaphysicae* etiam in initio, *ex causarum cognitione* dependet. Utroque autem modo possunt aliquid dupliciter definire: primo, *secundum se*, sicut definit verbigratia metaphysicus materiam primam nec esse quid nec quale, etc.; secundo, *in ordine ad aliud*, sicut definitur materia a physico, quae subiectum primum, etc.	Nota, primo, res naturales *posse dupliciter definiri*: primo *metaphysice, per genus et differentiam; secundo, physice, per* suas *causas*; tunc enim aliquid physice cognoscimus, cum eius causas scimus, *primo Physicae t. primo* et *octavo Metaphysicae*, ultimo; tertio, possunt definiri *secundum se*, vel *respective*.
Iuxta hos duos modos definiendi possumus *hoc modo definire elementum metaphysice: est corpus corruptible, simplex*; ubi corpus ponitur loco generis, *corruptibile* vero et *simplex* loco differentiae, *ut distinguatur* elementum *a mixtis*, corruptibile, et *a caelo corpore* simplici incorruptibili.	Totidem igitur modis definiri potest elementum. *Primo* enim *modo sic definitur: elementum est corpus corruptibile, simplex*. Dicitur *corruptible ad distinctionem caelorum*; dicitur *simplex, ad differentiam mixtorum*.
Physice, vero, secundum se *sic definiri potest: est corpus compositum ex materia et forma simplici a Deo productum ad* integritatem et *perfectionem universi*, incorruptibile secundum totum sed non secundum partes, in qua definitione sunt omnes quatuor causae.	*Secundo modo sic describitur: est corpus compositum ex materia prima et forma simplici, productum a Deo ad perfectionem universi.*
{Secunda definitio quae communiter afferi solet et explicari ab omnibus habetur *quinto Metaphysicae, t. 4*. Est: elementum *est illud ex quo aliquid primo* fit vel *componitur, inexistente, indivisibile specie in aliam speciem.*}	Tertio tandem modo varie definitur ab authoribus. Prima est Aristotelis, *quinto Metaphysicae, t. 4: est id ex quo aliquid componitur, primo, inexistente, indivisibili specie in aliam speciem.*

{Notandum tertio Aristotele variis in locis diverso modo *definire* elementa. *Septimo* enim *Metaphysicae*, t. *ultimo*, ita definit: *est id* in *quod existens* res *dividitur ut* in *materia.*}[45]

Secunda definitio est eiusdem, *septimo Metaphysicae, ultimo*:elementum *est id quod existens dividitur ut materia.*[46]

Note that Galileo starts by saying that natural things can be defined in two ways, metaphysically and physically (lines 1-6), and then goes on, inconsistently, to add a third way of defining natural entities, viz., absolutely or relatively (lines 8-11). Valla's parallel passage begins with the same twofold distinction (lines 1-7), but explains that in each of the two ways enumerated there is the possibility of a further twofold subdivision into absolutely or relatively (lines 7-13). Obviously Valla's way of presenting the matter is internally consistent whereas Galileo's is not; it thus appears that Galileo, in attempting to abbreviate a lengthier discussion of various definitions, adapted Valla's subdivision into a third way of defining but neglected to go back and change his *dupliciter* (line 1) to the *tripliciter* called for in his exposition. This likelihood is reinforced by the fact that Valla's treatment is much longer than Galileo's. A representative sample of Galileo's excerpting is provided in the parallel passage of lines 7-24. Finally, when Galileo comes to elaborate his third way of defining (line 27), he goes on to include under this seven different definitions drawn from Aristotle, Galen, Avicenna, and the Stoics (PQ5.2.5-11). Only the first two of these are given in the text shown (lines 28-35), and it is noteworthy that parallels for both (as for the others) can be located in Valla's notes, though a folio or two removed from the passages reproduced on lines 1-26 and so enclosed in braces.

A similar type of evidence occurs in the question on the forms of the elements (PQ5.4), to which both Valla and Galileo reply that these are substantial forms hidden from us but knowable through their qualities. Alessandro Achillini, to the contrary, taught in his *De elementis* that the forms of the elements are their motive qualities, *gravitas* and *levitas*, a solution rejected by both Valla and Galileo as part of their initial response. In developing their arguments against Achillini's teaching, Valla and Galileo then present objections drawn from Achillini and various counterarguments that can be used against them, as shown in the following parallel passages.

[45] Cod. APUG-FC 1710, Tractatus quintus, De elementis; Disputatio prima, De elementis in genere; Pars prima, De essentia elementorum; Quaestio quinta, De definitionibus elementi in genere.

[46] *Opere* 1:126.9-22; Wallace, *Galileo's Early Notebooks*, pars. Q2-Q6.

VALLA

Respondet primo *Achillinus . . . gravitatem et levitatem dupliciter sumi posse: primo, in actu primo, et hoc modo est forma substantialis; secundo, in actu secundo, et hoc modo est qualitas* et accidens elementi. *Contra:* quia . . . *operatio gravitatis est* motus vel *actio; ergo male ponitur* ab Achillino *in qualitate. . . . gravitas et levitas* aliquibus *sunt* accidentia, ut patet *in Sanctissimo Sacramento, ubi nulla est substantia* naturalis, ergo. *Respondet* Achillinus *ibi non esse gravitatem* in actu primo sed in actu secundo, *supplente Deo gravitatem* in actu primo. *Contra:* ibi est gravitatio, quae fieri non potest sine gravitate in actu primo . . . ; secundo, *non sunt multiplicanda miracula sine necessitate.*[47]

GALILEO

Respondet Achillinus, gravitatem et levitatem posse dupliciter considerari: vel in actu primo, et sic sunt formae substantiales elementorum; *vel in actu secundo,* idest ratione gravitationis et levitationis, *et sic dicuntur qualitates.* Sed *contra: gravitatio* et levitatio *sunt in praedicamento actionis* et passionis; *ergo non sunt qualitates.* Adde, quod *in Sanctissimo Sacramento est gravitas et levitas; et tamen ibi nulla est substantia. Respondet, Deum facere ibi gravitatem* et levitatem, *loco gravitatis* et levitatis *quae non adsunt. Contra: non sunt multiplicanda* nova *miracula sine necessitate.*[48]

Galileo's arguments here are obviously the same as Valla's, though expressed in slightly different terms and somewhat abbreviated. Here, as in the previous textual comparison, Galileo's attempt to be more concise leaves a telltale trace. In lines 9-10 Valla uses a theological argument against Achillini and states that there is no *substantia naturalis* in the Eucharist. Galileo contracts his parallel argument (lines 8-10) to say that there is no *substantia* there, apparently unaware that such a statement is theologically inaccurate in view of the Church's doctrine on transubstantiation. The remaining text (lines 10-16) provides another instance of contraction: Valla gives two refutations of Achillini's counterargument, whereas Galileo settles for his second, in terms well known to every student of scholastic theology.

The following question (PQ5.5) continues the discussion of the forms of the elements but takes up the special problem of whether or not they undergo intension and remission. In his exposition Galileo provides a brief prologue explaining the sense of the question and then gives the first position that is commonly defended, that of the Averroists, after which he takes up the second position, that of various scholastics and others. Valla's ordering is the opposite, for he begins with a lengthy explanation of scholastic differences on this topic and then presents the Averroist teaching. Apart from the ordering, however, Galileo's account is practically the same as Valla's, as can be seen from the following parallels:

[47] Cod. APUG-FC 1710, Tract. 5, Disp. 1, Pars 1, Q3, De forma et materia elementorum.

[48] *Opere* 1:131.31-132.6, emending Favaro's *praedicato* to read *praedicamento* (132.2); Wallace, *Galileo's Early Notebooks*, par. S12.

Valla

Secunda *sententia est Averrois*, tum alibi tum praecipue *tertio Caeli, com.* 67; Galeni, primo De elementis; Genuae, tertio Physicae, lect. 7; *Achillini, secundo De elementis, dubio tertio; Janduni, octavo Metaphysicae, quaes.* 5; *Contareni, tertio De elementis*, pag. 47 et 48; *Zimarae, propositione 20; Pauli Veneti et Suessani, primo* De *generatione, in fine*; Hieronymi *Tagliapetrae, libro secundo, tract.* 4; et communis Averroistarum, *qui asserunt formas elementorum intendi et remitti* et habere gradus sicut habent qualitates. *Eandem sententiam* multo magis *debent defendere Scotistae, ut Antonius Andreas, in Metaphysica, quaestione unica; Ioannes Canonicus, quinto Physicae, quaestione* ultimo; *et Pavesius in tractatu De accretione; qui dicunt omnem formam quae habet esse in materia, per quod excluditur animal rationalis*, posse intendi et remitti pro variatione suarum dispositionum. Eandem etiam sententiam debent defendere Graeci et hi omnes qui dicunt formas essentiales elementorum esse vel qualitates motivas vel activas. Eandem etiam defendit Nobilius, primo De generatione in fine in quaestione De mixtione, et indicit cap. 3, quaes. 10.

In hac tamen sententia *sunt varii modi dicendi.* Primus est *Contarenus*, qui *ait formas elementorum non induci successive ad* introductionem *qualitatum, sed fieri alterationem usque ad eum gradum ex quo sequitur formam mixti, et sub quo formae elementorum integrae esse non possunt; tunc autem, introducta forma mixti, introducuntur etiam formae elementorum in esse refracto.* . . . {Alii, tertio, ut *Achillinus*, dicunt *formas elementorum secundum partem aliquam minimam induci in instanti, facta prius debita dispositione; reliquum vero* formae induci *successive* eo modo quo alibi dixi.} *Alii dicunt formas elementorum intendi et remitti iuxta intensionem et remissionem qualitatum* . . .[49]

Galileo

Prima *opinio est Averrois, tertio Caeli, com.* 67; Niphi [i.e., *Suessani*] *et Pauli Veneti in fine primi De generatione; Zimarae propositione 20; Taiapetrae, libro secundo, tract.* 4; *Janduni, octavo Metaphysicae, quaes.* 5; *Achillini, libro secundo De elementis, articulo tertio; Contareni, libro tertio*; Alexandri octavo Metaphysicae, text. 10; *qui omnes dicunt formas* substantiales *elementorum intendi et remitti.Quibus addi potest Scotus*, octavo Metaphysicae, quaes. 3, *quem sequitur Antonius Andreas*, undecimo *Metaphysicae, quaestione prima; Pavesius, in libro De accretione; Ioannes Canonicus, quinta Physicae, quaestione* prima; *qua idem affirmant de quacunque forma* substantiali *quae educatur de potentia materiae, ut excludatur anima rationalis.*

In modo vero quo formae intendantur et remittantur, *non conveniunt. Contarenus* enim *dicit formas elementorum non introduci successive ad* intensionem *qualitatis, sed fieri alterationem* qualitatum *usque ad certum gradum ad quem sequitur forma mixti, et sub quo formae elementorum non possunt esse integrae; et, adveniente forma mixti*, tum primum *formae elementorum incipiunt refrangi.* Secundus modus est *Achillini*, qui videtur dicere *formam elementorum secundum aliquam partem minimam introduci in instanti, facta prius debita dispositione; postea vero*, intendi et remitti *successive. Alii* vero tertio communiter *dicunt formas elementorum intendi et remitti* successive, *ad intensionem et remissionem qualitatum.*[50]

[49] Cod. APUG-FC 1710, Tract. 5, Disp. 1, Pars 1, Q.4, An formae substantiales elementorum intendantur et remittantur.

[50] *Opere* 1:133.8-27; Wallace, *Galileo's Early Notebooks*, pars. T2 and T3.

All of the authorities cited by Galileo can be found in Valla with the exception of Alexander of Aphrodisias (line 9), who undoubtedly is covered in the *Graeci* to whom Valla refers in line 22. Note, however, that Galileo's *anima rationalis* (lines 19-20) is more correct than Valla's *animal rationalis* (line 19), since the rational soul is a form whereas a rational animal is not—a possible clue that Galileo used a revised version of the notes now extant from Valla's course. Otherwise the agreement for the remainder of the exposition (lines 28-44) is excellent, the only difference being that Galileo interchanges the second and third ways of undergoing intension and remission as these are found in Valla and again has a much abbreviated account of the differences between them.

The final question in the tractate on the elements is concerned with their number and quantity (PQ5.6). There are a number of passages within this question that show extensive parallels with Valla's treatment, and two of these will be reproduced below. The question lacks a title in Galileo's manuscript, and its structure is quite complex: first it inquires into the number of the elements, then into their size and shape, and finally whether they and other natural things have termini of largeness and smallness. The latter section is by far the longest of the question. It begins with eight notations that are necessary to formulate the problem, then gives four opinions or solutions that characterize the various schools, following which it formulates the preferred position, presumably Galileo's own, in ten conclusions, and finally ends with replies to arguments in support of the first solution. The four opinions as listed by Galileo are given below, set off against the corresponding passages in Valla's treatise:

VALLA

Prima sententia est eorum qui dicunt omnes res naturales praeter elementa habere terminos magnitudinis et parvitatis, eosque extrinsecos, elementa vero habere terminum intrinsecum parvitatis, non autem magnitudinis. Ita tenet D. Thomas, primo Physicae in text. 36 et 38, et secundo De anima, text. 41, *et Prima Parte, quaes. 7, art. 3; Capreolus, secundo* Sententiarum, *dist. 19; Sotus, primo Physicae, quaes. 4;* Ferrariensis, quaes. 7; *et communiter Thomistae;* Philoponus etiam, primo Physicae, loco citato. *Probatur primo auctoritate Aristotelis in text. illo 36; et ibidem text. 38 affert argumentum* quoddam *contra antiquos quod nihil valet nisi supponatur dari minimum. Ita enim argumentari contra Anaxagoram: si ex quolibet potest separari quodlibet, ergo datur minus minimo. Nec valet di-*

GALILEO

Prima opinio est dicentium omnes res naturales, praeter elementa, habere terminos magnitudinis et parvitatis intrinsecos; elementa vero habere terminum intrinsecum parvitatis, magnitudinis vero nullum. Ita D. Thomas in *primo Physicae, text 36, 38,* De generatione, text 41, *et Prima Parte, quaes. 7, art. 3; Capreoli, secundo* [Sententiarum], *dist. 19; Soti, primo Physicae, quaes. 4;*

et Thomistarum omnium.

Probatur primo authoritate Aristotelis, primo Physicae, 36, ubi hoc dicit; *et text. 38 affert argumentum contra antiquos quod, si non datur* maximum et *minimum, nihil valeret. Sic enim concludit contra Anaxagoram: si ex quolibet potest quodlibet separari, non datur minimum;* sed omnes res naturales habent minimum; ergo [etc.] *Nec dicas, Aristotelem supponere*

cere Aristoteles hoc supponere ad hominem, quia refert Simplicius ibidem, com. 34, Anaxagoram negasse minimum, et certe si loquebatur consequenter debuit ipse negare. *Confirmatur ex secundo De* [anima], *text. 41, omnium natura constantium certus est terminus magnitudinis et parvitatis*, quod idem est dicere omnes res naturales habent maximum et minimum. . . .

hoc contra Anaxagoram; nam refert ibi Simplicius, com. suo *34, Anaxagoram negasse minimum*; et hoc videtur verisimile, cum existimarit quodlibet ex quolibet posse separari. *Praeterea*, Aristoteles, *secundo* De anima, *41, docet omnium natura constantium esse certum terminum magnitudinis et parvitatis.* . . .

Secunda sententia est aliorum . . . Idque *videtur sentire Averroes*, comm. 37 et 28, *septimo Physicae com. 2; octavo Physicae, comm.* 62 et 64, *tertio Caeli, com. 9*, et secundo De anima, com. 41; *Themistius, primo Physicae in text. 37; Iandunus, quaes. 16*, Achillinus, tertio De elementis, dubio primo; Thiene, primo Physicae, quaes. 15; *Zimara, contradictione* 23, primi Physicae . . . {*dicentium omnes res naturales habere terminos magnitudinis et parvitatis, sed non explicant an illi sint intrinseci an vero extrinseci.*}

Secunda sententia est Averrois, in omnibus locis supra citatis ex Aristotele, et praeterea in sexto Physicae, 32, 91, et *octavo Physicae, 62, et septimo Physicae, 2, tertio Caeli, 9*, et alibi; *item Themistii, primo Physicae text.* 36 et *37; Ianduni, primo Physicae, quaes. 16; Zimarae in solutione contradictionis* 29, in text. 41 secundi De anima: *qui omnes dicunt res quascunque*, etiam elementa, *habere terminos magnitudinis et parvitatis; non tamen statuunt utrum intrinsecos an extrinsecos.*

Tertia sententia est Pauli Veneti, primo Physicae, text 38, qui ait illas res naturales *quae habent maximum vel minimum habere hos terminos extrinsecos*, non autem intrinsecos, quod probat *quia alioquin sequeretur substantiam desinere per ultimum sui esse. Si enim datur* maximum et minimum intrinsecum, datur ergo *maximus equus, et verberetur ita ut debeat tumefieri, toto tempore tumefactionis est non amplius forma equi. Tunc enim esset equus maior maximo et instanti ultimo esse talis formae; ergo*, illa forma *desinit* per instans *intrinsecum* et *per ultimum sui esse.* . . .

Tertia sententia est Pauli Veneti, primo Physicae, 38, dicentis, illa quae habent terminos, habere extrinsecos, tam quoad magnitudinem quam quoad parvitatem; quia alias sequeretur substantiam desinere per ultimum sui esse; quod est absurdum.

Et probatur sequela: *nam, si equus maximus verberibus intumescat, toto tempore tumefactionis*, quae incipit in tempore, *non est amplius forma equi, ne sit equus maior maximo; et tamen erat in instanti ante tumefactionem; ergo*, si habet terminum *intrinsecum, desiit per ultimum sui esse.* . . .

Quarta sententia . . . *Ita tenet Scotus, secundo* Sententiarum, dist. *2, quaes. 9, Occam, quaes. 8*, Burleus . . . Albertus autem primo Physicae, quaes. 9 et 10, ait excepto homine, in quo datur minimum via generationis, in ullo alio corpore naturali dari terminum ullum magnitudinis et parvitatis. {*Est viventia et heterogenea habere maximum et minimum intrinsecum, homogenea vero, qualia sunt mixta inanimata et elementa, nihil horum habere.*}[51]

Quarta sententia est Scoti in secundo [Sententiarum, dist.] *2, quaes. 9, Occam*, secundo, *quaes. 8*, Pererii, libro 10, cap. 23:

qui dicunt omnia heterogenea, ut viventia et aliqua mixta, habere terminos magnitudinis et parvitatis intrinsecos; homogenea vero, ut elementa et quaecunque *mixta homogenea, neque habere maximum neque minimum ullo modo.*[52]

[51] Cod. APUG-FC 1710, Tract. 5, Disp. 1, Pars 4, De quantitate elementorum; Q.1, An dentur maximum et minimum in elementis.

[52] *Opere* 1:144.15-29, 145.11-31; Wallace, *Galileo's Early Notebooks*, pars. U25, U30-U32.

The parallels throughout all four opinions, which follow the same ordering in both treatises, are quite striking. The first school treated in each is that of the Thomists, among whom Valla enumerates Capreolus, Soto, and Ferrariensis in addition to Aquinas, whereas Galileo omits the reference to Ferrariensis (line 10). Galileo also slips in explaining the first proof, which states that Aristotle's arguments against the ancients in text 38 of the first book of the *Physics* would be pointless if there were no minimum (lines 13-16); Galileo states in English translation, "if there were no maximum and minimum" (lines 13-14), and in this context the words "maximum and" make no sense. Either this was a *lapsus calami* or he failed to grasp the argument as properly formulated by Valla. The discussion of the second opinion is very similar in each, except that here too Valla enumerates more authorities than does Galileo. Valla also gives the school's teaching directly after identifying it as the second opinion, for his words encased in braces on lines 36-39 are taken from the ellipsis indicated in his line 28. The same inversion occurs in the fourth opinion, with his lines 60-63 being taken from the ellipsis in his line 53. Note that Galileo cites Pererius as holding the fourth opinion, whereas this name is missing in Valla. Otherwise the parallels are so close that one would have to admit a genetic connection between the two texts. If Valla's extant *Tractatus de elementis* is not the version of his notes that Galileo used, and this seems likely to the author, Valla probably changed very little in his enumeration of *sententiae* in successive revisions—a practice he himself acknowledges in the preface to the second volume of his *Logica* of 1622.[53]

The second passage to be compared occurs toward the end of the question, in Galileo's tenth conclusion, where he is maintaining that powers (*potentiae*) have a *maximum quod sic* but no minimum—a terminology that will be explained below in conjunction with an excerpt from Pererius's *De communibus*.[54] In this particular set of parallels the concept being discussed is lack of power or impotency (*impotentia*), and the respective texts read as follows:

VALLA

Est tamen hic difficultas *de impotentia* naturali, quia *Aristoteles, primo Caeli text. 116, ait impotentiam determinari per minimum quoad parvitatem, ut, verbi gratia, minima impotentia portativa est illa quae non potest portare aliquod minimum supra id quod potest maxima potentia*, ut si, verbi gratia, maxima potentia portativa potest portare centum *lib-*

GALILEO

De impotentia vero dicit Aristoteles et Averroes, *primo Caeli, 116, eam determinari per minimum quoad parvitatem; ut, verbi gratia,* dicant, *minimam impotentiam portativam esse illam quae ferre non potest aliquam minimam* particulam, verbi gratia, unam dragmam, *supra eam quam potest ferre maxima potentia.*

[53] See sec. 1.1.c *supra.*

[54] See sec. 2.3.d *infra.*

ras, minima impotentia erit illa quae non potest portare minimum quid supra centum.

Haec autem doctrina *videtur falsa, quia tunc sequeretur dari indivisibilia immediata. Maximum enim ad quod terminatur potentia consistit in indivisibili, ergo id quod superabitur* ab impotentia *debet esse etiam quid indivisibile.* Si enim sit divisibile, dividatur in duas partes, et una pars addatur obiecto maximae potentiae: vel potentia potest illud totum, vel non. *Si potest, ergo illud prius non erat maximum respectu potentiae; si autem non potest, ergo* illud quod erat assignatum *non erat minimum respectu impotentiae.*

Hoc tamen nihil concludit, quia *impotentia est* tantum *privatio, ergo debet habere tantum terminum* negativum et *extrinsecum, et ideo ille idem terminus qui est maximum potentiae et minimum impotentiae, et respectu potentiae est intrinsecus, respectu* vero *impotentiae extrinsecus.* Verbigratia, *potentia* portativa *potest* ferre *centum libras et nihil maius. centum librae sunt maximum potentiae huius, et minimum impotentiae* portativae. *Aristoteles autem* loquitur de maximo et minimo physice et secundum nos, et ideo *dixit impotentiam esse definiendum per minimum secundum nos* et physice, non autem mathematice, *ita ut terminus impotentiae* physicus sint *centum librae, verbigratia, cum uno minuto,* quamvis ille simpliciter non sit terminus talis impotentiae.[55]

Itaque *minima impotentia erit quae non potest ferre unam dragmam supra* 1000 *libras.*

Sed hoc videtur falsum. Nam terminus maximae potentiae, ut dixi, est intrinsecus et *indivisibilis; ergo quicquid est ultra ipsum, debet esse divisibile; alioquin duo indivisibilia erunt immediata;*

minima autem impotentia nil potest supra maximam potentiam; ergo non potest ulla ratione determinari per minimum; quia non datur minimum quod non potest. . . .

Respondeo: *impotentia est privatio; quare* non debet habere aliquos terminos extrinsecos, sed *debet terminare extrinsece,* per terminos potentiae cui adiungitur; *itaque minima potentia, quae est adiuncta maximae potentiae, habet eundem terminum cum maxima potentia; sed ille est intrinsecus potentiae, extrinsecus impotentiae. Cum ergo maxima potentia sit quae potest 100 libras, minima erit impotentia quae nihil potest supra 100 libras. Aristoteles autem, qui determinat impotentiam per minimum, debet intelligi quoad nos,* respectu quorum datur quantitas aliqua minima; *quare, secundum nos, minima impotentia est illa quae, supra 100 libras, ne dragmam quidem ferre potest.*[56]

Here attention should be directed to the example (*verbi gratia,* lines 3-10) in both texts, where Galileo is apparently abbreviating Valla's reasoning. Valla states that, if the maximum carrying power can carry 100 pounds, the minimum impotency would be one that could not carry a smallest amount (*minimum quid*) beyond 100 pounds. Galileo makes Valla's "smallest amount" more specific by rendering it as one dram (*unam dragmam*), and enlarges the maximum to 1000 pounds, but fails to state the antecedent on which his conclusion in lines 9-10 is based. His illustration should read, in English translation: "Thus, if the maximum power can

[55] Cod. APUG-FC 1710, Tract. 5, disp. 1, Pars 4, Q.1, An dentur maximum et minimum in elementis.

[56] *Opere* 1:155.22-33, 156.1-11, emending Favaro's *1000 libras* to read *100 libras* (156.7, 7-8, 11) as found in the manuscript; Wallace, *Galileo's Early Notebooks,* pars. U73-U75.

carry 1000 pounds, the minimum impotency would be one that cannot carry one dram beyond 1000 pounds." Lacking this antecedent, whose equivalent is in Valla, Galileo's reasoning does not make sense. Later on, at lines 31, 32, and 37, Galileo reverts to Valla's example of 100 pounds rather than the 1000 he had earlier substituted for it. Throughout this section of his manuscript, in fact, he vacillates between 100 and 1000 pounds in his examples; in the National Edition Favaro emended Galileo's numbers to read 1000 each time such an example occurs, thus making Galileo's exposition consistent but obscuring its relationship to the source on which it was based.[57]

b. TREATISE ON PRIMARY QUALITIES

The final tractate of Galileo's physical questions is concerned with the primary qualities of the elements, sometimes referred to as alterative qualities so as to distinguish them from motive qualities such as *gravitas* and *levitas*. The second question in this treatise (PQ6.2) treats a problem much discussed among sixteenth-century Aristotelians, namely, whether all primary qualities (hot, cold, wet, and dry) are positive, or whether some are merely privative. This is equivalent to inquiring whether coldness is a quality by itself or simply the absence of heat. (The same question was frequently raised, as will be seen in a later chapter, with regard to *levitas*, asking whether it is a positive entity in its own right or merely the absence of *gravitas*.[58]) Galileo begins this question by enumerating a series of opinions, three of which differ from his own solution, which he lists as a fourth *sententia*. Here, as in the previous enumeration of opinions from PQ5.6, Galileo's account correlates very well with Valla's. The agreement can be seen in the following parallels:

VALLA	GALILEO
Prima sententia {defenderunt aliqui ex antiquis, referente Plutarcho in opusculo De primo frigido} est Cardani . . . secundo De subtilitate, ubi ait duas tantum esse qualitates reales et positivas, calorem scilicet et humiditatem; *frigus vero et siccitatem esse harum qualitatum privationes . . .*	*Prima sententia fuit quorumdam antiquorum, apud Plurarchum libro De primo frigido, qui dixerunt frigiditatem esse privationem caloris.* Horum sententiam secutus est *Cardanus, libro secundo De subtilitate,* qui *idem affirmat etiam de siccitate, quam vult esse privationem humoris.*
Potest autem haec sententia *probare* his rationibus: *primo, quia Aristoteles saepe appellat frigus privationem caloris, et secundo De generatione, t. 32 . . .*	*Probatur, primo, ex Aristotele,* secundo Caeli 18, primo De generatione 18, *secundo De generatione 32,* duodecimo Metaphysices 22, *et alibi· ubi frigus appellat privationem.*

[57] *Opere* 1:154.24-156.11; Wallace, *Galileo's Early Notebooks*, pars. U68-U75, with their respective commentaries.

[58] See sec. 4.2.a *infra*.

Secundo: in febribus tertianis repetit quidam frigor qui secundum medicos *fit ex privatione caloris. Confirmatur in morte: statim corpus evadit frigidissimum, recedente anima et calore . . .*	*Secundo: febrium tertianarum rigor* non *fit* adveniente aliqua materia frigida, sed *per solam absentiam caloris;* et *idem dici potest de aliis multis, quae sola caloris absentia frigefieri videntur; et cum animal moritur,* cum ascendunt vapores ad mediam regionem aeris, etc.
Confirmatur: quia in libello De longitudine et brevitate vitae [Aristoteles] *docet omnes operationes fieri a calore,*	*Tertio: Aristoteles, in libro De longitudine et brevitate vitae, docet omnes operationes fieri a calore*: at, si frigiditas esse qualitas realis, aliquid efficeret; ergo. *Confirmatur: quia*
ob quam causam medici dicunt frigus non ingredi opus naturae .	*medici dicunt frigiditatem non ingredi opus naturae . . .*
{*Medici autem posuerunt duplex frigus: alterum reale* et positivum, *quod fit ab elementis, alterum vero privativum, quod fit ex recessu caloris . . .*}	Secunda *sententia est quorumdam medicorum, qui,* ob argumenta facta, *ponunt duplicem frigiditatem: alteram realem, ut est illa elementorum, privativam alteram, ut quae fit recedente calore.*
Secunda *sententia fuit aliorum: calores et frigus esse qualitates reales, humidum autem esse substantiam* quandam inseparabilem a corpore, ipso manente,	Tertia *sententia est aliorum dicentium omnes has qualitates reales esse et* positivas, *sed humorem* et siccitatem non *esse* qualitates, sed *substantias*; idest, humorem esse substantiam fluentem, siccitatem vero substan-
et ideo videmus experientia	tiam consistentem. Et probant: quia *nihil*
nihil posse humectare nisi per susreptionem alicuius corporis humidi . . nihil exiccatur nisi per evaporationem corporis illius humidi.[59]	*potest humefieri, nisi per receptionem substantiae humidae,* ut vapores, etc ; item, *nil posse exiccari, nisi per extractionem alicuius substantiae humidae . . .*[60]

This case is somewhat different from the preceding example in that the *sententiae* are not arranged and numbered in the same way by both authors, and, what is more unusual, Galileo's exposition seems expanded somewhat over that found in Valla, rather than excerpted from it as shown heretofore. This is noticeable in lines 8-24 and 30-39, and furnishes another clue that Galileo might have used a slightly later and more fully developed version of Valla's notes than those extant that provide the basis for the foregoing textual comparisons. Otherwise the resemblances are so close that there can be little doubt that the two texts are intimately related and thus that Valla is the likely source of Galileo's physical questions.

[59] Cod. APUG-FC 1710, Tract. 5, Disp. 1, Pars 2, Q.2, An omnes primae qualitates sint activae et positivae.

[60] *Opere* 1:160.13-161 6; Wallace, *Galileo's Early Notebooks*, pars. W1-W6.

c. CORRELATIONS WITH VALLA'S CONTEMPORARIES

Earlier the assertion was made that Valla's *Tractatus de elementis* is not as idiosyncratic as his *Logica* as exhibited in the *Additamenta*, and now evidence will be adduced to explain this further. One of the most interesting illustrations of the agreement to be found among all four of the Jesuits under study in this chapter, when compared with the notes in Galileo's MS 46, occurs in a question already discussed, namely, PQ5.4, where the topic of concern is the forms of the elements. Apart from Achillini's view that such forms are the motive qualities of the elements, another strong contender was the opinion of Alexander of Aphrodisias that they are the primary qualities of the elements, which is the third opinion listed by Galileo in his notes (Achillini being his second).[61] Galileo offers three proofs of Alexander's position, and following the third proof, given below, lists the fourth and final opinion, which turns out to be his own. The two paragraphs as written by Galileo are here shown on the left, with their English translation on the right:

Probatur, tertio: si qualitates non sunt formae elementorum, ergo sunt posteriores formae; sed hoc esse non potest, quia, cum prius conservetur sine posteriori, debebit forma conservari sine qualitate; sed hoc est contra experientiam; ergo [etc.]

Third proof: if qualities are not the forms of the elements, they must be subsequent forms; but this cannot be, because, since the prior is conserved without the subsequent, forms would have to be conserved without qualities; but this is contrary to experience; therefore . . .

Quarta opinio est dicentium formas elementorum esse formas substantiales [space for three words] qualitates nobis occultas. Est D. Thomae, Alberti, et Latinorum, secundo De generatione 16 et tertio Metaphysicorum 27; item Conciliatoris [space for one word] differentiae 13, Aegidii, primo De generatione, questione 19, Ianduni De sensu, quaestione 25 et quinto Physicorum, quaestione 4, Zimara in Tabula, Contareni, primo et septimo De elementis.

The fourth opinion is that of those saying that the forms of the elements are substantial forms hidden from us [but knowable through their] qualities. This is the position of St. Thomas, Albert, and the Latins, in the second *De generatione* 16 and the third *Metaphysics* 27; likewise the Conciliator [clarifying] difference 13; Giles, on the first *De generatione*, question 19; Jandun, *De sensu*, question 25, and the fifth *Physics*, question 4; Zimara in the *Table*; and Contarenus in the first and seventh *De elementis*.[62]

The first paragraph (lines 1-6) are unexceptional, but the second is unusual in that there are two *lacunae* in Galileo's composition, the first at lines 9-10 where space is left for about three words, and the second at lines

[61] *Opere* 1:130.5-20; Wallace, *Galileo's Early Notebooks*, par. S4.

[62] *Opere* 1:130.27-131.3; Wallace, *Galileo's Early Notebooks*, pars. S7-S8.

13-14 with space for about one word. The first paragraph has an obvious parallel in Vitelleschi's notes, where it is treated as an objection to his own position rather than as a proof in support of a different position, as can be seen in the juxtaposed texts:

GALILEO

Probatur, tertio: *si qualitates non sunt formae elementorum, ergo sunt posteriores formae; sed hoc esse non potest, quia, cum prius conservetur sine posteriori,* debebit *forma conservari sine qualitate*; sed hoc est contra experientiam; ergo [etc.][63]

VITELLESCHI (1590)

Obicies: *si qualitates non sunt formae* substantiales *elementorum, ergo sunt illis posteriores; sed hoc est falsum, quia prius conservatur sine posteriori; formae* autem elementorum *sine qualitatibus conservari* non possunt.[64]

The second paragraph, on the other hand, does not have such close agreement with any of the four Jesuits, but it shows some parallel with the notes of Menu, Valla, and Rugerius, as can be seen from the comparison below. Here Galileo's text is shown on the left with the same line numeration as above, and arranged around it the parallel texts of the others:

GALILEO

Quarta opinio est dicentium *formas elementorum esse formas substantiales* [space for three words] *qualitates* nobis *occultas.* Est *D. Thomae, Alberti,* et Latinorum, *secundo De generatione* 16 et tertio Metaphysicorum 27; item *Conciliatoris* [space for one word] *differentiae 13,* Aegidii, primo De generatione, questione 19, *Ianduni De sensu, quaestione 25 et quinto Physicorum, quaestione 4, Zimara in Tabula, Contareni, primo* et septimo *De elementis.*[65]

MENU (1578)

Tertia opinio est communis aliorum qui asserunt *formas elementorum esse* quasdam substantias *occultas* qui explicantur per *qualitates* motivas et alterativas. Ita *Albertus* Magnus, *secundo De generatione,* tr. 2, cap. 7, et *D. Thomas,* primo De generatione super t. 18, *Conciliator, differentia 13, Iandunus,* libro *De sensu* et sensili, *quaestione 25.*[66]

VALLA

Tertia sententia est communis omnium fere Peripateticorum, quia asserunt *formas substantiales elementorum* esse substantias quas-

RUGERIUS (1592)

Altera sententia est communis, differentias essentiales *elementorum esse* veras et proprias *formas substantiales, qualitates* vero

[63] *Opere* 1:130.27-31; Wallace, *Galileo's Early Notebooks,* par. S7.

[64] Cod. APUG-FC 392, [In libros De generatione disputationes]; Liber secundus, Tractatio de elementis ut sunt materia mixtionis; Disputatio [tertia], Quae sint formae elementorum.

[65] *Opere* 1:130.32-131.3; Wallace, *Galileo's Early Notebooks,* par. S8.

[66] Cod. Ueberlingen 138, [In libros Aristotelis De generatione]; [Tractatus quartus, De elementis]; Sectio prima, De elementis in genere; Disputatio prima, De existentia, quidditate, et causis elementorum; Caput quartum, Quae sint formae substantiales elementorum.

dam *occultas*, quae interdum explicant per *qualitates* motivas, interdum per alterativas. Ita tenent . . . *D. Thomas, secundo De generatione* in t. 24 et primo De generatione, lect. 8, *Conciliator, differentia 13, Albertus* Magnus, secundo De generatione, tr. 2, cap. 7, *Iandunus, De sensu* et sensili, *quaestione* 15 . . .[67]

tam motivas quam alterativas fluere ex illis tanquam passiones proprias. Haec sententia . . . *Alberti* hic, tr. 2, cap. 7, *D. Thomas,* primo De generatione 18 et *secundo De generatione* 6. Legite *Conciliatorem, differentia,* 13, *Iandunum, De sensu* et sensili, quaestione 28, et quinto Physicorum, questione 4, *Zimaram in Tabula . . . Contarenum* in *primo* libro *De elementis*, et alios . . . Propositio: *formae elementorum* sunt vere formae substantiales . . .[68]

Noteworthy is the fact that this is the fourth opinion for Galileo, whereas it is the third for both Menu and Valla, and the second (*altera*, as opposed to his *prima*) for Rugerius. Again, the word *occultas* occurs in Galileo (line 10), Menu (line 10), and Valla (line 22), but not in Rugerius. Rugerius, however, has the fullest citation of authorities, and in fact twenty-five of his words have corresponding terms in Galileo whereas only eighteen of Menu's and Valla's do so (all such terms have been italicized in the respective texts). From none of the three, moreover, can any words be extracted that would fit into the space left blank at lines 9-10 by Galileo, though there seems to be little doubt that all four writers maintain that the forms of the elements are substantial forms hidden from us and yet knowable through their qualities, as indicated in brackets in the English translation at lines 10-11. With regard to the space at lines 13-14, it is also noteworthy that the Jesuits cite the *Conciliator* (Pietro d'Abano) frequently, and in one of these citations (though not in the passage above) Rugerius writes out his reference as "Conciliator in dilucidario ad differentiam 10." It is quite likely that Galileo saw the abbreviation for "in dilucidario" and was unable to decipher it; thus he left a space, to be filled in later, for a term approximating that shown in brackets in the English translation at line 15.

A more striking example of Valla's agreement with another Jesuit's notes can be seen in conjunction with a passage already discussed, namely, that extracted from PQ5.5, where Galileo is discussing differences among the Averroists about how the forms of the elements undergo intension and remission.[69] Lines 28-44 of Galileo's text are again reproduced below on the right, only this time set opposite the parallel text in Menu's treatment of this question. As previously, words that are common or equivalent in both expositions are shown in italics.

[67] Cod. APUG-FC 1710, Tract. 5, Disp. 1, Pars 1, Q.3, De forma et materia elementorum.

[68] Cod. SB Bamberg 62-4, fol. 230v.

[69] See p. 74 *supra*.

MENU (1578)

Quo vero modo intenduntur et remittuntur dissentiunt inter se Averroistae. *Contarenus enim dicit formas elementorum non successive introduci* ad introductionem dispositionum, *sed fieri alterationem usque ad talem gradum ad quem sequitur forma mixti, sub quo forma elementi conservari non possit et introducta forma mixti tum primum incipiunt formae elementorum esse in actu refracto.* {Tertio modus ut *Achillinus decit formam elementorum secundum aliquam partem minimam induci in instanti, facta* tamen *prius dispositione, postea vero successive* introduci.} Secundus modus est *aliorum* qui *volunt formas elementorum intendi et remitti ad intensionem et remissionem* suarum dispositionum *successive*, . . .[70]

GALILEO

In *modo vero quo* formae *intendantur et remittantur, non conveniunt. Contarenus enim dicit formas elementorum non introduci successive* ad intensionem qualitatis, *sed fieri alterationem* qualitatum *usque ad certum gradum ad quem sequitur forma mixti,* et *sub quo formae elementorum non possunt esse integrae; et, adveniente forma mixti, tum primum formae elementorum incipiunt refrangi.* Secundus modus est *Achillini,* qui videtur dicere *formam elementorum secundum aliquam partem minimam introduci in instanti, facta prius debita dispositione; postea vero,* intendi et remitti *successive. Alii* vero tertio communiter dicunt *formas elementorum intendi et remitti successive, ad intensionem et remissionem qualitatum.*

If one compares carefully the above parallels with those already given, one can see that Menu and Valla, writing about a decade apart, have precisely the same degree of agreement with Galileo's composition: both authors have seventy-five words that are identical or synonymous with those used by Galileo in this passage. Also informative is the fact that they agree with each other slightly more than they agree with Galileo in the actual words used and in the way in which they order the variations in the opinion being dicussed. This strongly suggests that Valla's extant notes have a genetic connection with those written earlier by Menu, and thus with a tradition in natural philosophy that was already well established at the Collegio Romano by the time of his writing.

As a final example of how such a tradition persisted at least until the professorship of Rugerius, another passage that has been exhibited above will be set in parallel with its counterpart in that author. This occurs in the question dealing with the nature of primary qualities—specifically whether they are all positive qualities or some are merely privative (PQ6.2).[71] Lines 8-18 of Galileo's text are repeated below on the right, now set opposite the corresponding passages in Rugerius's notes:

RUGERIUS (1592)

Quod *probatur* tum *ex Aristotele,* quia *secundo Coeli 18* et *primo De generatione 18* et

GALILEO

Probatur, primo, *ex Aristotele, secundo Caeli 18, primo De generatione 18,* secundo De

[70] Cod. Ueberlingen 138, [De generatione], [Tract. 4], Sect. 1, Disp. 1, Cap. 5, An formae elementorum intendantur et remittantur.

[71] See sec. 2.3.b *supra.*

duodecimo Metaphysicae 22 frigiditatem appellat privationem, tum etiam probatur hoc signo: quod *rigor febrium tertianarum non fit adveniente aliqua materia frigida sed per solam abcessum caloris; et idem dici potest de aliis multis quae ex sola caloris absentia frigefiunt, ut cum animal moritur, cum ascendunt vapores ad mediam regionem aeris*, cum recedente sole terra frigefit, etc.[72]

generatione 32, *duodecimo Metaphysices 22*, et alibi: ubi *frigus appellat privationem*.

Secundo: *febrium tertianarum rigor non fit adveniente aliqua materia frigida, sed per solam absentiam caloris; et idem dici potest de aliis multis, quae sola caloris absentia frigefieri* videntur; et *cum animal moritur, cum ascendunt vapores ad mediam regionem aeris*, etc.

The same line numeration is given as before, for purposes of comparison. In the previous parallel, one can see at lines 8-18 that both Valla and Galileo present the above passages as two distinct proofs, whereas Rugerius presents them as related, the first on the authority of Aristotle (line 8) and the second based on a sign (line 12). Moreover, whereas Valla's expositon is briefer than Galileo's (lines 8-16), Rugerius's is fuller and contains Galileo's entire text (lines 8-18). This would seem to confirm the speculation already advanced in commenting on Valla's passage, namely, that Galileo worked from a version later than that extant in Valla's *Tractatus de elementis*—now seen to be more fully preserved in the lecture notes of Rugerius.

These illustrations should suffice to show the textual uniformities that can be discerned in Valla and his contemporaries and colleagues at the Collegio Romano. Obviously many more examples could be given, were this desirable. Since the author has fully indicated all cases of verbal agreement in the commentary provided in his *Galileo's Early Notebooks*, the reader who is interested in more details may refer to that work for complete documentation. In any event there can be no doubt that Valla's teaching throughout his *Tractatus de elementis* closely parallels the corresponding tracts in Menu, Vitelleschi, and Rugerius, and thus is quite representative of the tradition in natural philosophy that had developed at the Collegio Romano around 1590. Conversely, the portions of Valla's lectures of 1588-1589 and 1589-1590 that are no longer extant would seem to have been anticipated in the notes of his predecessors such as Menu or preserved in those of his successors such as Vitelleschi and Rugerius. On this basis, despite the uncertainty as to the exact exemplar used by Galileo when composing his physical questions, a body of knowledge is available for understanding their source quite analogous to that already identified, admittedly with greater precision, as the source of his logical questions.

[72] Cod. SB Bamberg 62-4, fol. 252r.

d. ADDITIONAL POSSIBILITIES

Before leaving the subject of textual comparisons, some additional examples will be given to show the range of parallels in Collegio lecture notes that extend from a decade before Valla to the end of the sixteenth century. An exceptional illustration of an earlier parallel may be pointed out in Pererius's textbook, *De communibus*, in the chapter where various definitions of *maxima* and *minima* are given. Mention has already been made of Galileo's teaching in PQ5.6 relating to the way these terms may be applied to powers or potencies and their related impotencies.[73] Galileo's terminology is that developed by the English Mertonians, who spoke of four different types of *maxima* and *minima*—two being the respective limits of which a power is capable (a *maximum quod sic* and a *minimum quod sic*) and two being those of which it is not (a *maximum quod non* and a *minimum quod non*). Galileo lays out these definitions in the tenth paragraph of PQ5.6. Oddly enough, none of the four Jesuits discussed thus far lists the four expressions in the same order as he, nor does any show close agreement with his explanation of them. Vitelleschi comes the closest, though the treatment in the manuscript containing his notes is brief and has been supplemented by marginal additions in the same hand. The excerpt below gives the corresponding passage in Vitelleschi, against which is juxtaposed the first sentence of Galileo's paragraph:

VITELLESCHI (1590)

Notandum sunt *hi quatuor termini* quibus utuntur philosophi in hac re: *maximum quod sic, maximum quod non, minimum quod sic, minimum quod non.* [M: Maximum quod sic et minimum quod non sunt *termini magnitudinis*, ille *intrinsecus et affirmativus*, hic *extrinsecus et negativus*. Minimum quod sic et maximum quod non *sunt termini parvitatis*, ille intrinsecus et affirmativus, hic extrinsecus et negativus.][74]

GALILEO

Nota, primo, explicationem *horum quatuor terminorum: maximum quod sic, minimum quod sic, maximum quod non*, et *minimum quod non*; quorum duo priores sunt *affirmativi* positivi *et intrinseci*, posteriores vero dicuntur *negativi et extrinseci*; item primus et quartus sunt *termini magnitudinis*, secundus et tertius *parvitatis sunt termini.*[75]

The teaching is obviously identical in both, though ordered differently. Galileo's order, it turns out, is actually that found in Pererius's *De communibus*, first published in 1576 but based on lectures given by Pererius at the Collegio in 1565-1566. To show this correspondence, the remainder

[73] See sec. 2.3.a, toward the end.

[74] Cod. APUG-FC 392, [De generatione]; Tractatio de augmentatione; Disputatio quinta, De maximo et minimo.

[75] *Opere* 1:139.5-10; Wallace, *Galileo's Early Notebooks*, par. U10.

of Galileo's paragraph is exhibited below, only now placed alongside the explanation as found in Pererius's text:

PERERIUS (1576)

Appelabant *maximum quod sic terminum quantitatis sub qua res* naturalis *esse potest, et non sub maiori, ita ut attingere quidem possit illum terminum transire autem non possit. Minimum quod sic,* dicebatur *terminus quantitatis sub qua res potest esse sed non sub* alia ulla *minori. Maximum quod non* vocabatur *terminus quantitatis sub qua res non potest esse sed bene sub quacunque maiori, modo non transeat modum magnitudinis* quem supra diximus *maximum quod sic,* et hoc modo . . . *Minimum quod non erat terminus quantitatis sub qua res non potest esse, sed bene sub minori, modo non excedat* terminum parvitatis quem supra vocavimus *minimum quod sic,* et . . .[76]

GALILEO

Maximum quod sic est *maxima quantitas* quam *res* potest attingere vel *sub qua potest esse, et sub maiore non* potest; *ita ut attingere quidem possit illum terminum, transire autem non possit. Minimum quod sic* est *minima quantitas* quam *res* potest attingere vel *sub qua potest esse, et sub minore non* potest. *Maximum quod non* est illa quantitas quam, ob parvitatem, *res* non possit attingere vel *sub qua non potest esse, et sub quacunque maiore* potest, *modo non transeat magnitudinem maximum quod sic. Minimum quod non* est *illa quantitas* quam *res,* ob magnitudinem, non potest attingere vel *sub qua non potest esse, et sub* quacunque *minori* potest, *modo non excedat minimum quod sic.*[77]

A comparison of the two sets of parallels will show that Galileo's composition is much closer to Pererius's than it is to Vitelleschi's. One might maintain, on this basis, that Galileo used Pererius's textbook when writing his notes; an alternative possibility, which seems more likely to the author, is that Valla had himself used Pererius when writing a revised version of his notes, and that Galileo appropriated these for his own use, thus basing himself on Pererius at second remove.[78]

An analogous instance, this time illustrating possible borrowing from Valla by a later professor at the Collegio, is found in the very last question of MS 46, PQ6.4, which is concerned with how primary qualities are involved in activity and resistance. Parallels for this question may be found in Vitelleschi, Valla, Rugerius, and Menu, in descending order of agreement, though none is close enough to suggest copying. It is therefore surprising that passages in another Jesuit who taught at the Collegio in 1597, Stephanus del Bufalo (de Cancellieri, Bubalus de Cancellariis, de Bubalis), are more similar to Galileo's text.[79] Two excerpts from

[76] *De communibus*, p. 355A.

[77] *Opere* 1:139.11-21; Wallace, *Galileo's Early Notebooks*, par. U10.

[78] This is argued in Wallace, *Prelude to Galileo*, pp. 200-217. Crombie and Carugo adopt the first possibility; see Crombie's "Sources of Galileo's Early Natural Philosophy," pp. 164-165, 170.

[79] For bibliographical information on Del Bufalo, see Lohr, "Renaissance Latin Aristotle Commentaries," *Studies in the Renaissance* 21 (1974): 282.

Galileo's treatment in PQ6.4 may serve to illustrate these correspondences:

Del Bufalo (1597)

Dicendum est resistentiam esse permanentiam rei in suo esse contra actionem contrarii . . . Etiam *videntur communiter intelligere per resistentiam quo res permanet in esse, licet corrumpere illam conetur contrariam* . . . {*Neque* etiam *bene* ipsum *Pomponatium dicere resistentiam* praedicamentaliter ac *formaliter esse impedimentum actionis contrarii et connotare permanentiam in proprio statu, quia impedimentum actionis videtur esse negatio actionis, et resistentia* non *accipitur tanquam quid* negativum sed *positivum.*} . . .

Ex his autem colligitur primo *non bene Flaminius* [Nobilius] *in capite septimo primi De generatione, quaestione undecima,* dicere *resistentiam in rebus inanimatis esse* ipsam naturalem potentiam, sed potius *impotentiam ad patiendum* a contrario; potentia enim *causa* est *resistentiae,* [non] ipsa actualis resistentia; unde *sublata actione contrarii,* sine qua non est resistentia, *adhuc remanet potentia illa naturalis.*[80]

Galileo

Dico tertio *resistentiam esse permanentiam in proprio statu contra actionem contrariam* . . . immo *resistentia* formaliter *dicit hanc permanentiam rei in suo statu,* et *connotat* impedimentum *actionis contrariae.* Et hoc est contra *Pomponatium,* sectione secunda De reactione, capite tertio, qui vult *resistentiam formaliter dicere impedimentum contrarie actionis, connotans* actionem contrarii et *permanentiam in proprio statu;* sed fallitur, quia *resistentia est aliquid positivium,* at vero illud *impedimentum est negatio actionis.* . . .

Ex quo apparet error Nobili qui, *primo De generatione dubio undecimo in capite septimo,* distinxit duplicem *resistentiam:* aliam animalium, quae consisteret in nixu quodam, qui est quaedam actio; aliam *in caeteris rebus,* quam reduxit ad *impotentiam ad patiendum;* ubi, ut videtis, accepit *causam* extrinsecam *resistentiae* pro resistentia formaliter, cum tamen distinguatur. Adde quod *naturalis potentia* vel impotentia *semper est in re;* et tamen res semper non resistit, *sed solum praesente actione contrarii.*[81]

The agreement in lines 1-12 is very close, even though a passage in Del Bufalo had to be rearranged to exhibit this; less striking is the agreement in lines 13-24, but still better than that with any of the other four Jesuit authors. This despite the fact that Del Bufalo wrote his notes in 1597, almost twenty years later than Menu, ten years after Valla, and five years after Rugerius. One might wish to maintain, on the basis of this type of evidence, that Galileo himself did not write the physical questions until after 1597, just as one might maintain that he did not write the logical questions until after 1597 because they too became available in Carbone's *Additamenta* of that year.[82] The more likely possibility is that Del Bufalo,

[80] Biblioteca Nacional Lisboa, Fundo Geral, Cod. 2382, fols. 129v-130r. This manuscript is anonymous, but there seems little doubt that it contains Del Bufalo's notes; see Wallace, *Galileo's Early Notebooks*, pp. 13, 20-21, and 308, note 19.

[81] *Opere* 1:171.1-11, 172.1-8; Wallace, *Galileo's Early Notebooks*, pars. Y4 and Y8.

[82] Crombie and Carugo have taken this position in their joint paper, "The Jesuits and Galileo's Idea of Science and Nature," presented at the International Galileo Conference, *Novità Celesti e Crisi del Sapere*, in Florence on March 24, 1983. For the author's evaluation

in preparing his lectures, made use of the same version of Valla's notes as was available to Galileo, and thus produced a set of notes in some respects similar to those contained in MS 46.

Finally, one might wonder if yet more textual comparisons might be exhibited between Galileo's notes and sources outside the Collegio Romano, including the writings of non-Jesuits. The answer to this is a decided affirmative, as can be seen from the parallels that have already been exhibited. The Collegio lectures were teaching notes that made full use of printed and manuscript works and frequently incorporated passages found in contemporary and earlier authors, with or without acknowledgment of source. Many of the resulting textual coincidences have been identified and noted in *Galileo's Early Notebooks*. For example, paragraphs PQ1.1.3, 7, 11, 14, 15, and 17 contain passages that seem to be excerpted from, or based upon, Jacopo Zabarella's *De naturalis scientiae constitutione*, printed at Venice in 1586. (The late date of this possible source is further evidence counting against Favaro's 1584 dating of the physical questions.) Parts of PQ2.3 and 4.3 contain teachings that are similar to those found in Francesco Buonamici's *De motu*, as do paragraphs PQ3.3.7 through 12. The distinctive views of Achillini are presented in PQ5.4, as already seen,[83] and those of Cardano in PQ6.2. A passage from Scaliger relating to impetus, to be noted later, was excerpted by Menu and incorporated into his notes, without identification of source; the same passage was subsequently used by Valla, and identified by him.[84] The remarkable thing, however, is that Galileo's style is unlike that of any of these non-Jesuit authors, and there are no extended passages in any of them that can be put side by side with his on the model of the parallels exhibited in this and the preceding chapter. The thought of many medieval and Renaissance commentators is preserved in the questions under discussion, but even when appropriated from them it is expressed in a terminology and an arrangement that are distinctive of the Collegio tradition and not of other authors or schools.

4. Provenance and Dating

The data presented in the foregoing sections of this chapter are so extensive and variegated that they are capable of generating many hy-

---

of their thesis, see his "Galileo's Sources: Manuscripts or Printed Works?" in *Print and Culture in the Renaissance*, Sylvia Wagonheim and Gerald Tyson, eds. (Newark, Delaware: The University of Delaware Press, forthcoming).

[83] Sec. 2.3.a.

[84] Sec. 4.3.a-b and chap. 4, note 94.

potheses about the date and purpose of Galileo's composition of the physical questions. Some of these possibilities have been investigated by the author in his previous studies.[85] Obviously the origin of the physical questions is more difficult to decide than is that of the logical questions, and so it is quite impossible to establish any dating for them with mathematical certitude. When considered together with the logical questions, however, they fit well into the pattern that has already been established in the preceding chapter; located in this context, the problem of their source and time of writing yields more readily to plausible reasoning.

The tactic to be pursued in this section, therefore, makes use of the evidence given in chapter 1. The physical questions of MS 46 were written *after* the logical questions of MS 27. The treatises based on Aristotle's *De caelo* in MS 46, moreover, were written *before* those based on Aristotle's *De generatione* and bound after them in the same codex. Assuming, as already argued, that the logical questions were not written until late 1588 or early 1589, and that the parts of the physical questions relating to the *De generatione* were not written until late 1590 or early 1591 (both based on the connection with Valla), the portions of the physical questions relating to the *De caelo* must have been written in 1589 or 1590. This conclusion agrees well with the notation of the curator who assembled the folios making up the first part of MS 46, for he entitled this codex "the examination of Aristotle's work *De caelo* made by Galileo around the year 1590."[86] And since the *De motu antiquiora* drafts of MS 71 have been consistently regarded as written around the same time, one may reasonably conclude that all three manuscripts, MSS 27, 46, and 71, date from the same period, which coincides with that of Galileo's first teaching post at the University of Pisa.

Good internal evidence indicates that Galileo wrote the logical questions before the physical questions. As noted in the last chapter, Galileo makes frequent reference to his treatment of the categories throughout both parts of his physical questions, and this treatment could have been contained in the missing portions of his notes on logic.[87] Apart from the missing portions, moreover, there are two references to material contained in the extant questions. One of these is explicit and mentions the *Posterior Analytics* as the source of the teaching: this occurs in PQ1.1.13, where Galileo states that there cannot be a science of individuals—a conclusion he establishes in LQ6.11.[88] Another makes use of a position

[85] Wallace, *Galileo's Early Notebooks*, pp. 21-24, 258-259; idem, *Prelude to Galileo*, pp. 135-138, 178-184, 217-228, 260-262, and 281-283.

[86] *Opere* 1:9: "L'esame dell'opera d'Aristotele 'De caelo' fatto da Galileo circa l'anno 1590."

[87] See sec. 1.2 *supra*.

[88] *Opere* 1:18.17-19; Wallace, *Galileo's Early Notebooks*, par. A13; MS 27, fol. 29r7-13.

defended in LQ3.3, namely, that the existence of the partial subject of a science can be demonstrated if it is not known to exist (PQ1.1.17), although Galileo does not there reference a previous treatment.[89] Throughout question PQ1.1, moreover, there are frequent indications of a good knowledge of the various subjects or objects of a science, whose foreknowledge is treated in LQ3.1 through 3.5, and which would also be considered in the treatise *De scientia*.[90] Such familiarity is a fairly good sign that Galileo had composed the logical treatises before he commenced his exposition of the *De caelo*. Confirmatory internal evidence would be the greater incidence of error in the Latin spelling and constructions of MS 27, already noted by Favaro, which would suggest that Galileo gained experience with them and so was more proficient when he came to write the materials contained in MSS 46 and 71.[91] External evidence providing additional support is the fact that, only shortly before Favaro's time, the logical notes had been bound together with, but before, those on natural philosophy, which leads one to believe that the curators who had examined them were convinced that the former was the earlier composition.[92]

If the foregoing dating procedure is accepted, the next problem is that of ascertaining how and why Galileo obtained these lecture notes from the Collegio Romano, in the center of Italy and fairly distant from his native Tuscany. Fortunately, an interchange of correspondence between Galileo and Clavius is extant that sheds light on how this might have occurred. Late in 1587 Galileo was apparently in Rome and left with Clavius a copy of some theorems he was developing on the center of gravity of solids. Details of the interchange that followed their meeting will be discussed in chapter 5 *infra*, but for purposes here it suffices to note that Clavius, on reading Galileo's treatise, experienced some difficulty with the logic of his proof.[93] So, too, did Guidobaldo del Monte, whose correspondence relating to the *Theoremata* is also extant. In letters written to Galileo on January 16, 1588, and again on March 5 of the same year, Clavius questioned whether Galileo's demonstration assumed what it attempted to prove, namely, that it was based on the fallacy of *petitio principii* and so might be defective.[94] In a letter dated June 17, 1588,

[89] *Opere* 1:19.27-30; Wallace, *Galileo's Early Notebooks*, par. A17; MS 27, fols. 9r24-10r12.

[90] *Opere* 1:16.15-19, 16.20-17.3, 17.14-20, 17.21-18.5, 19.15-26; Wallace, *Galileo's Early Notebooks*, pars. A6, A7, A9, A10, and A16. See also secs. 3.1.b, 3 4.a, and 4.1.a.

[91] *Opere* 9:282.

[92] The author is indebted to Raymond Fredette for this information, provided in his unpublished paper, "Bringing to Light the Order of Composition of Galileo Galilei's *De motu antiquiora*," delivered at the Workshop on Galileo, Virginia Polytechnic Institute and State University, Blacksburg, Virginia, October 1975.

[93] Sec. 5.1.b.

[94] *Opere* 10:24-25, 29-30.

Guidobaldo referred to the principle Galileo had employed as a *supposizione*, and in his reply to Guidobaldo of July 16 Galileo showed that he was still concerned over whether it was a *petitio* or not.[95] This problem, as will be elaborated more fully in parts 2 and 3 of this study, bears directly on the foreknowledge required for demonstration, the very topic that is treated in LQ2.1 through 4.2, all of which have counterparts in Valla's lecture notes—themselves not completed until August of 1588.[96] The coincidence here seems more than fortuitous, and indeed prompts the following speculation. Probably the letters of Clavius and Guidobaldo alerted Galileo to the fact that his knowledge of logic and methodology was not all it should be, and on this account he had recourse to Clavius, who had first raised the difficulty and was more expert in such matters, to assist him with its solution. No further correspondence between them on this matter survives, but it seems likely that Clavius, acceding to his request, obtained for Galileo a set of Valla's logic notes. These, as has been seen, have a very thorough treatment of this type of problem, and the fact that Valla distributed them to his students and that Carbone had also secured a set argues for their availability at precisely the time Galileo would have benefited from studying them.

If Clavius did Galileo this favor, once Galileo saw the thoroughness with which logic was taught at the Collegio, perhaps as contrasted with his own previous instruction in that subject at the University of Pisa, it would have been reasonable for him to seek additional lecture notes on the heavens, the elements, and the local motion of bodies.[97] These, after all, were topics in which Galileo was greatly interested, on whose mathematical treatment he would soon be, or already was, lecturing himself. Valla's notes on such subjects would be in continuity with his logic, and the time schedule on which they became available—the parts relating to *De caelo* late in 1589 and those on *De generatione* early in 1590—fits nicely into the dating of Galileo's MS 46 just proposed. Should this be the actual chronology, then the *Tractatus de elementis* of Valla that has been used for textual parallels throughout this chapter might well represent one of his earlier treatments of *De generatione*, probably that taught in 1586-1587, and later revised along the lines already conjectured.

The period "around 1590," moreover, was one during which Galileo had good reason to obtain knowledge of the universe and the heavens. In 1589 he began teaching the mathematics course at the University of

[95] *Opere* 10:34-36.

[96] Secs. 3.1, 3.2, 3.4.c, 4.4.a, 5.1.a, 5.2.b, 5.3.b-c, 5.4.a, 5.4.e, 6.3.d-e, and 6.4.a.

[97] Vincenzio Viviani, in his *Racconto istorico* of the life of Galileo, indicates that Galileo studied logic at the Monastery of Vallombrossa, but found the instruction there tedious and of little value; *Opere* 19:602.57-60.

Pisa, and continued doing so in 1590 and 1591.[98] His predecessor in the chair of mathematics at Pisa, Father Philippus Fantoni, had regularly been alternating his pure and applied mathematics on a two-year cycle, teaching the first book of Euclid's *Elements* together with the *Sphaera* of Sacrobosco in the first year, and then following this with the fifth book of Euclid's *Elements* and the *Theorica planetarum* in the second. When Galileo took on the course in 1589 he started the cycle anew. The *rotulus* of professors at Pisa lists him as teaching only the first book of Euclid in 1589, the fifth book in 1590, and the first book again, together with the "caelestium motuum hypotheses," in 1591. Since the *rotulus* frequently abbreviates teaching loads, it seems reasonable to assume that Galileo taught the *Sphaera* in 1589 along with the first book of Euclid, then the *Theorica planetarum* in 1590 along with the fifth book, and again the *Sphaera* under the title, "hypotheses of the celestial motions," along with the first book in 1591. In such a case, he would need knowledge of the heavens, and particularly the philosophical knowledge being covered in the parallel course in Aristotle's *De caelo*, to discharge his teaching responsibilities properly. It is noteworthy that during the period Galileo was teaching mathematics at Pisa the philosophers were covering Aristotle's *De caelo* and *Meteorology* in 1589, his *De anima* and *Parva naturalia* in 1590, and his *Physics* and *De generatione* in 1591.[99] (The philosophy cycle at Pisa was thus different from that at the Collegio Romano; if anything, it offered more instruction in natural philosophy than did the Collegio.) Therefore Galileo had good reason, in the years between 1589 and 1591, to acquaint himself with the best available philosophical thought on these subjects.

Another piece of evidence that supports the 1590 dating is a brief letter Galileo wrote to his father, dated November 15, 1590, wherein he asks the latter to send him a copy of the *Sfera* and some volumes of Galen.[100] The books requested, as the author has already argued, relate directly to the materials covered in the physical questions.[101] In the letter Galileo also mentions that he is applying himself "to study and learning from Signor Mazzoni," who sends his greeting—a statement that accords well with Mazzoni's published writings and the materials in the Jesuit lectures, as will be explained later in chapter 5.[102] Fortunately the letter has been

[98] For documentation, see Charles B. Schmitt, *Studies in Renaissance Philosophy and Science* (London: Variorum Reprints, 1981), especially Essays 9 and 10.

[99] The author has supplemented some of the information contained in Schmitt (ibid.) from his own studies of the materials contained in the Archivio di Stato, Pisa, Università G. 77, fols. 164v-194r.

[100] *Opere* 10:44-45.

[101] Wallace, *Prelude to Galileo*, p. 227.

[102] See sec. 5.1.c *infra*.

preserved in autograph, and this, of all available samples of Galileo's handwriting, shows closest agreement with the hand of the physical questions—again lending credence to 1590 as the year of composition for both.

If Galileo was thus determined to work through an entire set of notes on the universe and the heavens, one might wonder what his motivation could have been to do likewise with an extensive treatise on the elements and primary qualities, particularly when the former seems directly relevant to his courses in astronomy whereas the latter seems not. One answer might be that his effort was prompted by his desire for a firm foundation in the whole of natural philosophy, something he probably had not gotten during his own course work at Pisa. If Galileo's statements are to be taken on face value, moreover, he had completed questions on logic and on Aristotle's *Physics* by the time he composed his first set of questions on the *De caelo*; likewise in the second set on the *De generatione* he expresses his intention to treat each of the elements "in particulari," which was the normal way of signifying, in those days, that he was going to continue to the end of the *Meteorology*. Thus it would be natural to assume that, after finishing the questions on the *De caelo*, which introduced the problem of the elements, he would continue on through the *De generatione* and the *Meteorology* until he had completed the entire course.

Another plausible explanation would be based on the courses being taught by the philosophy professors at Pisa during the time that Galileo was preparing his own lectures on mathematics. As already noted, Aristotle's *De caelo* was being expounded in 1589 and his *De generatione* in 1591. If Galileo had prepared his *De caelo* questions in 1589 so as to be knowledgeable with respect to the matter his students were covering at that time, he could well have decided to do the same with the *De generatione* questions in 1591. This explanation gains credence from the fact that the teaching on the elements is also very important for understanding the astronomical theories of the time, and thus would have been directly relevant to Galileo's class preparation in the period between 1589 and 1591.

A more interesting and fruitful speculation, however, is the following. Just about the time Galileo was working on the tractate on the elements, according to the foregoing estimates, he was also beginning the various treatises on motion known as the *De motu antiquiora*. It was these studies, begun at Pisa but continued throughout Galileo's early years as a professor at Padua, that led ultimately to his *Two New Sciences*, for which he is celebrated as the Father of Modern Mechanics. Now, precisely at the point where Galileo's last question in the second set (PQ6.4) leaves off, most of the Jesuit *reportationes* begin with the study of the motive powers of the elements and the motion of heavy and light bodies, including projectiles. These, by what must surely be more than a coincidence, are the very topics in which Galileo became interested in his *De motu antiquiora.*[103] Thus the second set of questions (PQ4.1 through 6.4) need not be seen merely as a follow-up to the first set (PQ1.1 through 3.6); they may also be seen as in essential continuity with the next large project in which Galileo would become involved. The treatise on the elements could therefore be a stimulus for his further studies of motion and a parallel effort to understand more fully the composition of physical bodies, and especially their elemental components, so as to have a deeper philosophical basis for his ongoing investigations.

To conclude these remarks on the physical questions and their source, the extensive presentation of data in this chapter relating to MS 46 must not obscure one important fact. The notes contained in that codex embody a systematic approach to the problems of the heavens and the elements that was very much a part of the tradition of the Collegio Romano for a period of over twenty years, that is, from 1577 to 1597. The uniformity of these teachings complicates, as has been seen, the problem of precise dating. Yet there is another side to the coin. The very constancy of the teachings also means that, regardless of when and from what source Galileo composed these questions, they are but a cross-section of a body of knowledge that was generally accepted and taught at the Collegio in the last decades of the *cinquecento*. It was this natural philosophy that Galileo not only studied but carefully excerpted and attempted to make his own. It would seem that its understanding is indispensable for appreciating the background against which he worked, and which became the seedbed, as it were, from which his more important contributions would ultimately emerge.

With this, the historical analysis of Galileo's manuscripts recording his commentaries on Aristotle's *Posterior Analytics*, *De caelo*, and *De generatione*, and detailing the likely sources of these commentaries in lectures given at the Collegio Romano is complete. Contrary to practically all previous scholarship, this analysis reveals that Galileo was composing these commentaries while he was beginning his teaching career at the University of Pisa, contemporaneous with his initial attempts at elaborating a science of motion. This circumstance makes available a wealth of information for filling out the conceptual framework in which Galileo developed his thought, and for making intelligible not only his *De motu*

[103] Sec. 5.2.

antiquiora, but also—considering elements of continuity between this and *his later work—his more mature writings as well. Much of this infor*mation is in Galileo's own hand, and its value is hardly questionable on that account. A substantial additional amount is implied in what Galileo wrote: terms and expressions he uses and whose understanding he presumes without explicit definition; principles necessary for formulating the distinctions and arguments he employs; conclusions established in tracts contiguous with those he wrote out, as well as in those presupposed to their development or following from them as corollaries or further elaborations. In a word, there is an entire system of thought, well articulated and formulated in technical Latin, to which Galileo subscribed at this seminal period in his intellectual life. Thanks to the large body of literature generated at the Collegio Romano around this time—now recoverable in manuscript or printed form and clearly in continuity, as has been shown, with Galileo's Latin manuscripts—it is possible to reconstruct this philosophical background in its entirety. Such a task is the main burden of the second part of this study.

PART TWO
Science at the Collegio Romano

CHAPTER 3

Sciences and Demonstrative Methods

THE NOTION of science that was current when Galileo began his teaching career at the University of Pisa is different from that of the present day. While sharing in some characteristics, such as the publicly verifiable aspect of the knowledge it produced, the science of his day made more stringent claims to certitude and infallibility than does that of the modern era. The Aristotelian ideal of scientific knowledge is that of *cognitio certa per causas*, that is, knowledge that is certain through causes, or knowledge that cannot be otherwise because it is based on the causes that make things be the way they are. This knowledge is thus different from the product of probable or conjectural reasoning, which lacks certitude because it falls short of identifying true and proper causes or explanations, and so is revisable whenever more plausible reasons can be adduced in support of its conclusions. Now Galileo employed the terms *scientia* and *scienza* repeatedly throughout his entire life, from the earliest drafts of his first treatises to the masterful *Two New Sciences* that marked the summit of his intellectual achievement, but never once did he depart from this ideal of certain and irrevisable knowledge as the goal of his investigations. He had many arguments with the Aristotelians of his day, and often accused them of faulty and fallacious reasoning when their conclusions were at variance with his own. But never once did he question the ideal by which he judged their contributions. Where they purported to give a causal explanation that in his judgment was inadequate or downright erroneous, he did not hesitate to refute them by setting out the "true cause" as well as he could ascertain it. And whether he was arguing about the movement of the earth or about the laws that govern the motion of falling bodies, his claim was invariably for objectivity and certitude in the conclusions to which he came.

It should be quite clear that there was nothing novel in this ideal of science to which Galileo subscribed. Indeed it was the commonly accepted doctrine of the schools, expounded in every university, and regarded as a necessary propadeutic to all of the professions as well as to the world of learning generally. In the main it was an ideal set forth in

Aristotle's *Posterior Analytics*, led up to and supported by other works of the Stagyrite's *Organon* as well as the writings of logicians in classical and medieval times. Yet there were differences of interpretation with regard to the text of Aristotle and to the problems he bequeathed his followers—differences that led to a variety of schools, all of which could be labeled "Aristotelian" to a greater or lesser degree.[1] If Galileo had not written his logical questions, or if the manuscript containing them had been lost, it would have been practically impossible to ascertain the school in which he was educated and whose options he generally pursued throughout his later life. Fortunately his surviving materials bear directly on the *Posterior Analytics* and reveal him, as already noted, to have been strongly influenced by teachings at the Collegio Romano in the 1580s and 1590s, viz., by a somewhat eclectic Thomism containing elements deriving from Scotistic, Averroist, and nominalist thought. The aim of the present chapter is to lay out the elements of the teaching to which he subscribed as a consequence of such a commitment, particularly as this is relevant to his lifelong attempts to develop a science of motion.

Despite variations on particular points, sixteenth-century thinkers who pursued the ideal of *scientia* or *scienza* were agreed that its distinctive type of knowledge was the result of a reasoning process known as demonstration. A demonstration could best be seen as a rigorous argument when formulated as a categorical syllogism made up of two premises and a conclusion, the latter taking the form of a proposition containing subject and predicate. The premise stating the more universal truth was called the major premise, and it was composed of a middle term and the predicate of the conclusion. The other premise, known as the minor premise, was made up of the subject of the conclusion and the same middle term as occurred in the major premise. In virtue of this middle term functioning in both premises, one who assented to the truth of the premises would be led to affirm the predicate of the subject in the conclusion, and thus to assent to the truth of the conclusion itself. The middle term would invariably state the cause of, or the reason for, one's giving assent to the conclusion; in some types of demonstration, moreover, it would also formulate a definition of the subject of the conclusion expressed in causal terms. Thus definition and demonstration were closely interrelated in the process of reaching scientific conclusions. The various types of definitions and demonstrations also gave rise to different kinds

[1] For a survey of this development, see Charles B. Schmitt, "Towards a Reassessment of Renaissance Aristotelianism," in his *Studies in Renaissance Philosophy and Science*; also his *Critical Survey and Bibliography of Studies on Renaissance Aristotelianism* (Padua: Editrice Antenore, 1971), and *Aristotle and the Renaissance* (Cambridge, Mass.: Harvard University Press, 1983).

of sciences, since science was the effect of demonstration, which in turn was dependent on definition, and so all three—science, demonstration, and definition—would have to be understood thoroughly if one were to have a competent grasp of the scientific enterprise.

As explained in the first chapter, Galileo's extant logical treatises do not include those on science and definition, though they treat demonstration and its requirements in considerable detail. Fortunately, there are sufficient references to science and definition in the materials that have survived, however, to enable one to reconstruct, with the aid of his sources, Galileo's positions on these topics. Following the order of his logical questions, all three will now be discussed, beginning with demonstration, first in terms of the foreknowledge necessary to achieve it and second focusing on its nature and kinds, then considering definition and the process by which one can regress to it in a demonstrative context, and concluding with a treatment of science and its various kinds.

1. Prerequisites for Demonstration

The use of the term "foreknowledge" (*praecognitio*) when speaking of the requirements for demonstration arises from one of Aristotle's statements at the beginning of the *Posterior Analytics* wherein he asserts that one cannot communicate knowledge to another without presupposing that the latter knows something beforehand.[2] Such previous knowledge must be gained independently of the demonstration being proposed, since anyone who lacks a sufficient fund of information to assent to (or disagree with) the premises will be unable to pass from them to the conclusion they imply. Not only this, but at least some preexistent knowledge will have to be attained independently of any demonstration whatever, since the very nature of demonstration as generating true and necessary knowledge requires that not everything that is known can be demonstrated. Were this so, one could never demonstrate anything at all, since every premise would in turn have to be demonstrated, setting up an infinite regress, and nothing could ever be known as true. Since not everything that is knowable can be proved, therefore, it becomes extremely important to examine what can be known independently of demonstration, and in this sense be "foreknown."

Galileo's disputation on foreknowledge in general is missing from his logic notes, but in LQ3.5 he indicates that he has previously discussed how much must be foreknown in order for one to demonstrate, and in

[2] *Posterior Analytics*, bk. 1, chap. 1, 71a1-b8.

LQ4.1 he refers to Aristotle's teaching that only two general foreknowledges are required as something he has already explained.[3] In LQ4.1 he also refers to some types of preexistent knowing as "directing" (*dirigens*) one in the acquisition of knowledge, others as "acting" (*agens*) or immediately effecting such acquisition.[4] The basis for the lost treatise is fortunately contained in Valla-Carbone, and this furnishes an abundance of detail for filling out Galileo's views.[5] "Directing" foreknowledge assists one in attaining knowledge after the fashion of a condition *sine qua non*, whereas "acting" foreknowledge produces the knowledge desired after the fashion of an efficient cause. Examples of the former are knowledge of the meanings of the terms employed and of the existence of the entities referred to in the demonstration, whereas an example of the latter would be knowledge of the premises on which its conclusion is based. When Aristotle maintains that only two general foreknowledges are required for demonstration, he is indicating the principal "directing" knowledges, and these relate to the *quid nominis* and the *an sit*, that is, answers to questions respecting meaning and existence. For a fuller delineation of what must be foreknown, however, one must also include the "acting" foreknowledge of principles, whether these are the general axioms that govern all thought or the more particular principles, including definitions, that are used in the different sciences. Thus, in effect, for one to demonstrate a property (or predicate) of any subject, he must have some previous knowledge of that subject (at least that it exists), he must know something about the predicate (at least what the word means), and he must give prior assent to the premises or principles on which the reasoning is based. Galileo's detailed treatment of these kinds of foreknowledge is structured around such requirements: first, as obviously most important, he inquires into the foreknowledge of principles; then he examines in considerable detail the foreknowledge of the subject; and he concludes with a brief treatment of what must be foreknown about the property that is predicated of the subject in the conclusion.

a. FOREKNOWLEDGE OF PRINCIPLES

Galileo addresses the requisite knowledge of principles in four questions, the first (LQ2.1) inquiring whether all principles, including first principles, must be known to be true; the second (LQ2.2) asking whether one must know the meanings of the terms employed in first principles; the third (LQ2.3) whether the principles must be foreknown actually or

[3] MS 27, fols. 11r9-10, 11v15-17.

[4] MS 27, fol. 11v21-26.

[5] *Additamenta*, fols. 36r-40v.

merely as a matter of habit; and the fourth (LQ2.4) whether principles must be so evident that they cannot be subject to proof in any way.[6] His answers to LQ2.2 and 2.3 are straightforward and unexceptional: one obviously must know the meanings of the terms, and thus, where principles are concerned, one must grasp the *quid nominis* of their terms as well as be convinced of the *an sit* or truth of the principles themselves; and this must be actual knowledge of any principles or premises that enter directly into the demonstration, although for others that serve merely as background habitual knowledge suffices.[7] The answers to LQ2.1 and 2.4 are somewhat more interesting, for in these Galileo talks about how one comes to the knowledge of the requisite principles and about the possible role of proof in such acquisition, and thus they invite further examination.

As already explained in the first chapter, Galileo indicates a variety of routes to the knowledge of first principles, including one through "induction, division, and hypothetical syllogism," which he exemplifies with the principle stated by Aristotle at the outset of the *Posterior Analytics*, namely, that all teaching (*doctrina*) is based on preexistent knowledge.[8] Galileo merely states the example without explaining how it is related to induction, division, or hypothetical syllogism, and thus one wonders how he understood these various ways of acquiring universal knowledge. Here again the fuller discussion in Valla-Carbone can prove helpful for delineating the position he adopted. The teaching of the *Additamenta* is that several causes actually effect knowledge of principles, the more important being the power of intellect in the one acquiring the science and the teacher who leads him to acquire it.[9] Knowledge of principles is not innate in man, his mind being a *tabula rasa* at birth, and thus he requires sense knowledge and experience so as to have the material on which his intellect can work. When the latter extracts meanings it is able to grasp some principles from their terms alone, such as that the whole is greater than the part. Obviously the teacher is of lesser moment than the power of intellect, for all the teacher can do is tell the learner how to proceed if he is to gain knowledge, much like the doctor who tells the patient what to do if he wishes to get well. One can learn without a teacher, granted with some difficulty, but without power of intellect

[6] See table 3 *supra*, here and hereafter for the titles and locations of these questions in MS 27.

[7] Actual knowledge would be explicit awareness of the propositions used, whereas habitual knowledge would be that which could be recalled with little effort but was not in the knower's consciousness during the reasoning process.

[8] Sec. 1.3.a, especially pp. 33-34.

[9] *Additamenta*, fols. 40v-43v.

he can never attain an understanding of principles or of the science that flows from them.

Even so there is a difference of principles, some being principles of being (*principia essendi*) and others principles of knowing (*principia cognoscendi*) the former making things be what they are and the latter serving to make them intelligible to man. Galileo employs this distinction in LQ2.4, where he argues that principles of being are subject to proof, whereas principles of knowing usually are not, although they can be manifested "by some simple induction or division or hypothetical syllogism"—again using the same phrase as in LQ2.1, but this time without exemplification.[10] Examples of principles of being that can be demonstrated *a posteriori* are primary matter (*materia prima*) and the First Mover, both of which are proved to exist in Aristotle's *Physics*; the two examples are found in Galileo, in Valla-Carbone, and in Valla's later text.[11] The reason why such proofs are possible is that principles of being are generally hidden from the senses, whereas their effects are quite manifest; thus these principles can be demonstrated by reasoning from effect to cause. Principles of knowing, on the other hand, are denominated such because they are obvious by their very nature, and so can hardly be proved by something more manifest. Yet even here there are degrees of clarity and self-evidence: some principles of knowing are not immediately obvious, although once grasped they can be seen to be true on their own merits; they need only a brief explanation or justification to make themselves apparent. Such a principle is Aristotle's starting point in the treatise on demonstration, viz., all teaching must be based on preexistent knowledge. To see this all one need do is divide knowledge acquired by teaching into its various kinds (division), examine the different cases to show how some other knowledge is presupposed in each kind (induction), and then argue hypothetically: if this is true of each and every kind, it must be true of all (hypothetical syllogism). Thus some principles are capable of proof, either in the strong sense of *a posteriori* demonstration or in the weak sense of explanation and justification. That Galileo was convinced of this in more than an academic way is apparent, as will be seen, in his many attempts to prove or make manifest the principles on which his science was based.[12]

Galileo's treatment of the foreknowledge of principles in LQ2.1 through 2.4 is truncated with respect to that in the *Additamenta*, nor does it cover all of the points mentioned by Valla in his textbook on logic. Some of the material in these other authors is discussed by Galileo in his exposition

[10] MS 27, fol. 6v4-8.

[11] MS 27, fol. 6r23-25; *Additamenta*, fol. 44r; *Logica*, vol. 2, p. 151

[12] Secs. 5.2.a-b, 5.4.b, and 6.3.d.

of the principles required for demonstrative proof, for example, in LQ6.3, where he classifies various types of immediate propositions and includes *suppositiones* in their enumeration. Such propositions function as principles of demonstration, and are of some importance when viewing first principles from the perspective of teacher-student relationships. So, in his discussion of the foreknowledge of principles in his *Logica*, Valla notes that some principles are axioms (*dignitates*), and these are known to all and are indemonstrable; one cannot learn without them; there is no way of proving them by causal reasoning; and once the meaning of their terms is made clear, they immediately command assent.[13] Other principles are suppositions (*suppositiones*), and these are simply set down by the teacher; their function, according to Valla, is differently understood by different authors. Some would say that suppositions are of two types, depending on whether they are formulated as propositions in which something is said to exist or not, or to be true or false, in which case they are called petitions (*petitiones*); or, if this type of affirmation is not made, then they are called definitions (*definitiones*). Others would reserve the term suppositions for propositions that are mediate and demonstrable, but which are not demonstrated at the beginning of the science, and instead are presupposed (*supponuntur*) by the teacher. The teacher does this, either because such propositions are demonstrated in a subalternating science, or because it is not convenient for him to demonstrate them at the beginning, but rather when the science is further along. If such supposed propositions are not granted by the student and indeed are contrary to his own views, then they are called petitions, in the sense that the teacher asks the student to concede them for the time being until he proves them at their proper place—a procedure, Valla adds, that is often followed by mathematicians. A similar explanation, in even fuller detail, is given by Valla in his discussion of the principles of demonstration, to which attention will be directed in the following section when discussing techniques of demonstrating *ex suppositione*.

b. FOREKNOWLEDGE OF SUBJECTS

Galileo's account of the foreknowledge required with respect to the subject of a demonstration is quite full, being treated in five questions. The first of these (LQ3.1) inquires as to what Aristotle means by existence when he says that the existence of the subject must be foreknown; the second (LQ3.2), whether any science can demonstrate the *esse existentiae* of its adequate subject; the third (LQ3.3), whether it can demonstrate

[13] *Logica*, vol. 2, p. 145.

the existence of a partial subject; the fourth (LQ3.4), whether a science can manifest the nature of its subject and at the same time give a *propter quid* explanation for it; and the fifth (LQ3.5), what does Aristotle mean by the foreknowledge of the subject's nature (*quid est*) when he says that this must be foreknown. What seems characteristic of this particular disputation is that it employs a considerable number of technical expressions that are not explained by Galileo, but whose meaning he presupposes, either from his own general knowledge or from other parts of his logical notes that have not come down to us. Some of this material, such as the terminological distinctions pertaining to the subjects of a science, obviously is drawn from the treatise on science that is missing in Galileo's questions, to be reconstructed in the last section of this chapter.

The subject of a science, Valla informs us, is the same as the object of the science, and both become the subject of demonstration in that science, which is precisely the point of inquiry in these questions.[14] Here Galileo speaks of various ways in which this subject can be understood, and also of several ways in which the term "exists" can be taken when one says that the existence of the subject is presupposed for demonstrations within the science. Among the distinctions he employs are the total subject, the partial subject, the principal subject, the adequate subject, the *esse essentiae* of the subject, and the *esse existentiae* of the same. Fortunately all of these are defined in Valla's *Logica* or in the *Additamenta.*[15] The total subject contains everything that is treated in the science, either as its principles, its properties, or its various species or subjective parts; in the *Physics*, this is the natural body. A partial subject, on the other hand, designates any species or part of the total subject, and in the *Physics* could be applied to the elements, which are a kind of natural body. The principal subject, on the other hand, is the principal part of the total object; again in the *Physics* the heavens would qualify for most Aristotelians as fulfilling this designation. The adequate subject, finally, has somewhat the same connotation as the total subject, being the same as all the partial subjects taken together; in the *Additamenta*, nature is spoken of as the adequate subject of the *Physics*, and it is remarked that this cannot be demonstrated to exist, even though it has an effect—motion—that is more known than it is.

One of the points of contact between Galileo's logical questions and his physical questions is his use, in both manuscripts, of exactly this terminology. When discussing in the latter the subject of the *De caelo*, for example, he asserts that its total subject is the simple body;[16] the total

[14] *Additamenta*, fol. 45v; *Logica*, vol. 2, p. 409.

[15] *Logica*, vol. 2, pp. 160-161; *Additamenta*, fols. 45v-50r.

[16] *Opere* 1:16.5-8, 17.21-18.5; Wallace, *Galileo's Early Notebooks*, pars. A4 and A10.

subject of the *Physics*, on the other hand—and this he defines as that to which all things treated in that science are reduced—is the natural body.[17] A partial subject would be any part of the latter, such as the simple body and the composed body;[18] here it is noteworthy that the partial subject of the total science, *Physics*, is the same as the total subject of a part of that science, namely, the *De caelo*. The principal subject of the *De caelo* is obviously the heavens;[19] and finally the adequate subject of that same treatise is the universe, since "the adequate subject of the science is the thing whose parts and properties are considered in it."[20] These applications are completely consistent with those already noted in Valla and Valla-Carbone, and can serve to strengthen the case argued in part 1 of this study for the derivation of both sets of questions from courses offered around Valla's time at the Collegio Romano.

Foreknowledge of the existence of the subject of demonstration is the first topic treated by Galileo in LQ3.1, the entire text of which has been transcribed in chapter 1, wherein he employs the distinction between *esse essentiae* and *esse existentiae*.[21] The first is explained in the *Additamenta* as equivalent to a thing's essence, that which makes it have a determinate nature abstracting from existence, such as rational animal in man; as opposed to this, *esse existentiae* is its act of existing whereby it is placed outside its causes.[22] The latter in turn is twofold, actual and potential, the first when the thing is actually produced, such as man when fully existent; the second, when it is still contained in its causes, such as the animal in its seed. The difficulty of the question arises from problems relating to actual existence: must the subject of a demonstration be known to be here and now existing, particularly in the world of nature, where things continually come to be and pass away? What if God were to destroy the universe—would this negate the possibility of science, even of species that existed at one time? And what about subjects that are not known to exist at the beginning of a science—how can foreknowledge of their actual existence be required for demonstration? Galileo's replies, all explained at fuller length in the *Additamenta* and the *Logica*, are brief and to the point.[23] In any event the *esse essentiae*, or the nature of the subject,

[17] *Opere* 1:17.14-20; Wallace, *Galileo's Early Notebooks*, par. A9.

[18] Ibid.

[19] *Opere* 1:16.18, where Galileo identifies *caelum* as the *obiectum principalitatis*; Wallace, *Galileo's Early Notebooks*, par. A6.

[20] *Opere* 1:19.19-20, ". . . cum illud sit obiectum adaequatum alicuius scientiae cuius partes et proprietates in illa considerantur, . . ."; Wallace, *Galileo's Early Notebooks*, par. A16.

[21] Sec. 1.3.a.

[22] Fol. 46v.

[23] Fols. 45v-50v and vol. 2, pp. 158-167, respectively

must be foreknown for demonstration to occur. With regard to the *esse existentiae*, this need not be known for the subjects of logical demonstrations, since logical entities need exist only in the mind. With respect to the real sciences, on the other hand, the actual extramental existence of the subjects of demonstration must usually be foreknown. There are exceptions, however, for the existence of some partial subjects can be demonstrated *a posteriori* (the examples of primary matter and the First Mover are relevant here), and thus need not be presupposed. And finally, in the world of nature, where individual beings are contingent and not eternally existent, their existence must be foreknown in a qualified way, "for places and times and removing impediments," since, as Valla-Carbone explain, there are many kinds of things in nature that lack individual existents at particular times and places, and so these have to be supposed along with the removal of impediments that would prevent them from existing.[24] The case of God's willing to annihilate all existing species is handled in the same way: species are more important than individuals, and science is based on them, but demonstrations in this case have to be made on the supposition of there being a universe and removing the prospect of this being impeded by the divine will.[25]

Galileo's next two questions build on the basic reply to this first. One might wonder whether any science can demonstrate the existence of its total or adequate or principal subject (LQ3.2), or, alternatively, of one of its partial subjects (LQ3.3). The response to the first set of subjects is negative, for these are necessary to constitute the science and so there is no way of demonstrating them, either *a priori* or *a posteriori*, within the science itself. This does not rule out the possibility of their existence being demonstrated in a higher or subalternating science, to be sure, or of their being declared or manifested in the elaboration of the science whose subject they constitute. With partial subjects the situation is different, for not all of the parts of the subject are distinctly known at the outset of the science, and so there is the possibility of *a posteriori* proof of their existence; even here, however, *a priori* proof is ruled out, for one cannot argue from a thing's nature to its existence if he is totally unaware that it exists. Examples of partial subjects whose existence Aristotle demonstrated in his treatises, moreover, are fire in the *De caelo* and the agent intellect in the *De anima*, neither of which is immediately obvious as contained within the total subject of the science.[26]

The concluding two questions both respect the *esse essentiae* or the quiddity of the subject and inquire whether this can be manifested or

[24] The Latin text is given on p. 38 *supra*, lines 71-78.

[25] For the Latin, see p. 39 *supra*, lines 120-131, and p. 40, lines 196-202.

[26] MS 27, fol. 9v1-4

demonstrated (LQ3.4), and how it can be foreknown (LQ3.5). With regard to the first, Galileo maintains that it is possible to manifest and prove the quiddity (*quid sit*) of the partial subject of a science, since Aristotle clearly does this, establishing the nature of demonstration in the *Posterior Analytics* and the nature of the soul in the *De anima*.[27] When this is done, however, the demonstration must be *a posteriori*, since if the quiddity is unknown it cannot be used to demonstrate itself *propter quid*. With regard to the second query, Aristotle's statement that the *quid sit* of the subject must be foreknown is interpreted by some to mean that its real definition (*quid rei*) is prerequisite to any demonstration, but the better position is that a nominal definition (*quid nominis*) suffices by way of foreknowledge.[28]

c. FOREKNOWLEDGE OF PROPERTIES

In contrast to his account of the foreknowledge required for principles and subjects of demonstration, Galileo's treatment of property is brief, being confined to one question, which he considers along with another question relating to the conclusion in the final disputation of this treatise. There is nothing exceptional in either, since in the first question (LQ4.1) Galileo makes the common distinction between properties that are convertible with the subject and those that are not, and argues that the *quid nominis* of the property must be known in either case, whereas its existence must be foreknown in the second case but not in the first. The second question inquires whether the conclusion is known in the same act and at the same time as the premises (LQ4.2), and the response applies various distinctions to show in what way knowledge of the premises precedes that of the conclusion, and in what way both premises and conclusion can be said to be grasped in the same intellectual act.

This concludes the summary of Galileo's express treatment of the foreknowledge required for demonstration. Many of the points he makes may seem overly technical and even recondite to the modern reader, but a moment's reflection on his prolonged attempts to construct a science of motion will show that this is not the case. He was continually searching, for example, for cogent demonstrations that would manifest properties of various types of local motion, and was extremely painstaking in his efforts to supply all of the prerequisite information that would make such demonstrations intelligible to others. The formulation and justification of principles was a preoccupation with him, as was the precise delineation of the subjects with which he was concerned. In his

[27] MS 27, fol. 10r16-18.

[28] MS 27, fols. 10v30-11r8.

later work, the total subject of his *nuova scienza* was to be local motion, and under this would have to be included naturally accelerated motion as a species of such motion—in other words, as a partial subject. Must the existence of naturally accelerated motion as a partial subject be foreknown before one can demonstrate properties of it? How can such an entity, so contingent and so easily affected by impediments, be said to have actual existence? What *suppositiones* must be made to assure the validity of demonstrations concerning it? The answers to all these questions are contained in germ in this early treatise *De praecognitione*, and, as will be seen in the last two chapters of this work, were gradually worked out by Galileo in the elaboration of his "new science." But for the moment it may suffice to observe that these philosophical questions, over which he labored in his beginning years as a professor at the University of Pisa, were far from irrelevant to the concerns that dominated his later life.

2. Demonstration and Its Kinds

Galileo's treatise *De demonstratione* constitutes the larger part of his logical questions, and a good portion of it is concerned with difficulties relating to the text of the *Posterior Analytics* and its meaning—difficulties that occupied the commentatorial tradition from the early Greeks to professors who were his contemporaries. It is not necessary to rehearse all of this matter in the present survey, whose intent is to extract those items that relate to the ways in which one can formulate cogent demonstrations and be sure of the principles on which they are based. Except for one question, which was reordered by Carbone and therefore is not in its original place in Valla's 1588 notes, as discussed in the opening chapter, the treatise *De demonstratione* from which Galileo probably worked is no longer extant.[29] The fuller development of that treatise, to be sure, is preserved in Valla's text of 1622, and this is a most helpful source for filling out *lacunae* in Galileo's composition. In this connection, Galileo's treatise on demonstration is very compact, even cryptic in its expression, and thus its understanding requires some complement such as Valla provides. But the full exposition of all the matters contained in this treatise would expand this study far beyond reasonable length, and so greater selectivity will have to be employed here than in the previous section of this chapter.

Apart from the first question of the treatise (LQ5.1), concerned with

[29] Sec. 1.3.b.

the nature and importance of demonstration, most of Galileo's discussion of problems concerned with the principles on which it must be based is contained in the eleven questions that make up the second disputation of the treatise, viz., LQ6.1 through 6.11. These questions actually constitute a detailed exegesis of Aristotle's definitions of demonstration as these are given in LQ6.5, namely, that demonstration is a syllogism producing science, and that this syllogism consists of premises (or principles) that are true, first, and immediate, and that are more known than, prior to, and causes of the conclusion. The third and last disputation of the treatise is concerned with the various kinds of demonstration, in which Galileo has inserted his final question (LQ7.3), dealing with the demonstrative regress. This particular question, together with that in which he compares demonstration with definition (LQ5.2), is better left for discussion in the following section. The present section, therefore, will concentrate first on matters relating to the knowability of the principles or premises on which demonstration must be based, and then focus on the kinds of argument that Galileo is willing to regard as demonstrative.

a. PRINCIPLES OF DEMONSTRATION

The truth of the principles on which demonstration is erected (LQ6.1) would seem to offer little difficulty, since it seems obvious that such principles would have to be true in order for the argument to reach a true conclusion. In discussing this, however, Galileo follows the scholastic pattern of distinguishing all of the different senses in which the word "true" can be applied to principles of various kinds, whether these are principles of being or of knowing.[30] This entails a discussion of the truth or existence of complex principles, that is, those of knowing that are stated as premises, and then the truth or existence of noncomplex principles, that is, those of being that constitute the object of knowledge. Some would interpret Aristotle's statement to the effect that what is not, is not known, Galileo says, as applying to noncomplex principles, but this is not correct, since there can be science even of nonexistent things, and he gives the example of a rose. The example is mystifying, as thus given, until one consults the parallel passage in Valla, where one finds that Galileo means a rose in winter.[31] There can be science of natural things such as roses (and in LQ6.7 he adds the examples of eclipses and

[30] For a detailed study of the scholastic development, see Steven P. Marrone, *William of Auvergne and Robert Grosseteste: New Ideas of Truth in the Early Thirteenth Century* (Princeton: Princeton University Press, 1983).

[31] Cf. MS 27, fol. 18r26–27, and *Logica*, vol. 2, p. 223.

"hail and like things"),[32] even though such things are not always existent, provided they exist at some times and places. Even things that are regarded as totally nonexistent—and for these Galileo gives the examples in LQ6.1 of the vacuum and the infinite[33]—can be considered in a science, and truths can be stated about them in propositions that are true by complex truth. The theme of scientific statements that are true under the supposition of their times and places and "removing impediments" that he explained in LQ3.1 he further resumes in LQ6.7, where he takes up the problem of God's will changing the natural order and thus rendering knowledge of that order impossible for man. This is related to the necessity that attaches to propositions—some respect things that could not be made otherwise even by God himself, called absolute necessity, whereas others respect things that cannot be impeded according to the ordinary law of God but can be impeded by his absolute power, called natural necessity. As an example of the last, Galileo cites the necessity of the sun's rising and setting, unaccountably leaving out the word "sun" in doing so.[34] But his meaning is clear: natural science owes the very possibility of its existence to the supposition that God will not miraculously alter the natural order, so that the laws of nature will be continuously operative.

Galileo's discussion of Aristotle's condition that the premises of a demonstration must be immediate (LQ6.3) leads him to a further examination of the ways in which the term "immediate" can be understood and how this relates to the premises being grasped by the knower—another entry to the topic of *suppositio*. A proposition is said to be immediate when the predicate can be said of the subject directly, with no middle term intervening or without a cause having to be given for the predication. Examples of immediate propositions, and here Galileo follows scholastic usage, would be those in which a definition or part of a definition is predicated of the thing defined, or in which a first property is predicated of a subject, or in which one category is denied of another. Aristotle, he goes on, divides such propositions differently, saying that their two principal types are axioms and positions, which he defines as follows. Axioms (*dignitates*) are propositions that must be known necessarily by anyone who would learn the science, whereas positions (*positiones*) need not necessarily be foreknown to acquire the science.[35] Positions, in turn, are twofold: some are suppositions (*suppositiones*), wherein something is affirmed or denied of another, whereas others are principles or definitions

[32] MS 27, fol. 24r24-26

[33] MS 27, fol. 18r18-20

[34] MS 27, fol. 23v13-14: Galileo's text reads "Hac necessitate dicimus necessarium esse oriri et occidere." The word *solem* is obviously omitted between *esse* and *oriri*.

[35] MS 27, fol. 20r25-28.

(*principia seu definitiones*), wherein no affirmation or denial is made.[36] Galileo does not elaborate on this division here, merely mentioning it, but it is explained more fully, though in slightly different terms, in Valla's *Logica*.[37] Axioms in this context, Valla states, are neither suppositions nor positions, but they are propositions that are so easily known that no one can refuse assent to them internally, though he might deny them with his lips. Suppositions, on the other hand, are propositions that, although they might be able to be demonstrated, are accepted as true and are conceded to be such because they seem so to the learner. Petitions can also be demonstrated, but they are petitioned for the learner to concede them, either because he is not aware of their truth or because he himself holds a contrary opinion. A petition (*petitio*) is also called a postulate (*postulatum*), and both it and a supposition differ from a position (*positio*) only in that the latter has a wider meaning: it is a principle that is accepted without proof, even though it can be proved, or that is contrary to the opinion of another. Terms or definitions, finally—and note here that Valla speaks of "terms or definitions" rather than of "principles or definitions," as does Galileo (whose use of the word "principles" is so generic as to be meaningless in this context)—are not suppositions because they are not propositions and do not affirm being or nonbeing; rather, they express meanings that are grasped or not, and about which nothing is said, that is, whether they apply universally or particularly, to a whole or to a part, etc.[38] Valla here cites his earlier discussion, to which reference has been made in the first section of this chapter and which is consistent with this fuller exposition.[39]

In further examining whether immediate propositions must enter into every demonstration and how they can do so (LQ6.4), Galileo employs a distinction between propositions that are actually immediate and those that are only virtually so, and again between propositions that are actually indemonstrable and those that are virtually such, which he never fully explains.[40] The actual vs. virtual distinction apparently arose from the

[36] MS 27, fol. 20r28-31.

[37] Vol. 2, p. 218.

[38] Valla's text reads as follows: "Termini denique vel definitiones non sunt suppositiones, quia non dicunt esse vel non esse; suppositiones vero, cum sint propositiones, illud habent. Quare termini vel definitiones sola simplici apprehensione intelliguntur, quod non est in suppositione, nisi quis velit suppositionem esse ipsum audire verba. Suppositiones vero sunt ex quibus deducitur conclusio, eo quod ipsae sint." Galileo, on the other hand, has: "[Positiones] quaedam sunt suppositiones in quibus aliquid affirmatur vel negatur de alio; quaedam sunt principia seu definitiones in quibus nihil affirmatur aut negatur de alio."

[39] *Logica*, vol. 2, p. 145.

[40] MS 27, fol. 20v29-30: ". . . quod est dicere demonstrationem constare aut ex immediatis actu aut virtute . . ."; fol. 21r1-3: ". . . demonstratio in communi sumpta analogice dicitur de demonstratione constante ex indemonstrabilibus actu et virtute. . . ."

discussion of actual causes and virtual causes by the Thomistic commentator, Thomas de Vio Caietanus, and this is elaborated on by Valla in some detail.[41] Valla further applies the distinction to immediate and indemonstrable propositions, as does Galileo, but in so doing supplies a justification for this usage.[42] In the context of their employment in demonstration, some immediate propositions must enter actually into the demonstration, in the sense that they form its premises; others do not so enter, but are necessary to establish the premises that do. The latter can be said to enter virtually into the demonstration, even though they are not part of it. Again, some propositions are actually indemonstrable, in the sense that they cannot be demonstrated in any way but must be grasped immediately; others are only virtually indemonstrable, in the sense that they are incapable of demonstration in the particular science or in the part of it under consideration, though they are demonstrable elsewhere. Galileo takes the position that both virtually immediate and virtually indemonstrable premises are sufficient to constitute a valid demonstration (LQ6.4), although such a demonstration will be imperfect compared to those made from actually immediates and indemonstrables, insofar as it is made *ex suppositione* (LQ6.2, 6.4).[43] His, therefore, is the same view as Valla's, who holds that this type of demonstration, because really made from mediates rather than immediates, is not as perfect as the type made from immediates but must ultimately be reduced to the latter; insofar as it supposes these premises as true, moreover, it does not generate science *simpliciter* but only *ex suppositione*—a situation that is usually found in subalternated sciences (LQ6.4). Yet the argument it produces is truly scientific, and not merely probable, and so is capable of producing a conclusion that could not be otherwise.[44]

Galileo's discussion of principles being "more known than" the conclusion of a demonstration (LQ6.6) covers all of the distinctions com-

[41] *Logica*, vol. 2, p. 240ff.

[42] *Logica*, vol. 2, pp. 224-244.

[43] MS 27, fol. 19r4-6: ". . . at demonstratio quae procedit ex causis veris est huisusmodi, non illa quae procedit ex virtualibus, quae cum sint ex suppositione non possunt facere scire simpliciter"; fol. 20v13-21: "Dico primo, omnem demonstrationem debere aliquo modo constare ex immediatis. Probatur tum auctoritate Aristotelis in hoc secundo capite, tum ratione, quia scire nihil aliud est quam certo et evidenter assentiri conclusioni; at nemo potest certo et evidenter assentiri conclusioni nisi demonstratio constet ex immediatis; ergo omnis demonstratio debet constare ex immediatis. Dices, scientia subalternata habet perfectas demonstrationes; et tamen supponit sua principia immediata tanquam probata a subalternata; ergo, etc. Respondeo, scientiam subalternatam tanquam imperfectam non habere perfectas demonstrationes, cum prima principia supponat in superiori probata, ideoque gignat scientiam ex suppositione et secundum quid. . . ."

[44] Cf. *Logica*, vol. 2, p. 236. See also the author's "Aristotle and Galileo: The Uses of *Hupothesis* (*Suppositio*) in Scientific Reasoning," in *Studies in Aristotle*, D. J. O'Meara, ed. (Washington: The Catholic University of America Press, 1981), pp. 47-77, for a more

monly made by Aristotelians of the time, but is much condensed from the corresponding treatment in Valla.[45] More interesting, for purposes of this chapter, is his account of causality and causal connection and how these are related to the four modes of speaking *per se*, which is one of the celebrated doctrines of the *Posterior Analytics* (LQ6.7–6.9).[46] Here Galileo is unambiguous in his affirmation that there is a real connection between cause and effect, which, while not always present in the order of nature, does exist eternally in the divine mind (LQ6.7).[47] In the same context he argues that all four kinds of cause (final, efficient, formal, and material) are contained under Aristotle's fourth mode of speaking *per se* because of the intrinsic relationship that exists between cause and effect. This applies to the causal connections that are potential as well as to those that are actual. And even though many causes are contingent in their operation, and so permit only contingent predication in propositions, they are not contingent when considered in their operation and in the way they are connected with the effects they produce (LQ6.7).[48] He further argues that the fourth mode of predication applies to extrinsic causes as well as to intrinsic causes, and that there can be perfect demonstrations through extrinsic causes as well as through intrinsic ones (LQ6.9)—a topic that Valla develops at length in his *Logica*, while generally endorsing the teachings of Jacopo Zabarella on this disputed point.[49] Galileo's discussion of properties in his account of the first two modes (LQ6.8) is also of interest, since it shows a sophisticated awareness of the way in which white can be predicated of swan, for example, and uses this type of reasoning to show that Aristotle's arguments to prove the roundness of the heavens and the earth in the second book of the *De caelo* cannot be true and perfect demonstrations.[50] Galileo nonetheless holds that natural necessity alone is sufficient to guarantee the necessity

---

detailed account of Galileo's use of the expressions *suppositio* and *ex suppositione* in his various writings.

[45] *Logica*, vol. 2, pp. 248-253.

[46] A more systematic exposition of Galileo's knowledge of causes and their effects is given in the author's "The Problem of Causality in Galileo's Science," *The Review of Metaphysics* 36 (1983): 607-632.

[47] MS 27, fol. 24r19-22: "Respondeo, demonstrationem versari circa res aeternas, non quidem quae existerint ab omni aeternitate, cum nullae dentur tales uno excepto Deo; sed versari circa res aeternas, hoc est, circa res quae habuerunt connexionem verissimam ab aeterno in mente divina."

[48] MS 27, fol. 26r10-13: "Dices, causa ut respicit effectum in actu est contingens, ergo non poterit poni in hoc quarto modo. Respondeo, esse quidem contingentem huiusmodi propositionem, si praedicatio spectetur; non tamen, quod est ad rem nostram, si spectetur connexio praedicati cum subiecto, et habitudo causae inter haec duo."

[49] Vol. 2, pp. 265-267.

[50] MS 27, fol. 26v1, 27-29.

of propositions in these two modes (LQ6.8). Later, when discussing the requirements for perfect demonstration, he argues that circular motion is proper to the heavens, but that Aristotle's attempt to prove that the earth is at rest is not a true demonstration, since rest is not proper to earth (LQ6.11)[51]—a conviction he would maintain for the greater part of his intellectual life. He further uses these considerations to show that there cannot be demonstration, or science, of singular individuals—a conclusion to which he makes reference in his physical questions (PQ1.1.13) as one he has shown elsewhere, and so can be used to establish the temporal priority of the logical over the physical questions.[52] Again, all of this is covered more fully in Valla's *Logica*, though not in the same order and usually with a different selection of examples.[53]

b. KINDS OF DEMONSTRATION

Among the various kinds of demonstrations, Galileo enumerates five types in LQ5.1, namely, *ostensiva, ad impossibile, quia, propter quid*, and *potissima*;[54] but when he considers these kinds at some length in the last disputation of the treatise on demonstration, he effectively limits his treatment to the differences between demonstrations *quia* and those *propter quid* (LQ7.1 and 7.2). In the enumeration of LQ5.1 he defines an ostensive demonstration as one that proves something true from true premises (which would be applicable to the last three types), and a demonstration to the impossible as one that leads from the concession of one impossibility to another that is more known; Valla complements the latter definition by stating that the contradictory of the first impossibility is then known to be a true proposition.[55] For Galileo a demonstration of the fact (*quia*) is one that proves something from an effect or from a remote cause, whereas a demonstration of the reasoned fact (*propter quid*) is one that demonstrates one thing of another through true and proper principles. A demonstration that is most perfect (*potissima*), finally, manifests some first and universal property of an adequate subject through proper and proximate principles, and thus it represents a special type of *propter quid* demonstration.

In taking up the question of the number of kinds of demonstration (LQ7.1), Galileo immediately concedes the common view that demon-

[51] MS 27, fol. 28v16-18.

[52] MS 27, fol. 29r7-13; see sec. 2.4 *supra*.

[53] Vol. 2, pp. 255-285.

[54] MS 27, fol. 13r17-19.

[55] MS 27, fol. 13r20-21: ". . . ad impossibile est, quae ex aliquo impossibili concesso ducit ad aliud impossibile notius. . . ." Cf. *Logica*, vol. 2, p. 299.

stration to the impossible is not a true species because it argues from false premises. His major concern, therefore, is with the various forms of ostensive demonstration, namely, *quia, propter quid*, and *potissima*, to ascertain whether each of these constitutes a distinct species. Commentators were divided on this problem, some holding with Avicenna that there is only one true species of demonstration, namely, *propter quid*; others holding with Aquinas and others that there are really two species, *quia* and *propter quid*; and yet others holding with Averroes and his school that demonstration *potissima* constitutes a distinct species over and above *quia* and *propter quid*, thus making three in all. Galileo rejects the first and the third opinions and opts for the second. He elaborates his position under four conclusions, to wit, that demonstration *quia* is a true species of demonstration since it proceeds from necessaries and infers a necessary result, even though some authors refer to it as a topical or probable argument; that demonstration *potissima* does not constitute a species distinct from *propter quid*, since the latter fulfills Averroes's definition for most perfect demonstration under the required conditions; that there are therefore only two species of demonstration; and yet that *propter quid* demonstration can be said to contain under it two subspecies, one that argues from extrinsic causes, another that argues from intrinsic causes and manifests a property of its first and adequate subject through principles that are actually indemonstrable—which can with full justice be called most perfect or *potissima*.[56]

Question LQ7.2 resumes this discussion and inquires into the similarities and differences between demonstrations *quia* and *propter quid*. Galileo maintains that these two species are analogically the same insofar as both are based on true and necessary principles, whereas they are different in a variety of ways depending on the type of *quia* demonstration being considered. A demonstration *propter quid*, for example, employs a middle term that is prior both in being and in knowing (*in essendo* and *in cognoscendo*), while the middle in a *quia* demonstration is prior only in knowing (*in cognoscendo tantum*). Again, there are many different understandings of demonstration *quia*: the two basic types are those that proceed from a remote cause to prove an effect, in which Galileo professes no interest, and those that argue conversely from effect to cause. Some refer to the latter as demonstration *signi* (from a sign or indication), others as demonstration *evidentiae* (from evidence) or *existentiae* (of existence); yet others as demonstration *ab effectu* or *a posteriori* (from an effect or from what is later in being), or as *demonstratio coniecturalis* (one put together logically).[57] A demonstration *quia*, he further explains, is some-

[56] MS 27, fol. 30r1, 11, 21, 25-27.

[57] MS 27, fol. 30v32-37.

times constructed from convertible terms, as in the example: an eclipse is occurring, therefore the earth has come between the sun and the moon; sometimes from nonconvertible terms, as when one argues from the existence of heat to the presence of fire. And as far as existence is concerned, this type of demonstration can be used to prove either that a simple thing, an entity of some type hitherto unknown, exists, or that a complex thing exists, meaning by the latter expression a proposition whose truth is made manifest by *a posteriori* reasoning.[58]

Galileo concludes his discussion of demonstration from effect to cause with the remark that demonstration of this type is most useful in the sciences, because their principles are generally unknown and cannot be discovered without reasoning of this type.[59] This is a most telling observation, for it serves to explain many of Galileo's later attempts to prove principles on which he hoped to erect his new sciences. The force of this observation, it may be remarked, is easily missed by post-Humean philosophers of science, who adopt an agnostic stance with respect to causal connections in nature and effectively reject, in Avicennian fashion, the possibility of *a posteriori* demonstration. Such a position is not at all characteristic of Galileo, and so cannot be used as a basis for evaluating either his writings or the scientific conclusions he proposed to establish.

3. Definition and the Demonstrative Regress

As earlier remarked, Galileo does not have an explicit treatment of definition in his logical questions, though he does treat definition comparatively as it is related to demonstration when he inquires which of the two is the more noble instrument for acquiring knowledge (LQ5.2). This particular question has marked affinity to the problems treated in Valla-Carbone's *De instrumentis sciendi*, as explained in chapter 1, and was probably abstracted by Galileo from this treatise in the form it had prior to its appearance in the *Additamenta*. Since these materials have important methodological implications, some of which are implicit in Galileo's remaining logical questions, it will be helpful to summarize them here. This can be conveniently done by first explaining what is meant by an instrument of knowing and how this concept is related to definition and demonstration as well as to the general method of resolution and composition, and then addressing a special problem that arises from the latter method and was much debated among sixteenth-century Aristotelians, namely, the problem of the demonstrative regress.

[58] MS 27, fols. 30v43-31r2.

[59] MS 27, fol. 31r2-5.

a. INSTRUMENTS OF KNOWING

The ensemble of Aristotle's logical works has traditionally been referred to as the *Organon*, which means instrument, since these works were meant to describe the tools or instruments available to the human mind to achieve its natural goal, viz., knowledge in all its forms. Obviously many such instruments can be identified, some of which designate broader classes than others and so would include them as subspecies. When an attempt is made to eliminate duplication of this type, however, six instruments emerge as those most helpful in the acquisition of knowledge: definition, demonstration, division, proposition, argumentation, and method. All of these are mentioned by Galileo as serving to direct the operation of the human mind, though he notes that division and method assist the mind only mediately, that argumentation assists it imperfectly, and that demonstration and definition assist it immediately and perfectly (LQ5.2). He does not define any of these, but he does indicate that argumentation is a general term including under it the probable syllogism, induction, and enthymeme, and thus is not the same as demonstration.[60] Valla-Carbone supply the missing definitions, as follows: definition is an expression that clearly manifests the nature of a thing, meaning by this its essence or quiddity; division sets out the parts that make up any whole; proposition is a statement that expresses something true or false; demonstration is a syllogism that argues from necessary premises to necessary conclusions, employing either causes or effects; argumentation is a type of reasoning different from demonstration that establishes certain or probable conclusions, though not through proper causes and effects; and method is the order to be followed in pursuing the subject matter of a particular art or science.[61] In their schema each of these instruments serves a different act or function of the mind: definition directs the first act, which is concerned with the apprehension of meanings and the formation of concepts; division and proposition serve the second act by making the confused distinct and by promoting assent to the collated parts in the proposition; demonstration and argumentation regulate the third act, whose function is reasoning from one proposition to another; and method, finally, puts order into the entire process.

Galileo has a number of *obiter dicta* relating to the various instruments, most of which can be passed over as scaffolding to set up definition and demonstration as the most perfect instruments so that he can then adjudicate which is the more important of the two. Two of his statements, however, are worthy of note: one occurs when he is explaining that all other instruments of which people have spoken are reducible to these

[60] MS 27, fol. 14r26-28.

[61] *Additamenta*, fols. 25v-26r.

six, at which point he remarks that resolution and composition are reducible to demonstration;[62] the other, when locating method between proposition and division on a scale of increasing nobility, where he assigns this rank to method on the basis that it runs through everything.[63] Apart from these statements Galileo has no mention of the methodology of resolution and composition in his logical questions, despite the fact that most logicians of his period treated it at length. He does discuss, of course, the related problem of the demonstrative regress, about which more later. Fortunately the *Additamenta* contains a full discussion of resolution in its opening *Praeludia*,[64] and then takes up the special problems of how resolution is something distinct from demonstrations *propter quid* and *quia* and from the resolutive method when treating the various instruments in detail.[65] Thus one can turn to it for the needed background information.

Resolution, Valla-Carbone inform us, is nothing more than the Latin equivalent of the Greek *analysis* and as such is the primary concern of the two books of Aristotle's *Analytics*, known in Latin as the *Resolutoria priora* and *posteriora* respectively. The word itself means the act or operation by which something composed is reduced or returned to the simples from which it came. As applied to logic it can be explained in a number of ways, though usually it is seen as a regress to the principles from which a result comes or on which it depends. In the arts and sciences, moreover, it has been developed so that one can understand clearly and perfectly the parts, causes, and principles of the thing being studied. In view of these different usages, at least six different distinctions have been made in various meanings of the term, among which would be included real vs. rational, practical vs. speculative, metaphysical vs. mathematical, proper vs. improper, etc. The sixth distinction, which the authors state is noted by some, takes on special interest because of the focus of this study. It is called resolution *ex suppositione* and occurs when a conclusion is resolved to principles that have been supposed, but which need not have taken the form of suppositions because they are capable of proof; then the resolution continues until any further suppositions that might be required for their proof are uncovered, and each of these is proved in turn, until one comes finally to principles that are most easily grasped and that require no suppositions whatever.[66]

[62] MS 27, fol. 14r30: ". . . resolutio et compositio ad demonstrationem revocantur. . . ."

[63] MS 27, fol. 14v28-30: ". . . divisio nobilior est quam sit methodus et propositio, quia primae operationi intellectus omnium praestantissimae deservit, methodus autem propositione, quia per omnia vagatur, sicuti et discursus seu syllogismus. . . ."

[64] *Additamenta*, fols. 1r-2r.

[65] *Additamenta*, fol. 25r-v.

[66] *Additamenta*, fol. 2r.

The explicit treatment of resolution and composition in the *Additamenta* arises in reply to an objection as to why this instrument of knowing is not included in the six instruments already given.[67] The point of the difficulty is that resolution is not the same as the two species of demonstration that have been discussed (*propter quid* and *quia*), nor is it the same as resolutory method. It is not the same as *propter quid* argument, since it proceeds in the inverse way (from effect to cause, conclusion to premises) and has the opposite goal, namely, knowledge of principles and not of conclusions; nor is it the same as *quia* argument, because this is not found in some sciences, such as mathematics, whereas resolution is found in all, and *quia* argument proves a cause from an effect, whereas resolution uses an effect to inquire into its cause. Again, resolution is different from resolutory method, since in its practical understanding resolution is concerned with finding means to achieve preestablished goals, whereas resolutory method is concerned with ordering means among themselves to assure proper execution; in its speculative understanding, on the other hand, resolution is an inquiry into the causes of a proposed conclusion, whereas method is something quite different, namely, the order to be followed in treating the subject matter of a speculative science.

Valla-Carbone's response is that Aristotle himself is not completely consistent in his use of this terminology, but that generally speaking resolution is so pervasive in logic that it is part-and-parcel of its major instruments and so does not constitute a distinct instrument apart from those enumerated. As a general doctrine that is useful in all the sciences, it pertains to method; as one that aids in the analysis of the syllogism, it pertains to the teaching on demonstration. One might even ask if definition is known through resolution, and to this the reply is affirmative also: should definition be taken to mean the middle term of a demonstration, then obviously it is obtained by resolving the conclusion to its proper middle; should it be understood as an instrument of knowing itself, then it is resolved through the process of division. This explains why the two books of the *Posterior Analytics* are both well named *resolutoria*, for the first explains how to resolve demonstrations to their first principles, and the second, how to find definitions through a process of resolution that is appropriate to this task.

In LQ5.2 Galileo does not discourse about resolution, as already noted, but he is explicit that the two instruments just named, definition and demonstration, are the ones that direct the operation of the intellect perfectly and immediately, since perfect knowing consists in understand-

[67] *Additamenta*, fol. 25r.

ing the nature of a thing and its properties, and the nature is revealed by definition, the properties by demonstration.[68] Of these two instruments, however, it is difficult to decide which is the more important. Most Latin commentators, he notes, favor demonstration, while the Averroists favor definition, and some recent writers straddle the issue, saying that definition is the better in itself, but that demonstration is for us. Galileo's resolution is that of the *Additamenta*, as can be seen from the parallel texts furnished in the first chapter.[69] Definition is better than demonstration both in itself and for us; as instruments of knowing the two are related analogically, by an analogy of proportion and of attribution, wherein definition is the primary analogate; and definition is the goal of demonstration, and is the more perfect on that account, being concerned with substance rather than with the accidents that are found to be its properties through the process of demonstration.[70]

There is no further treatment of definition in Galileo's logical questions or in the *Additamenta*, but it is noteworthy that Valla's *Logica* devotes an entire disputation to this subject wherein all matters just discussed are examined at great length.[71] The first part of the disputation is concerned with the nature of definition and takes up the foreknowledge required for it, the act or operation of the mind that it perfects, its definition and division into various kinds, the laws for finding a good definition, the object of the defining process, and the methods to be used, including a brief treatment of resolution and composition.[72] The second and concluding part then takes up the questions whether definition and demonstration constitute distinct instruments of knowing, whether a definition can itself be demonstrated, and whether definition is better than demonstration, giving essentially the same answers as are found in Valla-Carbone and in Galileo, but arranged in more systematic form.

b. THE DEMONSTRATIVE REGRESS

The last of Galileo's logical questions (LQ7.3) addresses itself to the famous question: Is there such a thing as a demonstrative regress? The topic is not to be found in the *Additamenta*, though the same question is found in the *Logica* as the concluding part of the disputation on demonstration; the doctrine there is basically Galileo's, though expressed in

[68] MS 27, fol. 14v16-20: "Si autem considerentur instrumenta quae perfecte et immediate deserviunt directioni operationum intellectus, duo tantummodo sunt, definitio et demonstratio, quia quicquid perfecte scitur aut est natura rei aut est proprietas illius; sed natura per definitionem, proprietas per demonstrationem perfecte cognoscitur. . . ."

[69] Sec. 1.3.b.

[70] See pp. 49-50 *supra* for the Latin texts.

[71] *Logica*, vol. 2, pp. 377-409.

[72] *Logica*, vol. 2, pp. 396-397.

slightly different terms.[73] An earlier exposition of the question is contained in Toletus's logic text, in much abbreviated form, and practically all of the Jesuits who taught logic at the Collegio Romano devoted some time to it—each answering it, as does Galileo, in the affirmative.[74]

The problem of the regress arises from the proposal of ancient philosophers who wished the demonstrative process to exhibit perfect circularity, in the sense that the conclusion would be known perfectly through the premises, and the premises through the conclusion. The proposal was rejected by Aristotle, who offered in its stead an imperfect type of circle wherein premises could sometimes be inferred from a conclusion by a demonstration *quia*, and then the same conclusion deduced from the premises by a demonstration *propter quid*—a twofold *progressus* or two *progressiones* that came to be known as the demonstrative *regressus*.[75] His followers, unfortunately, were divided over this compromise: some, following Avicenna, rejected it because they could not accept the first *progressus* on which it was based, demonstration *quia*; yet others, following Pomponatius's teacher, Francis de Nardo or Neritonensis (Galileo writes this as Eritonensis, but the correct spelling is in Valla),[76] rejected it because the second *progressus* would be invalid if based on the first, since the conditions requisite for *propter quid* demonstration would be lacking in this event. Galileo and the Jesuits, having repudiated Avicenna's view of demonstration *quia*, were not prepared to follow his school on this matter; neither were they inclined to take up Nardo's position, for reasons that will be seen. Instead they followed a tradition developed by Aristotelians of Averroist leanings associated with the University of Padua, among whom Agostino Nifo and Jacopo Zabarella were the foremost proponents, which defended the possibility of the *regressus*.[77]

[73] *Logica*, vol. 2, pp. 340-350.

[74] Toletus, *Commentaria, una cum quaestionibus, in universam Aristotelis logicam* (Venice: Apud Georgium Angelerium, 1597), fol. 160r; Lorinus, *In universam Aristotelis logicam*, pp. 553-557; Rugerius, Cod. SB Bamberg 62-2, fols. 480r-490v. It is perhaps noteworthy that a copy of the Cologne 1596 edition of Toletus's logic was in Galileo's personal library; see A. Favaro, "La libreria di Galileo Galilei descritta ed illustrata," *Bulletino di Bibliografia e di Storia* 19 (1886): 245, n. 78.

[75] This is Galileo's terminology in MS 27, fol. 31r-v. The various occurrences of these and related terms are the following: regressus, 31r6, 11, 29, 33, 38, 31v7; circulus, 31r31, 47, 31v3; circulus perfectus, 31r7, 41; circulus imperfectus, 31r23; circulus improprie, 31v17; progressus, 31r33; primus progressus, 31v11, 19; secundus progressus, 31r16, 31v1, 20; duo progressiones, 31v7; processus, 31r21. Note also Galileo's use of *progressus* in the text of LQ3.1 cited *supra*, p. 37, line 24, which follows the usage of the *Additamenta*, same page, line 23. Ther term *progressio* reappears later in Galileo's discourse on floating bodies, *Opere* 4:67.23, as will be noted in chap. 6, p. 285 *infra*.

[76] MS 27, fol. 31r16; *Logica*, vol. 2, p. 346.

[77] A survey of the development of this teaching is contained in Giovanni Papuli, "La teoria del 'regressus' come metodo scientifico negli autori della Scuola di Padova," in

Galileo does not mention the source of this tradition, identifying the opinion merely as Aristotle's own, but Valla gives full credit to Zabarella, and there can be little doubt that the latter is the author from whom Galileo's arguments are ultimately drawn.[78]

Before offering his reasoning in favor of the *regressus*, Galileo gives two preambles that are necessary for its understanding. The first is that there has to be some connection between the proof being offered and the thing being proved, and that the former should come first and be more known than the latter. The second relates to various ways of looking at cause and effect, which he enumerates as three: (1) under the formal relationship of cause and effect; (2) as these are themselves disparate things; and (3) as cause is necessarily connected with the effect it produces. These understood, he proceeds to argue for three different conclusions related to the respective three ways, as follows: (1) if cause and effect are taken in the first way, there cannot be a demonstrative *regressus*; (2) if they are considered in the second way, there will be no circularity of argument; and (3) if the second way is taken in conjunction with the third way, a valid demonstrative *regressus* is possible. Each of these will be considered in turn.

The reasoning behind the first two conclusions is fairly straightforward. With regard to the first, if cause and effect are seen precisely as formally related to each other, then the two are correlatives, in the sense that the cause is a cause only insofar as it is producing the effect and the effect is an effect only insofar as it is being produced by the cause. But things that are correlatives are so interdependent in their being and thus in their being known that one cannot be more known than the other. Therefore the first condition requisite for the proof is violated, and there cannot be demonstration, let alone a demonstrative regress. With regard to the second condition, on the other hand, if cause and effect are regarded as totally disparate things, there is no possibility of circularity in demonstration, since demonstration requires that one thing be necessarily inferred from another, and this is possible only if there is some necessary relationship between the two. When viewed as totally disparate things, however, cause and effect lack this necessary relationship and thus are not capable of generating a circular argument.

Galileo's justification for the third conclusion is more difficult to explain, since there is a lacuna of five or six words at a crucial point in his

---

Aristotelismo Veneto e Scienza Moderna, 2 vols., Luigi Olivieri, ed. (Padua: Editrice Antenore, 1983), vol. 1, pp. 221-277. See also the author's *Causality and Scientific Explanation*, 2 vols. (Ann Arbor: The University of Michigan Press, 1972-1974; reprinted Washington: University Press of America, 1981), vol. 1, pp. 117-155.

[78] *Logica*, vol. 2, pp. 343-344

manuscript, and this requires reconstruction and interpretation.[79] The following, however, seems to be his argument. In the first *progressus* involved in the regress, cause and effect are understood disparately and not as formally related, and thus it is possible for the effect to be known without the cause; when this is the case, the existence of the effect can be used to prove the existence of the cause. Again, when one discovers a cause, one need not see this precisely as related to any effect; on consideration, however, one may come to the realization that it is necessarily connected with something not known or recognized before. When this occurs one is ready for the second *progressus* involved in the regress, for then the newly discovered cause can provide the basis for a *propter quid* demonstration. Thus the regress is possible if it occurs in two *progressiones*, one in which cause and effect are seen in the second way noted in the preamble to the proofs, the other in which they are seen in the third way.

Valla's reworked exposition does not follow the same order as Galileo's, and thus is not particularly helpful for clarifying the reasoning behind Galileo's third conclusion. Both he and Galileo follow up their arguments by enumerating the conditions that are necessary for the regress to occur, however, and here there is considerable similarity in the two texts—which can be used to confirm that the foregoing interpretation is essentially correct. For the regress to be possible, according to Galileo: (1) there must be two *progressiones* in the demonstration, one from effect to cause, the other from cause to effect; (2) the argument must begin with the demonstration *quia*; (3) the effect must be more known to us; (4) at the conclusion of the first step the cause is known only materially, and the second step cannot begin until the cause is recognized formally as such; (5) cause and effect must be convertible, that is, one cannot be of broader scope than the other; and (6) the reasoning must be in the first figure. Valla's conditions are not exactly the same; but they are comparable, as can be seen from the following enumeration: (1) both demonstrations cannot be *propter quid*; (2) neither can both be *quia*; (3) the terms employed must be convertible and reciprocal; (4) the argument must be in the first figure; and (5) the demonstration *quia* must come first, to be followed by the demonstration *propter quid*.[80] Galileo's fourth condition is lacking in Valla's list, but this is more than offset by the latter's devoting two entire chapters to a consideration of what must occur during the interval between the two steps of the regress.[81] At the end of the first step, Valla says, we do not have distinct knowledge of cause and effect, nor do we know them formally, but only materially.

[79] MS 27, fol. 31r38.

[80] *Logica*, vol. 2, pp. 343-344.

[81] *Logica*, vol. 2, pp. 345-346.

The interval between the steps is necessary to come to this formal recognition, and for this is required some activity of the intellect, variously named by different authors as *consideratio, negotiatio, mentale examen, meditatio, applicatio, intentio mentis*, and similar terms. However called, the effect of this consideration is to recognize cause and effect distinctly and formally as such, so that they can provide the basis for a *propter quid* demonstration. The best explanation of this, in Valla's estimation, is given by Zabarella in his treatise *De regressu*, but his mention of *negotiatio* and similar terms indicates that he is also familiar with the thought of Nifo and other Paduans who had addressed this very problem.[82]

Galileo's treatment of the demonstrative regress is understandably briefer than Valla's, but he does handle a few objections directed against the teaching. He also remarks that the regress is useful in all the sciences, but is of most frequent use in the physical sciences, for there causes are generally unknown to us; it is almost of no use, on the other hand, in the mathematical sciences, for in these disciplines causes that are most known in themselves are also most known to us.[83] These remarks, taken in conjunction with his later references to resolution and composition as scientific methods, show that Galileo was far from unfamiliar with the thought of the best logicians of his day. Methodologically, at least, he was well prepared to face the difficult problems he would early encounter in developing a science of motion.[84]

4. Sciences and Their Classification

From the foregoing it is clear that Galileo's logical questions contain many remarks about sciences and the differences between them. Such statements, like those on definition, usually are made in a context wherein they presuppose knowledge on the part of the reader and so do not spell out all that would be necessary for their understanding. As already noted, there is no treatise or disputation on science in the questions as they have come down to us, though this does not rule out the possibility of their having existed at one time, since some of Galileo's own references could have been to the missing tract—for example, his statement in the physical questions: "There cannot be a science of singulars, as we have shown

[82] Wallace, *Causality and Scientific Explanation*, vol. 1, pp. 140, 144-149.

[83] MS 27, fol. 31v3-6: "Respondeo, progressum demonstrativum esse utilem perfectioni omnium scientiarum; in physicis tamen esse frequentissimum, quia causae physicae ut plurimum nobis ignotae sunt; in mathematicis autem fere nullum, quia in talibus disciplinis causae sunt notiores et natura et nobis."

[84] See secs. 3.4.c, 5.2.b, 6.1.b, 6.2.a, 6.2.d, 6.3.d-e, and 6.4.a.

elsewhere."[85] Fortunate it is, therefore, that Carbone plagiarized such a treatise from Valla's lecture notes of 1588 and published it in his *Introductio in universam philosophiam* of 1599, as related in chapter 1. A similar treatise, reworked, of course, is also to be found in Valla's *Logica* of 1622. Recourse to these two works will prove helpful for reconstructing Galileo's early views on science and so for gaining a better understanding of his remarks.

Before proceeding with this task, it may be well to survey some of the statements about science and its kinds that are made in the logical questions. With regard to science in general, in LQ3.1 Galileo says that it abstracts from existence, that it is not concerned with contingencies, and that it does not consider singulars or individuals; the last remark he repeats in LQ6.11.[86] He also implicitly recognizes differences among sciences, for in LQ2.1 he asserts that the study of first principles pertains to metaphysics;[87] in LQ3.1 he recognizes logic as a rational science;[88] in LQ7.2 and 7.3 he points out the distinctive character of physics as employing *quia* demonstration and the demonstrative regress;[89] and throughout the questions he repeatedly characterizes mathematics as different from other sciences. Some of these particulars follow: in LQ3.1 he asserts that mathematics abstracts from being and goodness, that it abstracts from existence, and yet that it is a human science concerned with existence;[90] in LQ6.2 he characterizes mathematical principles as more known according to nature and to us, and he does the same in LQ6.6 and again in LQ7.3;[91] in LQ6.11 he claims that mathematical demonstrations are not perfect, or, if they are perfect, this is because they abstract from matter and use a superior method;[92] and in LQ6.5 he discusses the peculiar

[85] *Opere* 1:18.17-18; Wallace, *Galileo's Early Notebooks*, par. A13.

[86] For the Latin texts of LQ3.1 see pp. 39-40 *supra*, lines 176-177, 196-197, and 131-132; LQ6.11, MS 27, fol. 29r7-8: ". . . de individuis non posse esse demonstrationem neque scientiam. . . ."

[87] MS 27, fol. 4v11: ". . . prima principia in communi spectare ad metaphysicum. . . ."

[88] See p. 38 *supra*, lines 111-112.

[89] MS 27, fol. 31r2-4: "Nota autem huiusmodi demonstrationes [scil. a posteriori] in scientiis esse utillimas, quia principia illarum aliquando ignota sunt et non possunt probari nisi per huiusmodi demonstrationes." See also note 83 *supra*.

[90] For the Latin texts of LQ3.1, see pp. 38-40 *supra*, lines 168-169, 170-172, and 173-174; cf. lines 79-80 with 173-174.

[91] MS 27, fol. 19r6-8: ". . . aliquid posse esse notius . . . secundum naturam et secundum nos, qua ratione notiora sunt prima principia mathematica . . ."; fol. 22v1-2: ". . . patet hoc in demonstrationibus mathematicis, in quibus causae sunt notiores nobis et naturae . . ."; see also note 83 *supra*.

[92] MS 27, fol. 28v31-33: "Ex quo fit ut demonstrationes mathematicae, cum plerumque habeant unam ex praemissis communem, non esse perfectas. Quod si perfectae dicuntur, illud est aut quia, cum abstrahant a materia quae causa incertitudinis [est], certissimae sunt; tum propter praestantissimam methodum quam servant."

way in which axioms enter mathematical demonstrations, particularly those made to the impossible.[93] He also speaks of subalternated sciences, asserting in LQ6.3 that subalternation produces imperfect sciences that can be called such only *secundum quid* and *ex suppositione*, and in LQ6.6 that they show there can be a science with certitude but without evidence, so that certitude and evidence do not always go together.[94] In LQ6.7 he makes the cryptic remark that mathematicians demonstrate a true and real eclipse of the moon—cryptic because mathematics is usually regarded as a subalternating science with respect to astronomy or astrology.[95] Some of these assertions are proposed as difficulties or as objections to positions taken by others, but most are stated as truths Galileo himself accepts. Thus it is desirable to locate them in a larger framework in which their relative value can be more easily assessed.

To this end, a brief survey will first be given of the treatise on science in Valla-Carbone (referenced now as the *Introductio* rather than the *Additamenta* as heretofore), touching on the various kinds of science and special problems associated with their subalternation, followed by a more detailed treatment of mathematics as a science as found in these and other authors of interest for the further development of this study.

a. SCIENCES AND THEIR SUBALTERNATION

The *Introductio* begins its second book, *De scientia*, with a fairly lengthy discourse about the meaning of the term and the history of the origins of the sciences that have come down to us, preparatory to treating the nature of science itself.[96] If taken as the act of knowing *simpliciter*, as defined in the *Posterior Analytics*, then for Valla-Carbone science is nothing more than the conclusion of demonstration; more generally it is understood as a habit of mind that results from repeated acts of this kind and that disposes the one possessing it to know with facility and certitude.[97] The main problem that such a definition presents is that of delineating what makes for the unity of a science, and hence of deciding the number of independent disciplines that can be called sciences and how these are related to each other; it is this type of consideration that occupies the major part of the treatise.[98]

[93] MS 27, fol. 21v7-8: ". . . ut videre est in demonstrationibus mathematicis. . . ."

[94] MS 27, fol. 20v19-21; Latin text in note 43 *supra*; fol. 22v35: ". . . potest esse certitudo absque evidentia, ut patet et in subalternatis scientiis et. . . ."

[95] MS 27, fol. 24r27-29: ". . . mathematici veram realemque eclipsim [demonstrant] de luna, ut patet ex medio quod assumunt ad probandum."

[96] *Introductio in universam philosophiam* (Venice: Apud Marcum Antonium Zalterium, 1599), pp. 121-156; henceforth cited as *Introductio*.

[97] *Introductio*, pp. 157-171.

[98] *Introductio*, pp. 171-332.

The problem of the specification of the sciences divided scholastics and Aristotelians in many ways not mentioned by Valla-Carbone. Apparently for purposes of simplicity they restricted attention to two schools, that of the Thomists and that of the nominalists, to each of which they assimilate a few other writers. The Thomists, in their view, hold that each science constitutes a distinct habit of mind, which is acquired through the first demonstration made in a particular subject matter and then is strengthened and increased by subsequent demonstrations; the nominalists, on the other hand, say that each science is made up of many habits of mind generated by many different demonstrations, all of which go to make up one total science. The authors reject the first opinion and opt for the second on the grounds that a total science contains too many objects that are formally different to be grasped by a single habit of mind—citing the example of physics, which considers such diverse objects as the heavens, the elements, and compounds (*mista*). As to the precise number of habits that go to make up a total science, however, they are not sure: some would say that these equal the number of demonstrations made in the science, others the number of demonstrations made from different principles. To them, the second of these seems the more probable.[99]

The next problem is deciding what it is that makes a total science have its own unity so that it can be called one and distinguished from all others. Here again the authors invoke the two schools, characterizing the Thomists as holding that the three speculative sciences take their unity from the particular degree of abstraction that characterizes it—namely, physics from the fact that it abstracts only from singular matter; mathematics from the fact that it considers quantity as abstracted in the mind and not in reality; and metaphysics from the fact that it abstracts from all matter, not only in the mind but also in reality. The nominalists, on the other hand, maintain that the generic unity of a science comes from the formality under which it considers its object and from the intrinsic principles it uses in defining and demonstrating. Again Valla-Carbone decide against the Thomists, while admitting that the sciences, once constituted, can be located according to their indicated degrees of abstraction. Apart from this type of specification, they admit that there are many other ways of considering the diversity of the sciences, such as distinguishing human science from divine science, practical from speculative, and so forth. Restricting themselves to the human speculative disciplines, however, they seem to be concerned mainly with arguing against a proposal of Antonius Bernardus Mirandulanus, who wished to

[99] *Introductio*, pp. 171-182.

view physics, metaphysics, and mathematics as all parts of one large science rather than themselves constituting three different total sciences. His reasoning, as they report it, was that just as logic is made up of various tracts, and physics also, so the three speculative sciences have a unity in that they study various kinds of being and thus can be regarded as one total science made up of three parts: metaphysics, considering being in common; physics, considering natural being; and mathematics, considering quantified being. They do not reject this position outright, according it some probability, but they regard it as less probable than the traditional opinion, which sees each of these as a distinct total science.[100]

In the course of arguing out this thesis, Valla-Carbone run into difficulties with the science of metaphysics, which not only considers being in common but also abstract and immaterial beings such as God and intelligences, for according to their system of specification there seems no way of assigning the same formal object and the same mode of procedure to the study of things so diverse in their natures. They resolve this problem, in the end, by deciding that there are actually five total sciences—namely, a science of God, a science of intelligences, a science of being in common, a science of natural bodies, and a science of quantity as this is studied in mathematics.[101] The last division also presents its problems, because it would seem that, if the category of quantity can found a separate science, the other categories should be capable of doing so also, and, in any event, it would appear that arithmetic and geometry are sufficiently different to constitute distinct sciences. These considerations prompt a few observations about the nature of mathematics as a science that assume importance in what follows. Contrary to the view of some Scotists, Valla-Carbone maintain that mathematics is not concerned with corporeal substance but considers "nude quantity" (*nuda quantitas*) alone and so abstracts from being and goodness; in their demonstrations, moreover, mathematicians conclude as if there were no such thing as corporeal substance, and merely suppose quantity for their study.[102] Quantity, again, is different from other categories of being; it can be considered without substance, in complete abstraction from it, whereas

[100] *Introductio*, pp. 183-191, 238-240. It is perhaps noteworthy that Galileo considers Mirandulanus's positions in PQ1.2.1, 3.5.38, and 3.6.13, but is not favorable to them; see Wallace, *Galileo's Early Notebooks*, pp. 32, 112, 151, 256, 271, and 275.

[101] *Introductio*, pp. 249-250.

[102] *Introductio*, p. 242. "Primo, quia scientia mathematica non considerat substantiam corpoream, sed nudam quantitatem, et ita abstrahit ab ente bono et a fine. Secundo, quia mathematica numquam mentionem facit de substantia corporea, et eius demonstrationes ita concludunt ac si nulla esset substantia corporea, sed solam supponit quantitatem."

all of the other categories have an intrinsic order to sensible substance.[103] That is why quantity can be studied in three different sciences: in metaphysics, as it is a type of being; in physics, as it is a property of substance and according to its proper nature; and in mathematics, as it has certain properties that lack any order to substance.[104] And when mathematics considers nude quantity, it leaves aside natures and quiddities and does not proceed through causes and effects, but rather through necessary connections that are taken *ex suppositione.*[105] The resulting formal consideration and method, finally, are not different for arithmetic and geometry, and thus these two disciplines should be considered as but parts of the total science of mathematics.[106]

Having established the number of speculative sciences at five, and adding to these logic and ethics as the most important practical sciences, the next concern of the *Introductio* is to take up various matters relating to the comparison of all the sciences, ranking them according to nobility and certitude and then taking up the subject of their subalternation. In the order of nobility or dignity the proposed order is the following, beginning at the bottom of the scale: mathematics, whose mode of knowing is the least perfect since it lacks true definitions and demonstrations and true causes and effects; the science of God, because there can be no causes in God and thus no causal demonstrations concerning him; logic, because this does not consider real causes either; ethics, because it considers only extrinsic causes; physics, because, though treating of natures and causes, it arrives at definitions and demonstrations only with great difficulty; the science of intelligences, which is based on true causes and demonstrations; and metaphysics, which has perfect definitions and demonstrations. The ranking according to certitude is almost the reverse of this, for its order, again beginning lowest on the scale, is the following: the science of God, who is most remote from our minds; the science of intelligences, for the same reason; logic, whose subject is produced by the intellect; metaphysics, which is concerned with real being but not as this is grasped by the senses; ethics, which is based on experience and

[103] *Introductio*, p. 242: ". . . diversam esse rationem quantitatis et aliorum accidentium. Nam quantitas considerari potest sine substantia, ut ab ea abstrahit, alia vero accidentia intrinsece constituunt substantiam sensibilem."

[104] *Introductio*, p. 243: "Quo fit ut quantitas a tribus scientiis consideretur: a metaphysica ut est ens et unum genus generalissimum; a physica ut est proprietas substantiae et secundum propriam naturam; a mathematica, quatenus habet proprietates quasdam sine ordine ad substantiam."

[105] *Introductio*, p. 240: "Mathematica considerat nudam quantitatem et quantitatis accidentia extrinseca, relicta quidditate, nec procedit per causas et effectus, sed per ea quae habent quandam necessariam connexionem ex suppositione; igitur. . . ."

[106] *Introductio*, p. 241.

sense knowledge; physics, which is more firmly founded in sensible objects; and mathematics, which does not accept probable arguments and whose demonstrations completely convince the human mind. Thus mathematics, the least noble of the sciences, is also the one that enjoys the greatest certitude among all the human disciplines.[107]

Before considering the views of Valla-Carbone on the subalternation of the sciences, it may be well to remark that most of the foregoing material is taken up in Valla's *Logica* of 1622, with many of the points being argued in meticulous detail and with no major changes in the conclusions reached.[108] On the other hand, when one searches through the logic courses of Vitelleschi and Rugerius, the two professors who followed Valla at the Collegio Romano, one is surprised to find that Valla's positions on the specification of the sciences were apparently regarded as unorthodox and were not taken up by his successors.[109] The fivefold division of the speculative sciences is not mentioned, for example, nor is the same anti-Thomist bias in evidence. This would reinforce the thesis advanced earlier that Valla's treatment of science in the 1588 lectures may not have been completely representative of the ongoing tradition at the Collegio in the latter part of the sixteenth century.[110]

The discussion of subalternation in the *Introductio* is rather full, being concerned with its definition, the conditions for its occurrence, the relationships between subalternating and subalternated sciences, how subalternated sciences obtain their principles, and finally, an enumeration of the disciplines to which these terms apply. Subalternation is defined as the subordination of one science to another by reason of its dependence on the other by reason of principle, subject, or end. The first occurs when the higher science proves principles that are proper to the lower science and that enter into its demonstrations; the second, when the subject of the lower is contained under that of the higher, either as a species under a genus or as something essential that is contracted by an accidental difference or condition; and the third, when the end of the lower is included under the end of the higher. When subalternation in the strict sense occurs, moreover, the subalternating science will be more noble and more certain than the subalternated, the former will give *propter quid* explanations whereas the latter will give only *quia*, and the two will be related as the perfect to the imperfect. A subalternated science is

[107] *Introductio*, pp. 258-261.

[108] *Logica*, vol. 2, pp. 527-633.

[109] Probably because of teachings such as this Valla's *Logica* did not receive the complete endorsement of the Jesuit censors who approved it for publication; see chap. 1, note 69, *supra*.

[110] Secs. 1.3.c, 2.3.c, and 2.4.

imperfect, on this account, because it can never resolve its conclusions back to first principles without the aid of the subalternating science; it lacks principles that are *per se nota*; and it demonstrates properties of its adequate subject through principles that are only supposed from another science and thus not known in themselves.[111]

Within this context, a problem that was much debated in the sixteenth century was whether a subalternated science could ever lay claim to being a science in the strict sense, since it seemed to lack full knowledge of the principles on which demonstrative reasoning should be based. Possibly because St. Thomas Aquinas taught in the *Summa theologiae* that theology is a subalternated science and at the same time is fully scientific, Thomists generally answered this question in the affirmative. In doing so, they invoked a distinction as to the way in which the one possessing the subalternated science might know the principle involved: if he merely took it on faith and so did not grasp it by reason, he would not possess science; if his knowledge of it, on the other hand, was "in continuity" with the subalternating science, then he would truly know it, and his resulting science would be such in the accepted sense.[112] Valla-Carbone merely state this position without explaining the "in continuity" clause, and reject it along with the other Thomistic theses stated above.[113] In the *Logica*, however, Valla gives fuller consideration to the position and offers the following explanation of it. Suppose the case of optics (*perspectiva*) being subalternated to geometry, and the subalternating science (geometry) being in one person and the subalternated (optics) in another who knows no geometry. Can the latter prove a theorem in geometrical optics, and have scientific knowledge of his conclusion, if he is not certain of his geometrical principles but must accept them on faith? Some answer that he can, on the grounds that a reasonable faith is necessary for many scientific conclusions—for example, knowledge of the precession of the equinoxes, which depends on observations made over a period of many years and for which one must be dependent on others. Others reply in the negative, saying that in this case he has only opinion, not science. They do not rule out the possibility of his gaining scientific knowledge of the conclusion, however, provided that he can argue to the truth of the geometrical principle(s) he needs from experience and by *a posteriori* reasoning. If he does so, he will truly possess scientific knowledge of his

[111] *Introductio*, pp. 262-288.

[112] *Introductio*, p. 277: "Unde citatus auctor [Capreolus] ait, conclusionem ductam ex principiis fidei posse dupliciter considerari: primo, ut colligitur ex principiis inevidentibus, et sub hac ratione non est scita; secundo, ut continuatur cum scientia superiore quae illa principia evidenter cognoscit, et sub hac ratione est scientia."

[113] *Introductio*, pp. 276-281.

conclusion, but his science will be imperfect compared to that he would have had if based on *a priori* geometrical reasoning.[114]

To further elucidate the problem, Valla goes on to suppose that both subalternating science and subalternated science are possessed by the same person and thus in the same intellect, so that the subalternated science is conjoined with or "in continuity" with the subalternating. In such a case, can the optician derive or prove whatever geometrical principles he may need, and do so by geometry *a priori*, without exceeding the science of optics? Valla proposes two ways in which this might be done. First, in virtue of the subalternating science, or putting on its nature, as it were, the optician might demonstrate the theorem he needs, just as in the *Physics* Aristotle proves some principles that really pertain to metaphysics by adopting the stance of the metaphysician, since he cannot make progress in his science without the use of such principles. Second, in virtue of the subalternated science, or precisely as an optician, he might prove the principle(s) himself, though he is being helped by his knowledge of geometry, and in this way he would have an *a priori* proof without transcending the limits of the subordinated science. To explain the latter possibility, Valla gives the example of the optician who wishes to prove that things seen at a distance appear smaller than those seen close up. His proof might run like this: things seen under a smaller angle appear smaller; but things farther away are seen under a smaller angle; therefore things farther away are seen as smaller. Here both of the principles or premises invoked pertain to the science of optics, since they are concerned with visual lines or lines in sensible matter. Moreover they give the proximate cause or the *propter quid* explanation of the phenomenon the optician is trying to explain. Should he wish to resolve the demonstration back to first causes and completely indemonstrable principles, then he can invoke geometry to work out a proof of the minor premise. But even here he is thinking more as an optician than as a geometer, since geometry knows nothing about visual lines; therefore, even though helped by his knowledge of geometry in proving the minor premise, he does not transcend the limits of optical science and still has perfect scientific knowledge, *a priori* and *propter quid*, of his conclusion.[115]

The application Valla makes to theological science need not concern us here, for it suffices to note that in the *Logica* he accords some truth to the Thomistic position on subalternated sciences, even though this is rejected in the *Introductio*. The latter position is also consistent with some of the foregoing statements on suppositional reasoning, such as the possibility of resolution *ex suppositione*, and the way in which truly scientific

[114] *Logica*, vol. 2, pp. 649-651.

[115] *Logica*, vol. 2, pp. 651-652.

knowledge can result from its use. This topic will have to be resumed later because of Galileo's proposed demonstrations in the *Two New Sciences.*[116] For the moment it need only be observed that the view one takes of subalternation and subalternated sciences has important implications for mathematical physics, and thus that Galileo's apparently incidental remarks about them in the logical questions are deserving of the fullest concern.

To complete the examination of the *Introductio*, a brief account should be given of the sciences Valla-Carbone regard as either subalternating or subalternated.[117] Because of arguments in the scholasticism of their day, they are at pains to show first that neither physics nor mathematics are properly subalternated to metaphysics, and then that logic is not properly subalternated to any science. The major cases of subalternation, in their view, are provided by physics and mathematics. Physics or natural philosophy, as a speculative science, has two practical sciences subalternated to it: ethics, which depends on the part of natural philosophy that studies man's soul (later known as psychology); and medicine, which depends on the part that studies his body. They further divide medicine into two habits of mind, one theoretical and the other practical, the first of which has greater affinity with natural philosophy but is taken over by the physician because it is necessary for his science, and the second of which is more properly his own. By far the greater number of subalternated sciences, however, depend on mathematics, and these are called *scientiae mediae* or middle sciences because they consider a mathematical object that is applied to a physical thing; since the object is more essential than the condition of its application, they are more mathematical than physical. Subalternated to geometry are perspective or optics, concerned with visual lines; astronomy, concerned with the appearances of the heavens; and astrology, concerned with the movements of the stars. As speculative disciplines, these have practical sciences subalternated to them in turn: thus nautical science and agriculture are subalternated to astrology. Arithmetic takes under it the science of music, which, like medicine, has a speculative and a practical part. (In the *Logica*, Valla differentiates the two as follows: practical music proves its principles from experience, since it can tell by ear what consonances are good, whereas speculative music gives their cause "up to a point" [*usque ad certum limitum*] insofar as it is in continuity with the subalternating science.)[118] And finally, the part of geometry concerned with solids, or stereometry, has subalternated to it various arts of measuring, building, etc., usually referred to as the mechanical arts.

[116] Secs. 3.4.b-c; 4.4.a-b; 5.2; 5.3; 5.4, especially 5.4.e; 6.2; and 6.3.

[117] *Introductio*, pp. 282-288.

[118] *Logica*, vol. 2, p. 652.

The foregoing references to mathematics as a science furnish some idea of how this subject was seen at the Collegio Romano in the late 1580s. There is one brief section in the *Introductio*, however, that seems not completely representative but may have had an important influence on the way in which mathematics developed at the Collegio within the next few decades, and thus is worthy of consideration. The section is entitled *Dubitationes quaedam circa scientias mathematicas*.[119] Among the eleven doubts about mathematical sciences that are listed are several concerned with the type of abstraction that is characteristic of mathematics, wherein Valla-Carbone make the points that mathematicians consider nude quantity without any order to substance; that the intelligible matter at which they arrive when they leave aside sensible matter is merely fictive and cannot be defined in terms of true genus and difference; that mathematicians abstract from motion and from any natural forces that produce it; that they abstract also from goodness and being because these do not exist in mathematical forms apart from the world of nature; and that they abstract from all kinds of cause and so cannot use causal reasoning in any of their demonstrations. This section, it may be noted, is not reproduced in the *Logica* of 1622, although several of the points included above are touched on by Valla in the chapters devoted there to demonstration and its properties. Many similar remarks, moreover, are made by Pererius in his *De communibus* of 1576 mentioned in the first chapter, and they all add up to a view of mathematics that was advanced by Alessandro Piccolomini and was influential among philosophers, especially those of Averroist leanings, in the Italian universities of the time.[120]

b. CLAVIUS ON THE MATHEMATICAL SCIENCES

The professor of mathematics at the Collegio, Christopher Clavius, predictably was not sympathetic to this treatment of his discipline. In

[119] *Introductio*, pp. 288-302.

[120] See Paolo Galluzzi, "Il 'Platonismo' del tardo Cinquecento e la filosofia di Galileo," in *Ricerche sulla cultura dell'Italia moderna*, Paola Zambelli, ed. (Bari: Laterza, 1973), pp. 37-79, together with the more detailed studies of G. C. Giacobbe, "Il *Commentarium de certitudine mathematicarum disciplinarum* di Alessandro Piccolomini," *Physis* 14 (1972): 162-193; "Francesco Barozzi e la *Questio de certitudine mathematicarum*," *Physis* 14 (1972): 357-374; "La riflessione metamatematica di Pietro Catena," *Physis* 15 (1973): 178-196; "Epigoni nel Seicento della 'Quaestio de certitudine mathematicarum': Giuseppe Biancani," *Physis* 18 (1976): 5-40; and "Un gesuita progressista nella 'Questio de certitudine mathematicarum' rinascimentale: Benito Pereyra," *Physis* 19 (1977): 51-86; also Maria R. Davi Daniele, "Bernardino Tomitano e la 'Quaestio de certitudine mathematicarum,' " in *Aristotelismo Veneto e Scienza Moderna*, 2 vols., Luigi Olivieri, ed. (Padua: Editrice Antenore, 1983), vol. 2, pp. 607-621.

the 1580s, in fact, he prepared a disquisition for the Society of Jesus about the way in which the mathematical disciplines could be promoted in the schools of the Society.[121] Among the prescriptions he advanced were several relating to the choice of professors and students for advanced work in this field. Following this came a warning about professors of philosophy who gave an improper interpretation to passages in Aristotle and in other philosophers, which reads as follows:

> It will also contribute much to this if the teachers of philosophy abstained from those questions which do not help in the understanding of natural things and very much detract from the authority of the mathematical disciplines in the eyes of the students, such as those in which they teach that mathematical sciences are not sciences, do not have demonstrations, abstract from being and the good, etc.; for experience teaches that these questions are a great hindrance to pupils and of no service to them; especially since teachers can hardly teach them without bringing these sciences into ridicule (which I do not just know from hearsay).[122]

The parenthetical remark at the end is a sign of firsthand knowledge on Clavius's part of the effect of professors such as Pererius and Valla on their students' attitudes toward mathematics. Undoubtedly, through his influence in the Society, he was able to foster a much more positive view of his subject, beginning in the latter part of the 1580s.

In the preface to his second edition of Euclid's *Elements*, which was published in 1589, Clavius took pains to project a totally different appreciation of mathematics as a science.[123] The title page shows Euclid and Archimedes with instruments of their discipline standing in places of honor under the seal of the Society. The edition is dedicated to Charles Emmanuel I, Duke of Savoy, for whom Giovanni Battista Benedetti was then the court mathematician, and whom Clavius praises as most learned in things mathematical and appreciative of his patron—who, in turn, was eminently successful in applying such knowledge to practice. In the dedication Clavius mentions that the work is vastly improved over the first edition, with substantial additions from Archimedes, Apollonius,

[121] The background for understanding this is provided by Giuseppe Cosentino, "Le matematiche nella 'Ratio Studiorum' della Compagnia di Gesù," *Miscellanea Storica Ligure* II.2 (1970): 171-213, and "L'Insegnamento della matematiche nei collegi Gesuitici nell'Italia settentrionale: Nota introduttiva," *Physis* 13 (1971): 205-217 Additional details may be found in A. C. Crombie, "Mathematics and Platonism in the Sixteenth-Century Italian Universities and in Jesuit Educational Policy," in *Prismata: Naturwissenschaftsgeschichtliche Studien*, Festschrift für Willy Hartner, Y. Maeyama and W. G. Saltzer, eds. (Wiesbaden: Franz Steiner Verlag, 1977), pp. 63-94

[122] Cited from the English translation of his *Modus quo disciplinae mathematicae in scholis Societatis possent promoveri* in Crombie, "Mathematics and Platonism," p. 66.

[123] *Euclidis Elementorum Libri XV, accessit XVI De solidorum regularium cuiuslibet intra quodlibet comparatione, omnes perspicuis demonstrationibus, accuratisque scholiis illustrati* . . . (Romae: Apud Bartholomaeum Grassium, 1589).

Ptolemy, and others in order to connect their works with Euclid's theorems. The proofs are more rigorous also, and errors in a large number of axioms that were "mutilated" by Proclus have been repaired. Among the new matter to be found in the edition is an extensive treatise *De proportionibus* and a treatment of the five regular solids that can be inscribed in a sphere, taken from Pappus, much clearer than that given by Euclid.[124]

Of greater interest, however, are the Prolegomena to the edition, wherein Clavius provides his own introduction to the mathematical disciplines. With regard to the division of these sciences, he first gives that deriving from Pythagoras, which speaks of two pure sciences, arithmetic and geometry, the one concerned with discrete quantity and the other with continuous quantity, both in themselves and in comparison with the other; and then he speaks of two mixed sciences, music and astronomy. Music deals with discrete quantity as this is related to the blending and harmony of sounds, whereas astronomy is concerned with magnitude as this undergoes movement, such as is found in the heavenly bodies; to these, he adds, can be assimilated perspective, geography, and similar disciplines. Another division, he goes on, is that deriving from Geminus by way of Proclus, which holds for a twofold division into sciences that consider objects in the intellect that are separated from all matter and those that consider objects to be found in sensible matter. The first of these includes only arithmetic and geometry, whereas the second embraces six disciplines: astrology, perspective, geodesy, music or canonics, practical arithmetic or *supputatrix*, and mechanics. Each of the latter has many subdivisions. Illustrative of the others is that for mechanics, which Clavius says treats everything that has the power (*vis*) to move matter, including the study of instruments or machines capable of moving things, and the study of wonderful effects, such as those deriving from spirits, as examined by Ctesibius and Hero, those from weights, whose disequilibrium is regarded as the cause of motion and whose equilibrium the cause of rest, and those from spheres—much of which was pioneered by Archimedes.[125]

Other sections of the Prolegomena are concerned with the importance and usefulness of mathematics. The mathematical sciences, says Clavius, hold first place among all others because they are concerned with things that can be considered apart from sensible matter, although they themselves are immersed in such matter, thus being like both metaphysics and physics, each of which shares one of these modes of consideration. The preeminence of mathematics also derives from the certitude of its

[124] Ibid., fol. *2-*6.

[125] Ibid., pp. 9-12.

demonstrations, which remove all doubt and give birth to true science in the minds of its practitioners, and this can hardly be said of any other discipline, in most of which there are continual disputes and arguments that leave the student uncertain and in doubt. Here Clavius ridicules the philosophers, all of whom take their inspiration from Aristotle and call themselves peripatetics, but give different interpretations of him depending on whether they are Greeks, Arabs, or Latins, nominalists or realists. The theorems of Euclid and of other mathematicians, on the other hand, are as true today as they were centuries ago, and their demonstrations retain all of their forcefulness and certitude. With regard to the utility of mathematical disciplines, even apart from those concerned with arts and practice, they are essential for learning all the other sciences. Metaphysics, for example, has no path of entry except through mathematics, and physics itself gains much light from the use it makes of these disciplines. They are also valuable for the study of sacred sciences as well as for ethics, dialectics, and other subjects studied in the Academy of Plato, to which one would be refused entry if ignorant of mathematics.[126]

Clavius concludes his Prolegomena with brief sections describing the meanings of various terms used by mathematicians in proposing their demonstrations. These he divides into propositions and principles, including under the first category problems, theorems, and lemmas, and under the second, definitions, postulates, and axioms. The first are fairly straightforward, since a proposition is merely something proposed, a problem is a demonstration that requires some type of construction, a theorem one wherein a property of one or more quantities is manifested, and a lemma one that is assumed in another proof so as to make the latter simpler or easier. The second category Clavius relates to Aristotle's treatment of foreknowledge in the *Posterior Analytics*, saying that some principles are definitions of terms, formulated to remove ambiguity in the proof, and which are called *suppositiones* by Aristotle and Proclus. Others are postulates or *petitiones* that are so clear and require so little confirmation that they can be asked to be conceded by students. And yet others are axioms, which are common truths that are known to all and are used even outside the science, so that there is no reason why they cannot be accepted without proof.[127]

Such efforts on Clavius's part to improve the image of mathematics in the Society of Jesus are reflected in the Ratio Studiorum of the Society, which appeared along with his Euclid in 1589 and made adequate provision for mathematical instruction in Jesuit *studia* and universities.[128] At

[126] Ibid., pp. 14-17.
[127] Ibid., pp. 23-26.
[128] See the studies cited in note 121 *supra*.

the time of their appearance, Vitelleschi was teaching the course in natural philosophy at the Collegio Romano, and it is perhaps noteworthy that one does not find any invective against mathematics in his *reportationes*. Indeed, in his passage paralleling Galileo's PQ3.4.24 on the incorruptibility of the heavens, Vitelleschi rejects the possibility that the bright light that appeared in the heavens in 1572 was actually a comet in the upper reaches of the air, as some had said. Instead he states flatly that the luminous object was a star, and a new star (*stella nova*) at that, in the constellation Cassiopeia; Clavius, he says, "demonstrates" this in the first chapter of his *Sphaera*.[129] Again, when concluding his treatment on the elements as these are the matter of which compounds are formed, Vitelleschi refers his students for more particulars on the shape and quantity of the elements to the course on the *Sphaera*, which was being taught by Clavius that same year at the Collegio.[130] Both of these evidences manifest Vitelleschi's conviction that mathematicians do attain to reality and even provide demonstrations concerning the heavens and the elements that must be taken into account by natural philosophers.

Rugerius, teaching in 1590 under the newly promulgated Ratio Studiorum, is even more enthusiastic in his approach to mathematics than is Vitelleschi. He agrees with the latter in holding for a threefold division of the speculative sciences, and has a special *quaesitum* on the object of mathematics wherein he teaches that this is quantity as abstracted from substance, and so is different from the object of physics, which is the natural body, even though physical quantity and mathematical quantity are the same in the thing itself.[131] In discussing the subalternation of the sciences, moreover, he has a question on how subalternated sciences know their principles and treats the Thomistic position in a way similar to that found years later in Valla's *Logica*. His view of the "continuity" required between subalternating and subalternated sciences does not insist that both exist in the same intellect, for he allows that the person possessing the subalternated science can still have true science even though he merely "believes" the principles deriving from the subalternating science in another. In such a case his word "believes" refers to a firm and certain assent that is practically equivalent to sense knowledge, the way in which one trusts in astronomical observations made by others over a long period of time. Since these principles are supposed, however,

[129] Cod. APUG-FC 392, In libros De caelo, [Tract. 2], Disp. 6, An caelum sit incorruptibile; Wallace, *Galileo's Early Notebooks*, par. J24.

[130] Ibid., In libros De generatione, Lib. 2, Disp. [ult.], Cur aer in media regione sit frigidus.

[131] Cod. SB Bamberg 62-2, fols. 259r-260r.

they give rise to science that is not such *simpliciter* but rather *ex suppositione.*[132]

In the parts of his commentary on the *De caelo* that treat of the heavens, moreover, Rugerius introduces a few questions on the celestial orbs, advising his students that most of this material pertains to astronomy, "in which you have had an excellent course this year"—an obvious reference to the teaching of his colleague Clavius.[133] With regard to eccentrics and epicycles, he takes the position that their existence cannot be demonstrated mathematically, but that until someone thinks up a better way of saving the phenomena of the heavens he himself has no serious argument against them.[134] On the matter of the *nova* of 1572, on the other hand, he too asserts that this is a true star and that its location and distance from the earth have been established "by mathematical demonstrations." He still is not convinced on this account that the heavens are corruptible (nor was his predecessor Vitelleschi), but prefers to regard the star's appearance as either an event with supernatural significance or as something natural that a new system of epicycles and so forth would some day be able to explain.[135]

c. BLANCANUS'S DEVELOPMENT

The full effect of Clavius's program, however, was not seen until students whom he had prepared himself under the new Ratio Studiorum completed their studies and began to write on the nature of the mathematical disciplines. One such was Joseph Blancanus (Biancani), who entered the Society in 1592 and later enjoyed a considerable reputation as a mathematician, teaching mainly at the University of Parma, where Giovanni Battista Riccioli was in turn his student.[136] In 1615 Blancanus published at Bologna a compendious treatment of the references to mathematics contained in the Aristotelian corpus, to which he added a dissertation on the nature of mathematics and a chronological listing of famous mathematicians (*De mathematicarum natura dissertatio una cum clarorum mathematicorum chronologia*).[137] In the latter work he examined all

[132] Ibid., fol. 247r-v.

[133] Cod. SB Bamberg 62-4, fol. 61v: "Id totum astronomicum est, et hoc anno habuistis eleganter et egregie tractatum."

[134] Ibid., fol. 62r.

[135] Ibid., fols. 64r-66v; see Wallace, *Galileo's Early Notebooks*, pp. 268-269.

[136] Blancanus's teachings have already been noted by Paolo Galluzzi, "Il 'Platonismo' del tardo Cinquecento . . . ," and G. C. Giacobbe, "Epigoni nel Seicento della 'Quaestio de certitudine mathematicarum': Giuseppe Biancani."

[137] Authore eodem Iosepho Blancano e Societate Iesu, mathematicarum in Parmensi

the objections made against the scientific character of mathematics, whose origin he traced to Alessandro Piccolomini, but which he acknowledged were taken up by Pererius in his *De communibus* and by the Jesuits of Coimbra in their *Cursus philosophicus*. Piccolomini and his followers, Blancanus argues, either misread or misinterpreted many of the statements of Plato, Aristotle, and Proclus on the mathematical disciplines, and so he provides his own exegesis of these sources to show that they do not take the disparaging view of mathematics attributed to them. On the positive side, Blancanus maintains that mathematics is a strict science in the Aristotelian sense, that is, that it satisfies in a preeminent way the canons of the *Posterior Analytics*, for its demonstrations are *potissimae* and it is foremost among all the sciences in attaining truth with certitude. When these points are understood, and in their support he invokes the authority of Toletus and Zabarella, it becomes a simple matter for him to answer the "calumnies" brought by recent writers against his discipline.[138]

The subject matter of geometry and arithmetic, Blancanus begins, is intelligible matter (*materia intelligibilis*), since the quantity these sciences consider is abstracted from sensible matter. Such abstraction can be made, however, in a twofold way to yield either quantity considered in itself, as studied by the physicist and the metaphysician, or quantity considered as it is terminated in a particular way (continuous quantity by *figura* or shape, discrete quantity by number), as studied by the geometer and the arithmetician. Demonstrations of the latter two scientists are concerned with manifesting the properties of "terminated quantity"(*quantitas terminata*), and thus the latter is the proper subject of their investigations. By reason of its being abstracted such quantity acquires a type of perfection not found in terrestrial objects. And although mathematical entities do not exist in the physical world but only in the mind abstracting them, this does not mean that they are accidental, imperfect, and false beings, as some have alleged; rather they are *entia per se* and thus true beings. Their existence, moreover, is an *esse possibile*, and this suffices to constitute the object of a science, for science abstracts from the existence of its subject (*scientia enim abstrahit ab existentia subiecti*).[139] Intelligible matter, like other matter, is constituted of parts, and such parts satisfy all the requirements for being a true material cause, through which properties may be demonstrated. It is also possible to give essential definitions of mathematical objects, definitions that explain not only the

academia professore (Bononiae: Apud Bartholomaeum Cochium, 1615), henceforth cited as *Dissertatio*.

[138] *Dissertatio*, pp. 13, 19.

[139] *Dissertatio*, pp. 5-7.

meaning of the names applied to them but also the entire quiddity of the things themselves (*quae scilicet totam rei quidditatem explicent*).[140] Such definitions are both formal and causal, and thus are capable of furnishing middle terms from which *demonstrationes potissimae* can be formulated. Both formal and material causality therefore function in mathematical demonstrations, which as a consequence are not made through signs or effects, as frequently seen in the physical sciences, but rather through principles that are prior and more known than the conclusions drawn from them. Most of those who give a pejorative view of these demonstrations, Blancanus concludes, do so because they fail to recognize the precise formal or material cause on which the demonstrative inference is based, frequently confusing constructions that are prerequisite to a demonstration with the demonstration itself.[141] To make his point, he provides an appendix in which he analyzes forty-eight demonstrations contained in the first book of Euclid's *Elements* and shows the type of causality involved in each.[142]

The arguments brought against the position he has outlined are classified by Blancanus as twenty in number, the first three of a general nature and the remaining seventeen as formulated by various moderns (*recentiores*). The first is that the causes invoked by geometers are not true causes (*verae causae*) and are not sufficiently distinguished from their effects; his answer is that they satisfy the requirements for formal and material causality as contained in the tradition from Plato to the present, and that the distinction between cause and effect in these genera of causality is not merely a distinction of reason but is based on some reality in the entities considered, whatever theory of distinctions one may employ in characterizing it. The second is that geometers prove the same conclusion through different middle terms, and there can be only one proper and adequate cause of any particular effect; to this Blancanus replies that the existence of a single cause does not preclude one from viewing it in a variety of ways, and that frequently the same cause will be employed even though a different construction is used to make it evident. The third is that geometers do not consider the essence of quantity and the properties flowing from this essence; this objection, says Blancanus, comes from those who do not appreciate the different formalities under which quantity is considered by mathematicians and by physicists and metaphysicians. Once one sees that the proper subject of mathematics is *quantitas terminata*, and not merely quantity considered in itself, this difficulty disappears along with the others.[143]

[140] *Dissertatio*, p. 7.

[141] *Dissertatio*, pp. 10-18.

[142] *Dissertatio*, pp. 32-35.

[143] *Dissertatio*, pp. 18-19.

Of the remaining "calumnies," many are drawn from the writings of Piccolomini and his followers and from the interpretations they place on texts selected, usually arbitrarily and out of context, from Aristotle, Plato, and Proclus. One such argument is based on Plato, viz., that mathematicians only "dream" about quantity because their arguments are based on *suppositiones* and thus do not attain to its essence. Such an argument, Blancanus replies, must be understood in the context of Plato's philosophy as a whole, which admits only three types of discipline. The first is dialectics, or theology, which employs no *suppositiones* whatever and does not resort to reasoning; for Plato this alone deserves to be called a *scientia*. The second is mathematics, which does employ reasoning and on this account "supposes" its principles. The last is physics, or the study of natural things, which is located in the imagination and leads to nothing more than opinion. To say that mathematicians reason *ex suppositione*, in Platonic terms, is only to admit that their discipline lacks the certitude of theology, which has a direct intuition of its subject matter and thus need not even reason about it. Thus Plato's pejorative view is merely comparative—that is, compared to theology mathematics lacks the perfection of a science—but it is still superior to physics in this regard, for the latter never even rises above the level of a "likely story."[144] In other replies Blancanus shows how the ancients brought out the good to be found in mathematics, especially as a discipline that can exercise the minds of youths and bring them beyond the things of sense to a domain where they see reasons and proper explanations without having the experience requisite for more advanced studies.[145]

In another work published at Bologna in 1620, his *Sphaera mundi seu cosmographia demonstrativa*, Blancanus extends these considerations with an *Apparatus ad mathematicas addiscendas et promovendas*, that is, a preparation for learning and advancing the mathematical disciplines, which is patently addressed to his students.[146] In this he indicates that mathematics, like other sciences, can be divided into branches that are speculative and practical, pure and intermediate (*mediae*), and subalternating and subalternated. The speculative branches are six in number, two of which—geometry and arithmetic—are pure, and the remainder of which—viz., *perspectiva* or *optica* (including *catoptrica* and *dioptrica*), *mechanica*, *musica*, and *astronomia*—are intermediate. The pure sciences are also known as subalternating and the intermediate as subalternated, in his terminology. Each of the pure sciences, moreover, has a practical counterpart: *geometria practica*, concerned with the measurement of distances, surfaces, etc.;

[144] *Dissertatio*, pp. 20-23.

[145] *Dissertatio*, pp. 19-27.

[146] Bononiae: Typis Sebastiani Bonomii, 1620, pp. 387-414, henceforth cited as *Apparatus*.

arithmetic practica, concerned with calculation (*supputatrix* or *algebra*); *perspectiva practica*, used mainly by artists, sculptors, and architects; *mechanica practica*, whose experts are called engineers (*ingegnieri*); *musica practica*, concerned with the composition or performance of music; and *astronomia practica*, dealing with the calculation of astronomical tables, the prediction of celestial phenomena, and the making of clocks and other instruments such as globes, spheres, and astrolabes.[147] He then sketches a syllabus of readings for each of these branches, listing the authors and books indispensable for work in these fields, and ends with summary remarks on the method (*methodus*) to be followed to acquire competence in their pursuit.[148]

The concluding part of the *Apparatus* is devoted to special methodological problems encountered in geometry and contains Blancanus's specific instructions about how to secure demonstrations in that science, how to avoid fallacies and paralogisms, and how to employ the techniques of resolution and composition in its subject matter. The form of a geometrical demonstration, he says, can be set out in five steps, following Euclid and other mathematicians, each of which he explains by means of examples drawn from Clavius's works. First one proposes a proposition, which indicates either something to be proved, in which case it is a theorem, or something to be done, in which case it is a problem. Then a figure is drawn that explains the proposition, either supplying the data for the solution of the problem or, if a theorem is involved, exhibiting the subject matter of which a property is to be demonstrated. Third comes the construction, which is generally necessary, always in the case of a problem, sometimes not in the case of a theorem. The fourth step is the *discursus* based on the figure constructed, which is expressed in the form of an enthymeme and not syllogistically, although, as Clavius has shown, it can be reduced to syllogistic form—a process that is usually not worthwhile because it is tedious, repetitive, and adds little clarity to the reasoning. The *discursus* is itself the demonstration, and it will be one of two types: either *ostensiva* or *ad impossibile*. The *ostensiva* type offers a proof through a material cause, a formal cause, or a sign; the *ad impossibile*, on the other hand, leads to a result that contradicts either an accepted principle, or something already demonstrated, or a *suppositio* on which the reasoning was based. The fifth step, finally, is a restatement of the proposition to be proved and now demonstrated, to which the words are added *quod erat faciendum* in the case of a problem and *quod erat demonstrandum* in the case of a theorem.[149]

[147] *Apparatus*, pp. 388-390.

[148] *Apparatus*, pp. 391-404.

[149] *Apparatus*, pp. 406-408.

Blancanus's treatment of fallacies and paralogisms in mathematical reasoning is not of immediate interest for this study, but his account of resolution and composition assumes some importance in the light of what has already been said on that subject.[150] As used in mathematics, following Euclid, Blancanus maintains that resolution is the adoption of something inquired about as conceded, through consequences that yield some truth [likewise] conceded (*resolutio est sumptio quaesiti tamquam concessi, per ea quae consequuntur in aliquod verum concessum*). This cryptic expression he explains as follows. Resolution is a *discursus* in which we investigate the truth of a theorem or the solution of a problem by first supposing that the theorem is true or the problem solved; on the basis of this supposition we then proceed to deduce consequences until we come to a conclusion that is recognizable as false or as a truth already conceded. If we come to a true conclusion, this is a sign that our initial supposition was either a truth or a solution, on the basis of the logical principle that truth can come only from truth if the matter and form of the reasoning process are correct. Composition, on the other hand, is the acceptance of what is conceded through consequences that yield the conclusion or a grasp of the thing inquired about (*compositio est sumptio concessi per ea quae consequuntur in quaesiti conclusionem sive deprehensionem*). This process, Blancanus similarly explains, follows an order that is the inverse of that pursued in resolution. Once the truth has been discovered, he says, the demonstration of what is inquired about proceeds compositively, that is, it reasons back from the truth discovered to the conclusion previously sought. If something false or impossible results in the process, this is an evident sign that the conclusion is itself false or impossible. Here the logical principle involved is that falsehood can come only from falsehood if the matter and form of the deduction are correct. Blancanus adds that both processes are much easier to see through examples than they are through prescriptions such as these, and refers the student to appropriate places in Apollonius, Archimedes, and Pappus for their illustration. For purposes here, however, his explanation has a twofold value: it shows what he means by mathematicians employing an ostensive proof through a sign, rather than through a cause as he has already explained, and it shows how different are the processes of resolution and composition as used by mathematicians from those employed in the physical sciences.[151]

Before leaving Blancanus it may be well to note the favorable notice he gives to both Galileo and his father in his listings of famous mathematicians and in his syllabus of recommended readings. In the first, published in 1615, he calls attention to the two volumes of Josephus

[150] Sec. 3.3.

[151] *Apparatus*, pp. 411-412.

Zarlinus on music and then cites the five *Dialogi de musica veteri et nova* of Vincentius Galilaeus, noting that the latter points out the absurd errors of the *contrapuntistae,* as they call themselves.[152] Later, after extolling the merits of Clavius (*praeceptor meus*), he describes Galileo's discoveries with the telescope and lists his *Sydereus nuncius*, his *De maculis solaribus*, and his "very intelligent treatise" *De iis quae natant aut moventur in aqua*; Galileo is still alive, he writes, and is working on a new system of the universe.[153] In the syllabus of 1620 he again calls attention to the *Dialogi della musica antica et moderna* and stresses how essential this is "for correcting and restoring the music of our time."[154] He also notes the work of Copernicus, "which now can be read with the Church's permission," and those of Tycho Brahe, Kepler, Galileo, and Scheiner, among others.[155] In the title of the *Sphaera mundi* of the same year he even advertises that it has been brought up to date with the latest discoveries of the latter (*Sphaera mundi seu cosmographia demonstrativa, in qua totius mundi fabrica una cum novis Tychonis, Kepleri, Galilaei, aliorumque astronomorum adinventis continetur*).[156]

This completes what need be said here about the science of mathematics in the context of Galileo's logical questions.[157] If the extracts from those questions relating to mathematics that are given at the beginning of this section are compared with the teachings of Clavius and of philosophy professors from Valla to Rugerius, it will be seen that they all become intelligible in the setting of the Collegio Romano around 1589. Whether Galileo had access to a complete set of Valla's logic notes and worked through the treatise *De scientia* contained in them, even making an abstract of them that was subsequently lost, need not be decided at the moment. What seems clear is that the portions of his questions that have survived are derivative from the treatises on foreknowledge and demonstration contained in Valla's lectures of 1588, and that these implied views of demonstration and science that are quite consistent with Galileo's remarks on these subjects. The more pejorative statements about the status of mathematics as a science that occur in Valla-Carbone certainly would not have been accepted by Galileo at that period of his life, committed

[152] *Dissertatio*, p. 61, pp. 64-65.

[153] *Dissertatio*, pp. 62-64.

[154] *Apparatus*, p. 396.

[155] *Apparatus*, p. 397. For additional details on Blancanus, see secs. 4.4.b and 5.4.c *infra*.

[156] In a private communication Dr. Ugo Baldini has informed the author that documents exist in the Jesuit Archives in Rome censoring Blancanus's writings on Galileo's behalf and prohibiting their further publication.

[157] Of related interest is the essay of Adriano Carugo, "Giuseppe Moleto: Mathematics and the Aristotelian Theory of Science at Padua in the Second Half of the [*sic*] 16th-Century Italy," in *Aristotelismo Veneto e Scienza Moderna*, Luigi Olivieri, ed., vol. 1, pp. 509-517.

as he then already was to the pursuit of mathematics as a professor of that discipline. The correctives suggested by Clavius and worked out in detail by Blancanus, probably on the basis of Clavius's teaching in the 1590s, undoubtedly reflect more accurately Galileo's personal views on this subject at that time, were one to attempt their reconstruction.

In his later life, of course, Galileo was to be completely absorbed in the study of mechanics, which at this earlier period was regarded, according to both Valla and Clavius, as a *scientia media*. Such a hybrid type of science, the medieval ancestor of mathematical physics, was described by Valla-Carbone as more mathematical than physical in nature, and indeed an attempt was made by them to remove it completely from the sphere of physics. Others around that time regarded the *scientia media* somewhat differently, for among the Thomist commentators were many who saw it as more physical than mathematical in nature—an interpretation that Valla himself viewed more favorably in 1622 than he did in 1588, as can be seen from his more considered remarks on the scientific character of a subalternate science.[158] Galileo apparently went through a similar change as his own career developed, becoming more and more concerned with the physical problems of local motion than with the pure and applied mathematics that first attracted him to Euclid and Archimedes. More will therefore have to be said about the status of mechanics as a science that does fuller justice to both its physical and its mathematical aspects—a subject that will be broached anew in the latter part of the following chapter.[159] But in the period between 1589 and 1591, when Galileo was writing his logical and physical questions and making his first attempts at constructing a science of motion, the materials presented in this chapter delineate, better than anything heretofore available, his general understanding of what a science was, and of the methodology one would have to employ to attain the true and certain knowledge that was its goal.

[158] Sec. 3.4.a.

[159] Sec. 4.4.

CHAPTER 4

The Study of Local Motion

MOST ATTEMPTS that have been made to fill out the background of Galileo's study of local motion have focused on the writings of philosophers in the universities of northern Italy or on those of applied mathematicians in the Archimedean tradition working largely outside the universities. Jesuit mathematicians such as Clavius have attracted attention mainly because of their astronomical writings, but they seemingly left no works dealing with the quantitative aspects of local motion, nor was this a field in which courses were offered at the Collegio Romano or other Jesuit *studia*. The striking parallels between Galileo's Latin compositions and those of Jesuit professors displayed in part 1 of this study do not deal *ex professo* with problems of local motion; they were concerned, as has been seen, mainly with logical methodology and with questions concerning the heavens and the elements that terminate before the discussion of the movement of elemental bodies has begun. This perhaps explains why those who have studied Galileo's early Latin writings have tended to ignore the logical and the physical questions, or to view them in any event as in no way connected with his first attempts to construct a science of local motion. And if such a connection were lacking in Galileo's own compositions, *a fortiori* there would be no inclination to search in *reportationes* of lectures given at the Collegio Romano for teachings that could have influenced such a development.

Yet, as has been intimated, there are reasons to suspect that Galileo's interest in local motion was *not* unconnected with the logic and the natural philosophy he had carefully studied and excerpted from the notes of Jesuit professors. The problems he discusses in his *De motu antiquiora* and that surface in his later writings have many counterparts in Jesuit lecture notes when these continue on from the point at which Galileo's extant physical questions leave off. Again, the methodological questions he raises in his early notes on the *Posterior Analytics* are still being asked by him until after his definitive treatment of local motion in the *Two New Sciences*. It is hard to believe that such parallels would be completely fortuitous, and in any event they should be investigated for the light they may shed on the development of the *nuova scienza*.

The main purpose of this chapter is to set out teachings on motion that were being advanced by the Jesuit professors whose writings have been reported and analyzed in the three previous chapters, preparatory to evaluating, in the chapters that lie ahead, their possible influences on Galileo. These teachings continue to show the same type of uniformity as is reflected in the physical questions, though, on a few matters, there are variations that can be examined with profit. The principal topics that require investigation are those dealing with the possibility of a science of nature and of motion, difficulties associated with motion in a void and with natural and violent motion, and special problems relating to the motion of heavy and light bodies and of projectiles. The first three sections of this chapter will focus on the ways in which the natural philosophers of the Collegio Romano from Menu to Rugerius dealt with these topics. The final section will complement this survey by detailing developments in the science of mechanics that took place outside the Collegio, but that have methodological similarities with materials covered in the earlier chapters and thus could be part of the tradition that influenced Galileo's final drafting of the *Two New Sciences*.

1. Nature and Motion in General

The study of motion as it occurs in nature is obviously only a part of a broader science that considers nature in general, commonly referred to as *scientia naturalis* or natural philosophy. Because of difficulties arising from the variability of nature and the ideal of scientific knowledge as being itself invariable and necessary, those who taught natural philosophy usually began by inquiring whether it was possible to have a true science of nature—it being understood that their students would already have completed the course on the *Posterior Analytics* and so would be acquainted with the requirements for a strict *scientia*. The Jesuit professors were no exception in this regard, and in the period of interest in this study their mode of treating the possibility of a science of nature was fairly uniform. Antonius Menu is a convenient person with whom to begin, since he is the earliest author whose physical questions show any substantial similarity with Galileo's, and since, as will be seen in the sequel, he appears to stand at the head of a tradition that continues throughout the entire period of the correspondences noted in the first two chapters of this inquiry.

a. THE POSSIBILITY OF A SCIENCE OF NATURE

Menu's basic answer to the problem is that there can be a science of nature because natural things have principles, parts (or elements), and properties, and so they satisfy the requirements for an object of scientific inquiry. Natural things are variable, he admits, but this does not preclude there being some necessity associated with them; to make this clear, he distinguishes between a necessity of existence and a necessity of essence, and maintains that the first is not generally found in the objects of physics, whereas the second is. Some physical entities, such as the heavens and the elements, may exist necessarily, but most species that come to be and pass away do not. Regardless of what their mode of existence may be, however, their essences are invariable, and it is under the aspect of their essential characteristics that they are studied in natural science. So, one may object that accidents make such a science impossible, and the objection has some force, but not when one leaves aside accidental considerations and focuses on what is essential. Again, the fact that a natural thing can corrupt and become other than it is poses a difficulty for its being an object of scientific inquiry; but here, too, the corruptibility and variability are in the order of existence, not in that of essence. Even though natural things come to be and pass away, while they exist they are what they are, and it is only when considered in relation to what makes them be such—that is, their essences—that they are capable of generating scientific knowledge.[1]

This understood, Menu goes on to delineate the object of natural science in more detail. As he sees it, and this becomes common teaching at the Collegio Romano, the adequate object of physics is the natural body or the natural substance precisely as natural. The qualifications are required because the term "body" can be taken in two ways: in a mathematical sense, as indicating the pure dimensionality of length, width, and depth; and in a substantial sense, as indicating the substance that can be extended through quantity and so underlies such dimensionality. The consideration of the naturalist or physicist differs from that of the mathematician in that the latter studies the first type of body, whereas the former studies the second. This distinction made, it is further necessary to state that the physicist considers the substantial body as it is natural, that is, as something having a nature. Now nature, as Aristotle makes clear, is a principle of motion and rest, and so the natural body has within it an order or aptitude for particular types of activities or operations;

[1] Cod. Ueberlingen 138, Quaestiones in philosophiam naturalem, Prolegomena in eandem, Cap. 1. An de rebus naturalibus detur scientia.

these are not its essence but they are related to it, for it is the essence considered as a principle of characteristic activity (and reactivity) that is known as the nature. Finally, to consider such a natural body precisely as natural, one must abstract from, or leave aside, particular or individual matter (called "signate matter") with all its accidental modifications, and focus rather on its universal or common characteristics as a body of a given nature or kind, for this is required to assure the essential type of knowledge that is sought in natural science.[2]

Natural bodies, as classified by Menu and his successors at the Collegio, fall into five different species: the heavens, which are incorruptible; the elements, which are corruptible; imperfect compounds, which contain two or three elements and are not essentially different from elements; perfect inanimate compounds, which contain all four elements and are essentially different from them; and animate compounds, which are further divided into plants and animals. All of these share in some common properties, such as having quantitative dimensions and shapes, being in different places and configurations, undergoing movements and changes, and being in time.[3] The foundational science of nature, contained in the eight books of the *Physics*, considers the natural body in a general way: in its principles, causes, and elements, and in the properties that are common to all these species. Subsequent books of Aristotle's *naturalia*, Menu goes on, then take up partial sciences whose adequate objects are one or another species. Thus the heavens are studied in the first two books of *De caelo*; the elements in the last two books of *De caelo*, the two of *De generatione*, and the fourth of the *Meteors* (which Menu feels is really the third book of *De generatione*); imperfect compounds in the first three books of the *Meteors*; animate compounds in the *De anima* and related books; and so on. And the proper order for working through all of this material is to begin with the more basic, common, and universal features of natural things and proceed from this to the investigation of the simpler types, then to the more complicated, following the order of the Aristotelian corpus.[4] The knowledge that is gained from such a procedure, Menu concludes, is both useful and necessary, for without it one could never advance to metaphysics, nor would one be in a position to do medicine (a science subordinated to physics, in his view), or astrology,

[2] Ibid., Cap. 2. De obiecto philosophiae naturalis.

[3] Ibid., Cap. 3. De principiis, causis, et proprietatibus subiecti in communi.

[4] Ibid., Cap. 4. De partibus philosophiae naturalis, earum ordine ac methodo quae in his libris observatur.

perspective, music, and ethics, to say nothing of all the factive arts that are concerned with the manipulation of natural things.[5]

b. NATURE AND ITS CAUSALITY

Aristotle's statement that nature is a principle of motion and rest was the subject of much comment among peripatetics, particularly in the context of discussions of causality, for the text of the *Physics* is not at all clear as to the way in which nature might be regarded as the cause of motions that are generally regarded as natural. This particular problem assumes considerable importance in Galileo's *Two New Sciences*, for in the discussions of the Third Day, Galileo invokes the concept of nature to explain naturally accelerated (or falling) motion, and then offers some *obiter dicta* on the general subject of causality that have been elaborated by some authors into a completely acausal philosophy of science, which they propose as unambiguously his.[6] Fortunately the Jesuit professors whose teachings have already been expounded devoted attention to this problem, and their views on it can now be examined with profit. Menu and Vitelleschi have the most extensive treatments of nature and its causality, and since their expositions are representative and the second complements the first, they will be considered in that order.

Menu approaches the difficulty by way of the commentaries of two Neoplatonic Aristotelians, Philoponus and Simplicius, the first of whom saw in Aristotle's definition some universal principle that enlivens all natural things, and the second who saw a variety of accidental principles that seem to account for particular types of motion. Instead of defining nature simply as a principle of motion and rest, as Aristotle had done, Philoponus characterized it as a life (*vita*) or kind of force (*vis*) that is diffused through all natural bodies and is a formative and governing principle of their motion and rest. In Philoponus's view, this force proceeds from the First Cause and is communicated to bodies through the motion of the heavens; it was this that Aristotle described when he attempted to define nature through an order to various operations. Simplicius, on the other hand, saw nature as being more connected with the instrumental principles within a body that account for its aptitudes and propensities, and that germinate and produce its characteristic activities. Menu rejects both interpretations, first that of Philoponus because nature is an intrinsic principle and as such is within things, whereas no universal

[5] Ibid., Cap. 5. De utilitate et necessitate physicae et de quibusdam servandis in tota philosophia.

[6] See sec. 6.3.c; also 5.4.d-e.

principle is within in this sense; clearly the heavens and God can count as universal principles of such motions, but both of these are extrinsic principles, and so do not fit the definition of nature. Simplicius's interpretation is more acceptable in that instrumental principles are at least within things, but it is wrong if it sees nature as realized in the accidents from which their activities proximately proceed. Accidents are secondary principles, whereas nature is a primary principle in the order of substance. Thus, if Simplicius's position identifies nature with matter and form insofar as these have an aptitude for operation and can cause motion either actively or passively, it is correct; if it sees aptitude or inclination as types of accident, then it is an improper interpretation of Aristotle's definition.[7]

This leaves a difficulty that still requires solution, however, for one can inquire as to whether the motion of vapor upward is natural or not, and if so, what is the cause of its upward motion. It would appear that it is moved upward by reason of some quality inherent within it, say, its *levitas*, and if so, then this is its nature. Menu's answer to this is somewhat involved, but may be summarized as follows. For such a motion to be called natural, it does not suffice that there be a quality within the body that actively causes the motion, for over and above this such a quality must also exist naturally in the body by reason of its substantial form. In order for any quality to be operative, it must exist in a substance by which it is sustained in being; similarly, for it to act in an enduring way it must exist in a subject to which it belongs naturally and for which it can serve as a natural instrument, and this cannot be the case unless the quality is in the body by reason of its substantial form. Therefore the upward motion of water vapor is not said to be natural on the part of an active principle such as *levitas*, because *levitas* is not a natural instrument of the substantial form of water. (*Gravitas*, on the other hand, would be such a natural instrument, for water is normally heavy in air and therefore its falling motion in air is regarded as natural.) Menu goes on to state that the motion whereby iron moves toward a magnet is not natural for the same reason—the quality that effects the movement is not a natural instrument of the iron, but rather is an instrument of the attracting substance alone.[8]

Another question whose answer serves to illuminate how nature is the cause of motion addresses the precise formality under which substantial form can be regarded as a principle of operation in other bodies as well as in the body of which it is the form. Menu equates this question to

[7] Cod. Ueberlingen 138, Tract. 2 [*bis*], Disp. 1, De principiis operativis, Cap. 5. Quale principium sit illud quod definitur in descriptione naturae.

[8] Ibid.

the following: Can nature (as this is identified with form) be called an efficient cause, understanding efficient cause to mean something that causes motion in another? Many commentators, he says, take an affirmative position on this, pointing out, for example, that when fire heats wood it actually acts on the wood, that such heating action is as much a natural motion for fire as its moving upward, and therefore that fire's being a natural agent qualifies its form to be denominated as an efficient cause. Menu, however, disagrees with this way of speaking, for in his view nature is a principle of operation *within* the thing of which it is a substantial principle and not a principle of its acting on something extrinsic to itself, which would warrant its being called an efficient cause. Many non-natural things contain active principles within them that enable them to act on another, and here Menu cites intelligences and the movers of the heavens, but these are distinguished from natural things precisely in that the latter have a principle of physical change within themselves, whereas the others do not. So, while admitting that the term "nature" can have many meanings, in the precise sense in which this word is used and defined in the second book of the *Physics* nature is not a principle of motion in another, which is the way one usually thinks of an efficient cause. To clarify the differences between the two, Menu goes on to specify the following: nature includes both active and passive principles, whereas an efficient cause is an active principle only, not in any way passive; nature is a principle of motion in that in which it is, whereas an efficient cause is a principle of motion either in itself or in another; and nature is merely a principle of a substantial body, that is, its matter or its form, whereas an efficient cause is the substantial body itself (technically a *suppositum* from which actions proceed, according to the scholastic adage that *actiones sunt suppositorum*).[9]

Vitelleschi, as has been said, addresses these same questions and provides similar answers, although in a slightly more nuanced way than Menu. Vitelleschi formulates the difficulty of the vapor rising upward somewhat differently: the vapor seems to go upward naturally, because it does so from an intrinsic principle, and this cannot be anything other than *levitas*, an accidental principle, since the substantial form of water inclines it to move downward; therefore it would appear that the term "nature" could be applied to an accidental principle of motion, provided this is within the thing that undergoes the movement. To this Vitelleschi replies that vapor is not necessarily the same as water, since the two might differ essentially, and if this be the case then the motion upward is natural insofar as it proceeds from the substantial form of vapor as its

[9] Ibid., Cap. 7, Quaes. 3. An forma sit natura ut est principium operationis non in se sed in alio.

principal cause. If vapor and water are essentially the same, on the other hand, then one would have to admit that the upward motion of the vapor is violent, if speaking in an unqualified way, even though it proceeds from an internal principle—for the principle would have to be connatural to the thing in order for the motion to qualify as natural in a proper sense. Even here, however, one could still speak of the vapor's upward motion as natural in a qualified way, since the vapor moves upward in virtue of an instrumental principle, namely, *levitas*, and it is natural for *levitas* to effect this kind of motion.[10]

Vitelleschi's discussion of the way in which nature can be called an efficient cause similarly makes greater use of distinctions than does Menu's, although both agree on the final result: Aristotle's definition of nature does not employ cause in the sense of an efficient cause, precisely as this causes an effect in another object. Vitelleschi, however, offers a fuller explanation of the difficulty in the following terms. There are two kinds of movement: one wherein the motion remains in the thing moved, as when fire moves upward; the other wherein it passes to another, as when the foot moves the body. In both these cases one would have to admit that the motion comes from the nature of the mover and therefore that it is natural in its principle—meaning by natural whatever is in accord with the nature of its principle. But we usually do not say that the essence of a natural thing is nature with respect to the motions it effects in things extrinsic to itself: thus motion in another does not come from nature precisely as it is nature, although it can come from nature as it is an efficient cause. The reason for this is that whenever motions are imposed on things from without in ways that are clearly evident, we attribute them to the cause we recognize, and so we attribute the heating of the wood to the fire that heats it. When motions take place in things themselves, on the other hand, and there is no extrinsic cause that is obvious, we attribute these to the nature of the thing and we say that such motions are natural. Again, we distinguish between things that are natural and those that are artificial or abstract, an example of the latter being the movers of the heavens. Natural things seem to have a principle within themselves whereby they produce motions that are perfective of them and at the same time cause motion in other things. Abstract things, on the other hand, are able to effect motion in other things but do not produce motion in themselves, since they are actually unmoved movers. Artificial things, finally, have no principle for effecting motion within

[10] Cod SB Bamberg 70, fol. 112r-v. This disputation, the second in Vitelleschi's exposition of the second book of Aristotle's *Physics*, has been transcribed by the author in his *Prelude to Galileo*, pp. 294-297. A discussion of the teaching will be found in the same, pp. 286-299, and, in a broader context, on pp. 110-126. See also sec. 4.2.a.

themselves, nor can they initiate it in another, although they can modify such motion once it is initiated. It is difficult, states Vitelleschi, to define nature in such a way as to take account of all these differences, as well as of the way in which the natural differs from the violent, but it is with such an end in view that Aristotle defines nature the way he does in the second book of the *Physics*.[11]

A better insight into Vitelleschi's position on nature as an efficient cause, which will be elaborated more fully in the next section when discussing the causality involved in falling motion,[12] is given by his treatment of the case mentioned by Menu. This he presents in the form of an objection: not only does fire go upward naturally but it also heats naturally, and with respect to both motions the nature is the form of fire; in the second case, however, the form functions as an efficient cause; therefore in the first case, the upward motion, the form must be an efficient cause also, and so nature (as identified with form) is properly called an efficient cause. Vitelleschi replies to this by distinguishing four different ways in which one can use the term "natural": (1) as opposed to violent, as that which follows the inclination of a thing and is not contrary to it; (2) as opposed to voluntary, as that which follows necessarily and determinately from that inclination; (3) as including everything that is attributable to a natural agent when acting naturally; and (4) as opposed to violent, as that which has a principle of activity within itself. These distinctions understood, one can say that the motion of fire upward is natural in all four ways, whereas the heating action of the fire is natural in the first three ways but not in the fourth. This being the case, when using the term "natural" in a completely exclusive and proper sense, one must say that the upward motion of fire is truly natural, whereas the heating action wherein nature is an efficient cause is not completely so, and thus one cannot say that nature, precisely as nature, includes the notion of being an efficient cause.[13] Neither Vitelleschi's nor Menu's resolution of this difficulty, it should be noted, excludes the possibility of nature's being an efficient cause in certain cases, as will become clear later, despite the insistence of both that nature and efficient cause are not reciprocal concepts.

c. MOTION IN A VOID

Apart from common problems such as the possibility of a science of nature and the way in which nature itself can be said to be a cause of

[11] Cod. SB Bamberg 70, fols. 112v-113v; Wallace, *Prelude to Galileo*, pp. 290-292, 295-296.

[12] Sec. 4.2.d.

[13] Cod. SB Bamberg 70, fols. 113v-114r.

natural motions, the Jesuit philosophers discussed a number of problems relating to local motion in general, abstracting from its particular types. Among these, one of the most interesting from the viewpoint of Galileo's science is the possibility of there being local motion in a void (*vacuum*). On this matter the Averroists and the conservative Aristotelians in the universities of northern Italy took a strong negative position, and it is noteworthy that Pererius adopted this stance in his textbook of 1576, and so did Vitelleschi and Rugerius in their lecture notes of 1590 and 1591, respectively. In the intervening years, however, it appears that the more scholastic position, generally adopted by the Thomists, to the effect that successive motion in a void would be possible, was being taught at the Collegio Romano. Certainly Menu gave strong support to the affirmative, and since there is reason to believe that his influence extended to Valla, it will be well to review the arguments he offered.

Menu begins by noting that both sides of the issue are defensible; that Averroes and his followers thought that if there were motion in a void it would have to be instantaneous; that Philoponus, Avicenna, St. Thomas, Scotus, and the Thomists taught that local motion in a void would be successive and not instantaneous; and that some recent moderns (*recentiores*) have taken the position that not only would the motion of elements in a void not be successive, but neither would the motion of compounds. His own view is that both motion and mutation can take place in a void, and that on the point at issue relating to local motion, the succession that characterizes it arises not only from the plenum—that is, the medium—through which it ordinarily passes, but also from other factors, namely, the quantity of the thing moved, the power of the mover, and the distance across which the thing moves. If one admits this, then it follows that the motion of any body, whether it be simple (that is, an element) or compound, would be successive even if it took place in a void, for granted that there would be no extrinsic resistance arising from the medium, intrinsic resistance would still be present, and this would suffice to make the motion successive.[14]

The arguments Menu offers in support of his conclusions are drawn from Thomistic and nominalist sources, and will be easily recognized by students of medieval science. Whether the thing moved passes through a vacuum or a plenum, he says, it will always encounter some resistance, and thus the motion will be successive. To see this, consider a three-foot-long body moving through a void; in such a case it is impossible for the first footlength to be in the space occupied by the last unless it

[14] Cod. Ueberlingen 138, Tract. 4, De passionibus rerum naturalium in communi, Disp. 2, De vacuo, Cap. 10. An posito vacuo possit in eo fieri motus vel mutatio.

previously is in the space occupied by the middle, and this can only take place in time. Similarly, the ultimate sphere moves successively and in time, and yet it encounters no extrinsic resistance. One may object that the heavens move successively because of an intrinsic resistance arising from the proportion of the thing moved to the mover that determines its velocity, but against this, let the entire substance of the heavens be annihilated by the power of God, leaving only the circular quantity. In such an event, will the sphere rotate successively or not? One cannot say that it will not, and therefore succession can come from quantity alone. Again, suppose that the entire space between heaven and earth is void, and let God create fire on earth in an instant; will the fire come to exist in the heavens in the same instant? If not, this will be because the same thing cannot be simultaneously in the terminus *a quo* and in the terminus *ad quem*. But then, if different instants are involved, this means that motion through the void must take place in time. Yet again, let a very long lance extend from the concavity of the moon's orb through the intervening void all the way to the earth. On this supposition, let mobile object *A* move from the middle of the lance to the earth while mobile object *B*, of the same motive power, moves from the upper end of the lance to its middle. Obviously *A* will reach earth more quickly than *B*, which shows that the motion of both must be in time. More, the fact that a mobile object has opposite dispositions requires that it move in time, just as it takes time for a cold body to become hot. But when a body moves through a void it always has opposite dispositions insofar as it is always in different locations, whether real or imaginary. Such locations are opposites because they are contraries that cannot exist in the same subject at the same time. And finally, just as in the motion of growth, succession results from the quantity of food that is gradually added to the living thing, so in local motion a similar succession arises from the quantity of the moving thing, since it has part outside of part that must succeed each other at least in imaginary space.[15]

In addition to such arguments, Menu also attempts to show the flaws in various proofs that are adduced to establish that motion in a void would be instantaneous. Among these is the admission on the part of both Thomists and Scotists that Aristotle's arguments relating to the void in the *Physics* are merely *ad hominem*; if this is so, they are not to be taken seriously as stating what takes place in the real world. Here Menu would abstract from the value of Aristotle's particular arguments and still maintain that, in truth, motion in a void would be successive. His reason is that all arguments based on the medium through which

[15] Ibid.

the object passes are based on extrinsic resistance, and none of these apply to intrinsic existence. He admits that there is always some ratio between motion that takes place in a plenum and that which takes place in a vacuum, since both are in time, but the ratio does not arise from the extrinsic resistance of the plenum, which is clearly lacking in the vacuum; rather it comes from the intrinsic resistance that is found in both. Moreover, granted that there can be no ratio between the vacuum (as non-being) and the plenum (as being) when considered under the formality of being, there can be a ratio between them when considered under the formality of space, for it is this that prohibits one part (of space) to be occupied simultaneously with another.

Again, Menu is aware that one can construct many arguments that lead to apparent absurdities, such as the following based on text 72 of the fourth book of the *Physics*. Let a weight of four units move through a vacuum in a quarter hour and through a plenum in a half hour; if it does this, then a weight of eight units would move through a plenum in a quarter hour, whereas a weight of four units would move through a medium that is twice as yielding (i.e., offers only half the resistance) in the same quarter hour, which is absurd when compared to the fact that the identical weight takes the same time to move through a vacuum. To this Menu replies that should the heaviest body move through a plenum and the lightest body move through a vacuum, there would be no absurdity in their taking the same time to cover an interval, since the conditions are not the same. All things being equal, however, a heavy object and a light object will not move through a plenum and through a vacuum in the same time, because there will always be more resistance in the plenum, no matter how yielding one wishes to make it, even by allowing its subtlety to approach infinity in the process.

Finally, in defense of his use of intrinsic resistance, Menu distinguishes two kinds of resistance, one that takes place with action, the other without. The resistance that arises from the quantity of an element as it moves through a void is of the latter kind, and for this it is not necessary that the quantity act in any way; rather it suffices that one part of the quantity not be able to be in the same place as another part in the same time. And this argument applies even to imaginary space, for granted that such space does not have positive resistance in the way just described, it does have privative resistance, insofar as the object's being in one part excludes it from being in another at the same time.[16]

Vitelleschi and Rugerius, as already noted, take a position opposite to Menu's in this matter, but there is evidence that they considered Menu's

[16] Ibid.

arguments carefully, even though they were not prepared to accept them. Vitelleschi expresses his convictions in three conclusions that make this rather clear: (1) if there were a vacuum, there is no way in which the motion of heavy and light bodies would take place through it; (2) similarly, if there were a vacuum there would be no motion of the heavens; and (3) if one allows that the influence of the heavens is not necessarily required for a motion on earth, the progressive motion of animals might be possible, provided that there would be something on which the animal could move, but certainly not for an appreciable interval of time. The arguments Vitelleschi offers in support of these theses need not be entered into here; suffice it, perhaps, to note why he rejects the position favored by Menu. The latter position, he states, supposes that there is a kind of resistance that is intrinsic to the thing moved and is different from the resistance of the medium, but this is false. Granted that the slowness and fastness of motion can arise from variations in the motive power of the thing moving, such variation arises from the motive power's ordination to the medium, namely, precisely because a larger or smaller power overcomes the resisting medium with more or less ease; therefore, should the medium be removed, the motive power would have no resistance to overcome and thus would not effect any succession. (This view of velocity variation, as will be seen, is intimately connected with Vitelleschi's explanation of acceleration in falling motion, to be gone into in detail in the following section.)[17] Even though one can admit, he goes on, that motive power is a *per se* cause of local motion, it is not such a cause, nor can it be such a cause, except in a plenum, and thus it is by reason of extrinsic resistance that motion is successive. And, as to the argument based on quantity, this supposes that there are various parts and various locations in a vacuum, so that when the moving object traverses the void it really passes over them and touches them, but in reality there is nothing there, only in one's imagination.[18] Apparently Rugerius was impressed by this last refutation, for he summarizes his entire treatment of the problem in two somewhat enigmatic propositions: (1) Aristotle did not deny that there could be any motion in a void, if by void one means three dimensionality that is real and devoid of all bodies, usually called separated space; but (2) Aristotle absolutely denied that there could be motion in a void understood in the sense of a space that is completely imaginary and not real, and in such a space there would be no successive motion, either natural or violent.[19]

[17] Sec. 4.2.d.

[18] Cod. SB Bamberg 70, fols. 210v-220r.

[19] Cod. SB Bamberg 62.3, fols. 306v-308r.

d. OTHER PROBLEMS WITH LOCAL MOTION

Among the remaining general problems relating to local motion, three are particularly noteworthy for this study because they enter into Galileo's earliest essays on motion, his *De motu antiquiora*.[20] The first concerns the possibility of motions intermediate between the natural and the violent, the second concerns the requirements for the continuity of motions, especially for the cases where a natural motion follows after a violent motion and vice versa, and the third, directly related to the second, concerns the existence of a point of rest at the moment of reflection, for example, when there is a turning point such as occurs when motion upward is succeeded immediately by motion downward. On all these difficulties Galileo adopted a position not to be found in the text of Aristotle, and consequently rejected by the conservative Aristotelians teaching in the universities of his time. Scholastic positions, on the other hand, allowed some latitude in their solution, and this attitude is generally reflected in the lectures given at the Collegio Romano. Unfortunately Menu's notes are quite schematic in the portions of the *Physics* where such problems are raised, and Valla's notes are missing entirely—although one can infer from his treatment of projectile motion, to be discussed in the third section of this chapter, that the solutions he offers were quite similar to Galileo's.[21] Vitelleschi, however, treats all of them in some detail, and his solutions are of interest not only in their own right but also for their probably having conserved vestiges of the teachings of Menu, Valla, and others who preceded Vitelleschi at the Collegio.

The dichotomous division of local motions into natural and violent that is implied in Aristotle's *Physics* poses a difficulty for Vitelleschi, and on this account he explicitly raises the question whether it is possible to have a motion that is neither natural nor violent but actually intermediate (*medius*) between the two. It seems that such is the case, he says, because fire moves circularly in its proper place (i.e., when at the concavity of the moon's orb), and such circular motion is not natural for fire, which can only have one natural motion, namely, upward; nor is it violent, because it is perpetual and nothing contrary to nature can be sempiternal; therefore this must be a type of intermediate motion. On the other hand there are many statements of Aristotle that seem to rule out the possibility of any motion being neither natural nor violent, and these can be urged against an affirmative answer. When one examines these texts, however, one can see that the negative reply interprets Aristotle's expressions "beyond nature" (*praeter naturam*) and "contrary to nature" (*contra naturam*) as meaning exactly the same thing as "violent" (*violentum*). Actually

[20] Sec. 5.2.

[21] Sec. 4.3.a.

violence has two senses for Aristotle: (1) sometimes it refers to what is preternatural, that is, to something that is over and above the nature of the thing, in the sense that the nature has neither a repugnance nor a positive inclination to it; and (2) sometimes it refers to what is contranatural, that is, to something that opposes the nature of the thing and does violence to it in a strict sense. In the broader first meaning would be included anything that would not be "according to nature" (*secundum naturam*), whereas in the narrower second meaning the focus would be only on what is "contrary to nature."

This distinction understood, Vitelleschi states that if "violent" is taken in the first sense, then the dichotomous division is correct, for every motion will have to be either according to nature or not, and there is no possibility of a third alternative. If "violent" is taken in the second sense, however, then there is room for a type of intermediate motion that would be either "beyond nature" (*praeter naturam*) or "above nature" (*supra naturam*), but would not be related to nature positively (*secundum naturam*) or negatively (*contra naturam*). And it is obviously the second sense that is taken when one asserts that nothing violent can be perpetual or sempiternal, for the contranatural will clearly take away the force of nature (*vim naturae*) and ultimately destroy it. But when one allows the possibility of the preternatural and the supernatural, and takes "violent" in the first sense, there are cases where the adage "nothing violent can be perpetual" does not apply, such as that of fire circulating perpetually at the concavity of the moon. Thus an intermediate motion is possible, for not every motion need be according to nature or violent in the strict sense.[22] This conclusion has obvious implications for the development of a concept of circular inertia, as will be discussed in following chapters.[23]

The question of the continuity of motion is raised by Vitelleschi in a quite general way: Can motions that are specifically diverse be in continuity? As a rule, he replies, the answer must be in the negative, for such motions cannot be intrinsically in continuity; yet an exception should be made for local motions, for in some cases these can be. To explain this, Vitelleschi invokes a distinction based on ways of looking at local motion: such motion can be considered precisely as it is local, namely, as involving changes of location, or precisely as it is natural, namely, as coming from nature. Now succession and continuity are not accompaniments of local motion precisely as it is natural, but rather as it is local, for it is this that requires a medium and some real space through which or around which it takes place, and from which its succession and tem-

[22] Cod. SB Bamberg 70, fols. 257r-259v; for a summary of this teaching and further reference to texts, see Wallace, *Prelude to Galileo*, pp. 111-114, 313-314, 334-335.

[23] Secs. 5.2.b and 6.1.c.

poral character arise. When one speaks of local motions that are specifically diverse, however, such specification is taken from their termini, not as these are local motions, but rather as they are natural motions, that is, insofar as they have from nature certain prescribed termini from which they are denominated and specified. Therefore it is possible that local motions, even when specifically diverse from the viewpoint of their being natural, can still be in continuity. The reason is that they are not specifically different under the formality of their being local motions, and it is under this formality that succession and continuity are proper to them. Granted, therefore, that one is considering the same moving object and the same uninterrupted change of location to any terminus whatever, the local motion will continue to be of the same species precisely as local, even though as natural it would pertain to a different species if made to a different terminus.[24]

It is this type of consideration that supplies the background for Vitelleschi's discussion of simple and reflex motions, and in particular for the position he adopts on whether rest always interrupts a reflex motion when it reaches the point of reflection. Based on his general view of succession and continuity, Vitelleschi maintains that it is not always necessary for rest to intervene at this point, and that in particular there is no interval when a violent motion follows a natural one or when one violent motion follows another. In the other cases one can think of, however, that is, when a natural motion follows a violent one or one natural motion follows another, he is of the opinion that some rest intervenes. The reason he gives is important because of its similarity to an argument advanced by Galileo in his early *De motu*, as will be seen in its proper place.[25] The problem is posed in the context of a stone being thrown upward, reaching its point of reflection at the summit of its travel, and then falling downward under the influence of its *gravitas* or gravity. Vitelleschi observes that the force impelling the stone upward, whatever this might be, always begins to weaken gradually. (There is a marginal note in the manuscript here stating that whether this force is a *virtus impressa* will be addressed in its place,[26] about which more will be said when treating projectile motion in the third section of this chapter.)[27] As a consequence of this weakening, the upward motion is faster at the beginning and slower toward the end, and the force that impels the stone upward finally reaches the point where it is equal to the gravity that draws it downward. Since there can be no action from a ratio of equality,

[24] Cod. SB Bamberg 70, fols. 253v-255r.

[25] Sec. 5.2.

[26] Cod. SB Bamberg 70, fol. 281v.

[27] Sec. 4.3.c.

the force at this point is unable to propel the stone further, nor is the gravity effective, with the result that the stone remains at rest until the gravity can take over, and then the downward motion begins. As a confirmation of this argument, Vitelleschi observes that it is easier for a small force to impede the descent of the stone than it is for such a force to move it upward; therefore it can happen that the force reaches a point where it cannot move the stone farther and yet can prevent it from descending, although it does this in such a very small interval of time that the rest can hardly be perceived by a human being.[28]

Valla, as will be seen, apparently took a different view of this particular case, for he held that no point of rest would interrupt the upward and downward motion of the stone,[29] just as did Galileo in his early writings.[30] Rugerius, on the other hand, returned closer to the conservative Aristotelian position, maintaining that in all cases where the movement of a single body is being considered there will be a moment of rest at the point of reflection.[31]

e. THE STRUCTURE OF THE CONTINUUM

Before concluding this discussion of the continuity of motion there is another problem relating to the constitution of the continuum generally that is treated by Menu, Vitelleschi, and Rugerius, and that is relevant to Galileo's later writings, and so should be mentioned here.[32] This has to do with the presence of indivisibles such as points in extensive continua such as lines, a problem which leads naturally to a consideration of the parts of lines and of the local motions that traverse them. Of the three authors Menu has the briefest treatment, which is amplified considerably by Vitelleschi, and then taken over with a slightly different emphasis by Rugerius.

Menu's main thesis is that the continuum is not composed of indivisibles, and one of the strongest arguments he is able to bring against this thesis is the mathematician's proof that a sphere touches a plane surface at a point, which seems to prove that the point really exists in the surface. Menu does not deny the mathematician's conclusion, and thus is faced with the problem of how to admit indivisibles into the continuum and

[28] Cod. SB Bamberg 70, fols. 280v-282v. The Latin of the argument just cited reads as follows: "Confirmatur: quia facilius est et minoris virtutis impedire lapidis descensum quam promovere ascensum; potest ergo eo devenire virtus illa sursum tendens ut non possit amplius promovere lapidem; possit tamen exiguo aliquo tempore quod a nobis vix percipitur, illum interdum detinere contra propriam naturam ne descendat"—fols. 281v-282r

[29] See sec. 4.3.b *infra*.

[30] Sec. 5.2.a-b.

[31] Cod. SB Bamberg 62-3, fols. 338v-340.

[32] See sec. 6.3.a.

yet avoid the extreme position of Ockham and the nominalists.[33] Some indivisibles, he writes, are real and positive entities *in actu*, since they are considered in both physics and mathematics, which are real sciences and so must be concerned with real things. He introduces a distinction, however, between indivisibles that are terminating (*terminantia*) and those that are joining (*copulativa*) or continuing (*continuativa*), and says that only the first are actual, although even these are not really distinct from the quantity they terminate. Since the sphere touches the plane at a point that is clearly not terminating, but rather is joining or continuing, Menu proposes a distinctive way out of the difficulty: the sphere's contact of the plane makes the point where it touches actual in some way, but not really, only through assignation (*aliquo modo actu, non reale sed secundum assignationem*).[34]

This teaching is rather cryptic, but it is clarified somewhat in Vitelleschi's exposition and in his solution of a similar difficulty. Vitelleschi first inquires whether indivisibles are something positive or not; this question is answered in the negative by the nominalists, he notes, and in the affirmative by Scotus, the *Parisienses*, and Soto and other Thomists. A point cannot be a mere negation, he goes on, since mathematicians demonstrate that a sphere contacts a plane at a point. Therefore it must be something positive, although it also has a negative connotation, namely, that it completely lacks extension. Although a positive entity, moreover, it is not really distinct from the quantity of which it is a part; rather it differs from it *formaliter ex natura rei*—the famous Scotist distinction. Does this mean that the continuum is composed of indivisibles? Vitelleschi replies in the negative, and gives both physical and mathematical proofs to show that indivisibles cannot be either essential or integral parts of the continuum. Yet this is not to deny that indivisibles are in the continuum in some way, as the example of the sphere touching the plane makes clear. Vitelleschi's explanation is that they are not there actually (*actu*) in the sense of being separated *res a re*, but they are present potentially and so can be made actual by division, designation, or contact. The last mode is the one that applies in the example of the sphere touching the plane. When the sphere touches the plane, the contact makes the point at which it touches something actual, and it can do this because the point itself is a positive entity, though not in virtue of its negative connotation.[35]

[33] For a general characterization of nominalist thought and its relation to other medieval and Renaissance schools, see Wallace, *Prelude to Galileo*, pp. 18-26 and 341-348.

[34] Cod. Ueberlingen 138, Tract. 4, Disp. 1, De quantitate, Cap. 3. An et quomodo in quantitate continua sint admittenda indivisibilia.

[35] Cod. SB Bamberg 70, fols. 260r-269r.

What emerges from this discussion is, in effect, the conclusion that both divisibles and indivisibles go to make up the structure of the continuum, the first as integral or quantitative parts and the second as positive terminating or continuing entities that serve to terminate or join the extended elements. Rugerius has a fuller discussion of these components and how they can be identified in successive continua such as motion as well as in permanent continua such as a line. In this context he goes beyond Menu and Vitelleschi and speaks of the instants of a motion being its indivisible "parts." In his terminology motion has *partes intermediae* that are of two types: one is a *pars divisibilis*, which is extended and quantified, and the other is a *pars indivisibilis*, which is not.[36] Galileo, as will be seen, adopts a similar way of speaking when he talks of *parti quante* and *parti non quante* in his discussions of the continuous character of motion.[37]

2. Natural Motion

The two most important topics in the study of local motion, from the viewpoint of natural science, are those that deal with natural motion, that is, motion that proceeds from an internal principle such as heaviness and lightness, and those that deal with violent motion, that is, motion that is caused by an external principle as in the case of projectiles. Both types of motion were examined by Galileo in considerable detail, and without doubt there was an evolution in his thought concerning them.[38] Similarly, both types were investigated by the Jesuit philosophers at the Collegio Romano, with results that were not too different from those propounded by Galileo in the drafts of his early essays on motion. So as to furnish an appropriate background for appreciating the relationships that may have obtained between these efforts, the present section will be devoted to Jesuit teachings on the natural motion of heavy and light objects, and the following section to their views on the motion of projectiles. Particularly worthy of note in the first category are their definitions of gravity and levity, their search for the mover or movers involved in the natural motion of the elements, their speculations as to whether or not elements gravitate when in their natural places, and their explanations as to why heavy bodies accelerate when they fall, each of which will be considered in turn.

[36] Cod. SB Bamberg 62-3, fols. 347r-348r; cf. fols. 241r-259r.

[37] Secs. 5.4.b, 5.4.d, and 6.3.a.

[38] Secs. 5.2.b-c and 6.3.d-e.

a *GRAVITAS* AND *LEVITAS*

All four professors whose teachings have been the focus of interest in the foregoing, namely, Menu, Valla, Vitelleschi, and Rugerius, devote some consideration to problems concerning the motive qualities of *gravitas* and *levitas*, with little variation in the conclusions to which they come. Since both Menu and Vitelleschi have already received a fair share of attention in the previous section, and since Valla and Rugerius recapitulate their teachings and advance them to some degree, the latter pair will provide the main source for the information that follows.

Valla's analysis of gravity and levity is contained in a series of notations wherein he defines these qualities and explains how they have come to be recognized and differentiated one from the other. He points out that Aristotle never defines gravity and levity, but rather heavy and light, both of which are described in a variety of ways. Sometimes they are defined through upward and downward, as when the heavy is said to be what moves downward and the light what moves upward; sometimes through the center and the extremities, as when the heavy is said to be what moves toward the center and the light what moves away from it; and sometimes through both, as when the heavy is said to move downward and to the center, the light, upward and from the center. All of these are definitions through motion. Alternatively, sometimes Aristotle defines both heavy and light through rest, as when he states that the heaviest is what stands under all things and the lightest what stands above all things. From these characterizations, Valla says, one can gather the definitions of gravity and levity, for these are nothing more than the principles that make things heavy and light. Thus gravity is a principle whereby a body tends downward to the center and comes to rest under light objects, whereas levity, conversely, is a principle whereby a body tends upward away from the center and comes to stand above heavy objects. Whether principle here is understood as primary or instrumental, says Valla, is immaterial for these purposes.[39]

Other notations clarify what is implied in these definitions. The motion that is involved, for instance, is rectilinear motion alone, for light and heavy objects move up and down in straight lines; therefore such definitions do not apply to the heavens, even though these are above all the elements, because they move circularly. Yet they do apply to the elements, since it is through such qualities that they attain their natural places. Even here, however, it should be noted that they apply to the

[39] Cod. APUG-FC 1710, Tract. 5, De elementis, Disp. 1, De elementis in genere, Pars 5, De qualitatibus motivis, Quaes. 1. Quid sit gravitas et levitas. Notandum primo. . . .

elements considered only as a whole (*totum elementum*), since it is not necessary that every particle of a particular element be in its natural place. Again, the definitions apply strictly to the absolutely light and heavy, namely, fire and earth, and not to the relatively light and heavy (air and water), since the relatively heavy (water, say) would have to be defined as what stands below some things (fire and air) and above others (earth). And finally, the definitions of light and heavy that are given in terms of rest, that is, standing above and below, are to be preferred to those given in terms of motion, since they better characterize the nature of the elements, whose perfection consists more in rest than in motion.[40] (Vitelleschi, it may be observed, takes the opposite position on this last point. He concedes that rest is more natural in itself, but from the viewpoint of the science under consideration motion offers a better way of explaining a nature than does rest. Thus, although motive qualities are causes of both motion to a proper place and rest in that place, they are called "motive" because the motion they initiate better reveals the nature of the elements.)[41]

The source of gravity and levity is considered by Valla in relation to their priority when compared to other qualities that are found in elemental bodies as well as to the substantial forms of such bodies. The mere presence of the primary qualities (hot and cold, wet and dry) does not suffice to explain gravity and levity, he says, nor do rarity and density, nor does the substantial form alone. Yet all of these are required in one way or another, as is clear in the cases of rarity and density: all light things are rare and all heavy things are dense, which shows that there is some type of precedence here. Valla concludes that the fundamental requirement is the substantial form, although both the primary qualities and rarity and density are prerequisite also; his reason is that motive qualities are instruments of the substantial forms of the elements, since these are necessary to produce the motions proper to them. Therefore, if the motive qualities are in their connatural mode, they should flow from the substantial forms of the elements. Valla notes that he includes the qualification, "if they are in their connatural modes," to take care of the cases of vapors and exhalations; in such cases, he holds, gravity and levity can be present without their appropriate substantial forms provided that the primary qualities, together with density and rarity, are there also.[42] (This hints at his probable answer to the way in which the upward

[40] Ibid., Notandum secundo . . . , tertio . . . , quarto. . . .

[41] Cod. SB Bamberg 70, fols. 359r-361r; see Wallace, *Prelude to Galileo*, p. 114.

[42] Cod. APUG-FC 1710, Tract. 5, Disp. 1, Pars 5, Quaes. 2. Unde proveniant levitas et gravitas.

motion of water vapor can be called natural: not absolutely, but in a qualified way, to account for the presence of *levitas* without the substantial form that usually accompanies it.)[43]

Vitelleschi has a parallel discussion of the source of gravity and levity in the context of inquiring which qualities are more primary, the active and alterative qualities or the motive qualities, when primary is understood to mean that which does not follow on or arise from others. He notes that one thing can follow on another in two ways: the first may be the cause of the second, and in this way rarity follows on heat; or the first may be a disposition for the second, and in this way the substantial form of fire follows on heat and dryness. A precedence of the first kind, he says, is absolutely such, whereas that of the second kind is not. This understood, motive qualities are not primary in an absolute sense but arise from other qualities. Rarity and density by themselves do not suffice to explain levity and gravity, although it is necessary for light things to be rarefied and for heavy things to be dense; his reason is that levity and gravity are operative qualities and principles that determine motion, and as such they cannot arise from matter and quantity, which in themselves are passive and indeterminate. Again, gravity and levity do not follow on active qualities alone, although they do follow on them when density and rarity are also present, for gravity seems especially to follow on cold, levity on heat. More properly, however, pure gravity and pure levity come primarily and fundamentally from the substantial forms of the elements, presupposing density and rarity as material dispositions.[44]

An even more important question relating to motive qualities, from the viewpoint of Galileo's early writings, is that concerning the existence of both gravity and levity as dual principles of natural motion. Is it necessary for both to exist as positive qualities, or does it suffice to posit one principle alone and account for phenomena attributed to the other by saying that it is not a positive quality but merely the privation and imperfection of the other? Galileo vascillated on this problem in successive drafts of his *De motu antiquiora*,[45] and it is difficult to ascertain the background out of which such vascillation arose. As it turns out, the question is discussed by Valla, Vitelleschi, and Rugerius, with all three maintaining that both motive qualities, *gravitas* and *levitas*, are real and positive in the order of nature. Since Rugerius's treatment is chronologically last and incorporates materials found in the others', it can conveniently be followed to set out the arguments used by the Jesuits to resolve the issue.

[43] Cf. sec. 4.1.b.

[44] Cod. SB Bamberg 70, fols. 361r-363v; see Wallace, *Prelude to Galileo*, pp. 114-115.

[45] Sec. 5.2.

One might propose, begins Rugerius, that all elements are actually heavy but to greater or less degree; if this were so, then the heaviest elements would go downward and in the process would propel those less heavy and force the latter to go upward. On this accounting, such elements would ascend not because they were light but because they were less heavy. And this seems consonant with experience, for if one inserts his hand into a jar full of quicksilver he senses it being propelled upward and buoyed up by the quicksilver, which tends downward by its own gravity. Similarly one may experiment with oil and note that it rises when water is poured in upon it, for it is obviously propelled upward as the water descends around it. In both cases the descent of the heavier material is *per se,* whereas the rise of the less heavy is *per accidens,* being brought about only incidentally and extrinsically by the motion of another body.[46]

To decide the question, Rugerius goes on, one must consider the goals or ends intended by nature through the motions of elemental bodies—ends intended not merely incidentally and *per accidens* but directly and *per se.* There are three that he can think of: (1) the distinction and separation of bodies; (2) the filling of the entire space below the heavens, making the universe a contiguous whole; and (3) the generation and conservation of compounds. On the supposition of two places, one up and the other down, all of these goals or ends are attainable. Thus all bodies do not tend to the same place, but some go up and others down, permitting them to be differentiated and at the same time integrated into the universe as a whole. Again, the intermediate space is filled orderly and quickly, without any intervening void. Finally, earth and the heavier elements are brought to the center, making the earth itself a place suitable for the generation and conservation of compounds as well as a habitat for animals and man; fire, on the other hand, is brought to the upper regions and spread out so as to be in contact with the concave orb of the moon and prevent the action of the heavens from burning up the terrestrial regions. To achieve these goals, moreover, it seems necessary to admit two different motive qualities, for one alone would not suffice even if it admitted of greater or lesser intensity. Consider, for example, the case of a less intense gravity: in what direction would it *per se* impel the element or body in which it was found? Should one reply downward, then any motion upward attributable to the body would be *per accidens*; certainly its ascent could not be *per se* intended by nature. Should one reply upward, on the other hand, then he would have to admit either that one and the same motive quality can incline in contrary directions,

[46] Cod. SB Bamberg 62-4, fol. 86v.

which seems absurd, or that the less intense gravity differs essentially from the more intense, which is to say that it really is lightness only called by another name.[47]

Following this line of reasoning, Rugerius comes to the conclusion that gravity and levity are both distinct positive qualities and thus that one is not the privation and imperfection of the other. He offers two proofs, one *a priori* and the other *a posteriori*. The first goes as follows: the elements are natural bodies that are essentially different from each other; therefore they have natures that are essentially distinct, and similarly, motions that are essentially distinct, since nature is a principle of motion. If this is the case, it is not true that all things are carried downward by the force of nature, for some are carried upward and others downward, and so they must have motive qualities that serve as instruments to effect motions compatible with these natures. The second argument, proceeding from effect to cause, is based on the observation that when air is quiet, flame ascends rapidly and earth descends with considerable velocity, and that neither of these ascents and descents can be explained by their being propelled by another element, with the result that both must be attributed to the natural levity and gravity of their respective bodies. In confirmation of this, if either of these motions resulted from the impulse of another body, the greater the quantity of fire or of earth the slower they would ascend or descend, for the greater the size of a body the more slowly it can be moved by another.[48] (Valla adds to this that fire goes up even when air is at rest, and so it is not pushed up by the air; again, there is a greater supply of fire than of air in the universe, and it still goes upward even when there is only a modest supply of air—in itself insufficient to propel the greater quantity of fire upward.)[49] All of these consequences are contrary to experience, and thus the proposal of a single motive quality is to be rejected.

Rugerius concludes by returning to the experimental evidence offered in support of the contrary opinion. With regard to the experiment of the hand in quicksilver, this proves nothing more than that a heavy body in its descent can propel another body that is less heavy, just as a light body that is ascending can propel one that is less light, for we see that when a bladder full of air is held under water it impels the hand holding it upward. Neither of these experiments prove, however, that the ascent or descent of *every* body results from its being propelled by another, which is the point of the theory being proposed.[50]

[47] Ibid., fols. 86v-87r. [48] Ibid., fol. 87r.

[49] Cod. APUG-FC 1710, Tract. 5, Disp. 1, Pars 5, Quaes. 1, Notandum quinto. . . .

[50] Cod. SB Bamberg 62-4, fol. 87r-v.

b. THE MOVER OF THE ELEMENTS

What moves the elements in their natural motions is a topic discussed by all four Jesuits under consideration, with more or less unanimity in the conclusions to which they come. The problem has a long history, with a variety of solutions being offered by Aristotle, by his Greek commentators, by the Arabs and the Latins, and by contemporary professors in the Italian universities. Valla offers a representative classification of the different solutions as five in number, which he describes as follows: the first is that the elements are moved by the heavens; the second, by the places to which they tend; the third, by their gravity and levity but through the intermediary of the medium through which they pass; the fourth, by the generator and whatever removes the restraints that might be holding them away from their natural places; and the fifth, by themselves in some way but without requiring also the agency of the medium. The third opinion Valla identifies as that of the Averroists, the fourth as that of the Thomists, and the fifth as that of more recent thinkers, including a number of nominalists. Menu, Valla, and Vitelleschi all adopt some modification of the fifth position, while Rugerius, though presenting much the same teaching as the others, is explicitly more favorable to the fourth. It is noteworthy that Valla, after presenting his conclusions, adds qualifications to the effect that natural motion is helped by the motion of the medium, which is why such motion is faster at the end than at the beginning, and that succession in such motion does not arise from the medium alone, which is why motion of this type can take place in a void. When stating these reservations, he adds that he has explained them elsewhere, and this is a clear indication that he sided with Menu and against Vitelleschi and Rugerius when treating the possibility, already discussed, of local motion in a void.[51] On the present topic, however, Vitelleschi has the clearest and fullest statement of the Jesuit position, and thus he will be the main source of the explanation that follows.

Aristotle's treatment of this problem in the eighth book of the *Physics*, notes Vitelleschi, occurs in the context of his defense of the principle that whatever is moved is moved by another (*omne quod movetur ab alio movetur*). It would seem that, if this principle were taken literally, everything would be moved violently, and yet Aristotle's meaning is that among things that are moved by another, some are moved naturally, others violently. The definition of motion, he goes on, is given in terms of potency and act, and much of the difficulty of understanding the principle is associated with different meanings of the term potency. For,

[51] Cod. APUG-FC 1710, Tract. 5, Disp. 1, Pars 5, Quaes. 5. An elementa moveantur a seipsis an vero a generante vel removente prohibens.

says Vitelleschi, just as there are two kinds of act, so also are there two kinds of potency. First act is the form of a thing, and second act is the operation that follows on the form; similarly, first potency is that to first act, that is, to form, and second potency, to second act, that is, to the operation that comes after the form is acquired. In the case of the natural motion of heavy bodies Aristotle maintained that the element is led from first potency to act by the generator that gives the element its form; and it is led from second potency to second act by the agent that removes the restraints holding it outside its proper place. Further light on the question comes from inquiring what Aristotle might mean by saying that something is moved by itself. One meaning would be that the thing has within itself a principle of motion, not of any kind whatever, but of beginning motion and of terminating it when begun, in such a way as not to require anything extrinsic. Now Aristotle says that heavy and light bodies have within themselves a principle of their motion, but not such that they move themselves. By this he apparently means that they do not have a principle of beginning motion or of causing it to cease when they wish; animals have such a principle within them, whereas elemental bodies do not. And it is for this reason that elements need to be moved by another.[52]

These matters understood, Vitelleschi continues, the principle whatever is moved is moved by another, can be taken in four different ways: (1) that in motion there must always be two different bodies, one the mover and the other the thing moved, and this is not true; (2) in every motion the mover must be spatially separated from the thing moved, and this also is false; (3) in every motion there must be, in the thing moved, a moving principle that is distinct from the thing moved, and this is true; and (4) in some things that move (a) a moving part can be distinguished from a moved part, as with animals, whereas in others (b) the two parts cannot be distinguished, as with inanimate objects. With the former (4a), although these are moved by themselves, nonetheless they are so moved that one part, the body, is moved by another, the soul, and so the principle of movement by another is verified. With the latter (4b), even though they do have a principle of their motion in themselves and are moved by their own forms, since their principle cannot go into act unless it is moved from without, they are said to be moved not by themselves but by another, and this is the sense of Aristotle's principle.[53]

With this as a background, Vitelleschi seeks to determine more pre-

[52] Cod. SB Bamberg 70, fols. 277v-278v.

[53] Ibid., fol. 278v.

cisely what it is that moves the elements. There are two principal opinions on this, he says, the first being that the elements do not have an active principle of motion within them, but only a passive principle, and that they are actively moved by an external mover, namely, by the generator *per se* and by the remover of restraints *per accidens*. The second opinion holds that the elements are moved actively by their proper forms. To decide between them one must in effect reply to the following questions: (1) do the elements have within themselves a first active principle of their motion, or must this be an extrinsic principle; and (2) if they have such a principle within themselves, how can they be said to be moved from without? Vitelleschi's answer is that it seems certain to him that the elements have not merely a passive principle of their motion within themselves, but an active principle also. He concedes, moreover, that the elements must first be moved by something extrinsic to them. But since this extrinsic mover is not, nor can it be, conjoined to the elements, it must impress on them some quality that inheres in them and by which they are moved as by an instrument, just as the magnet attracts iron by impressing some quality on it. This is the motive quality that is in the element, and so one must concede that elements have within themselves at least an instrumental active principle of their motion. Moreover, since whatever is done instrumentally by an instrument can be said to be done principally by the agent of which it is the instrument, and motive qualities are the instruments of the substantial forms from which they flow, the substantial forms of the elements are also active principles of their motion in a primary sense. Therefore, if one wishes to distinguish in the elements their matter, their form, and their motive quality, the passive remote principle is primary matter alone, the passive proximate principle is the same matter actuated by a form that puts it in potency to a particular type of motion, and the first and fundamental principle is the substantial form itself.[54]

To see how elements are moved both by themselves and from without, Vitelleschi goes on, we must see how many potencies are in the elements and then identify the agent that reduces each one from potency to act. He has already mentioned two potencies, the first to the form and the second to the motion that follows on the form; to these he now adds a third, which the element has while it is actually being moved, and this is its potency to a particular place or terminus. For each potency there must be a corresponding agent, and there are two types of agents—and for these two types Vitelleschi here acknowledges his debt to Zabarella. One effects motion truly and produces it in something distinct from

[54] Ibid., fols. 373r-374r.

itself, whereas the other acts through emanation and does not result in an external operation, merely causing, for example, the properties that flow from a particular form. The first type of agent requires that the patient be really distinct from it, whereas the second obviously does not. To illustrate the two types Vitelleschi reverts to his teaching on the causality of nature and cites the example of fire. In the first type, the form of fire acts as a true agent and produces heat in something different from itself, whereas by the second type it produces a motion upward that remains within itself—and in both cases the form is a true efficient cause.[55]

To sum up the various potencies and agencies, Vitelleschi continues, the motion of the elements involves a threefold potency and a twofold agency: to the first and second potencies correspond agents of the first kind, to the third potency an agent of the second. With respect to the third potency, Vitelleschi feels confident that the elements are reduced to act by their proper forms; with respect to the second, they are reduced to act by the remover of restraints; and with respect to the first, they are reduced to act by the generator. Aristotle in the eighth book of the *Physics*, he says, speaks only of the two potencies that precede motion, and not of the third potency with which he is concerned. And just as the first two agents are efficient causes in the first sense of the term agent, so the substantial form is not only the formal cause of the element but it is also the efficient cause of its motion, understanding this in the broad sense of being the cause of properties and of accidents that are proper to the element through emanation. Aristotle, moreover, is interested in finding the first principle of motion in each thing, and when discussing projectile motion mentions only the projector, although he later discusses the medium when investigating the proximate principle that sustains the motion. Similarly, if one passes from the first principle of the natural motion of the elements to seek a principle that is proximate and coexists with the motion, one must say that this is the proper form of the element.[56]

In another way of looking at it, Vitelleschi goes on, the elements can be said to be the true efficient cause of their motion, although this is *per accidens* and not *per se*. For in the motion of a heavy body, for example, the sequence of events is the following: (1) when restraints are removed the element is moved by its substantial form as by an agent, and its motion emanates from its motive quality; (2) not only the form but the entire element moves the surrounding air, and in this action the element is truly and properly an efficient cause, just as when it heats or cools something; and (3) as the medium is moved, the motion of the element

[55] Ibid., fol. 374r-v.

[56] Ibid., fol. 375r-v.

is assisted by the motion of the medium, because it effectively lessens the resistance the element would otherwise encounter, and in this sense the element effects its own motion *per accidens*. Yet the element is moved by its substantial form *per se*, through the agency of emanation, in the way just explained.[57]

All of these factors may now be brought together, Vitelleschi concludes, to supply a complete doctrine on the motion of the elements in the following propositions: (1) The elements are not moved by the heavens. (2) Nor are they moved by their proper places, in the sense of being drawn to them efficiently by some occult force. (3) They are not moved *per accidens* with the motion of the medium, as some Averroists hold. (4) The elements are not moved by the generator in the sense that the generator elicits the movement immediately and *per se*, or even through the agency of some quality. (5) The elements are moved actively and principally by their substantial forms, instrumentally by gravity and levity. The main reason for this is that there must be some active principle that coexists with the element throughout the entire motion, and this cannot be other than its form. (6) The generator or the remover of restraints is the efficient cause of the motion, taking efficient cause in the narrow sense of that which produces an effect in another; neither, however, is a proximate principle of the motion, since both are remote. (7) The proximate cause of the motion of the element itself is its form, acting *per emanationem* and thus as an efficient cause in the broad sense; if one concentrates, however, on the motion the element produces in another body such as the medium, then it is not the generator or the remover of restraints that is the true efficient cause but rather the element itself, and here the same argument applies to local motion as to heating and cooling, as already explained. (8) The element can be said to be the efficient cause of its own motion, but not *per se*, only *per accidens*. In the light of these statements all of the difficulties surrounding the natural motion of the elements can be solved. Not only this, and here Vitelleschi adds what seems to be an afterthought, but a better understanding of the phenomenon whereby water cools itself can be attained than that offered by Scotus.[58]

This teaching, as previously mentioned, is already contained in Menu and Valla, though not in the developed form found in Vitelleschi. Rugerius also has a rather full treatment, adding a few insights into how the elements might be moved by the heavens (Capreolus, he says, identifies the heavens with the generator, whereas Achillini and others identify them with the universal cause), but focusing his problematic around

[57] Ibid., fol. 376r.

[58] Ibid., fols. 376r-380v.

whether the elements are moved by the generator or not. The Greeks, Averroes, and Aquinas all answer this in the affirmative, he notes, whereas Scotus, Burley, and Gregory of Rimini hold the negative position. Rugerius himself takes the affirmative, but not in such a way as to deny that the elements are moved by their forms also. Peripatetically speaking, he concludes, it cannot be said that the form is the true and proper efficient cause of the motion of the element, and yet it cannot be denied that the element has within itself an active principle of its motion just as it has a passive principle. The remainder of his explanation is essentially that of Vitelleschi, though not proposed, as in the latter's case, as a position opposed to the teaching of St. Thomas Aquinas.[59]

c. MOTION IN A NATURAL PLACE

Another problem, no less interesting, and in fact even more relevant because of its being discussed explicitly by Galileo in his early *De motu*, is whether the elements undergo motions by reason of their gravity or levity when they are in their proper spheres, that is, in their natural places.[60] Valla has a good treatment of this question, and the main lines of his solution are taken up and developed by Vitelleschi and Rugerius. The topic is especially important because the principles employed by Archimedes to deal with bodies immersed in water seem contrary to a number of statements contained in the text of Aristotle. Thus anyone interested in applying Archimedes' teachings to problems of local motion, and particularly in reconciling Archimedes with Aristotle, would be obliged to address himself to such questions as whether air has weight in air, water has weight in water, etc., thus whether water will gravitate when in its natural place, and similarly for the other elements.

Valla poses the question in terms of the three elements that are regarded as in some way heavy, namely, air, water, and earth: do these gravitate in their own spheres? The opinions on this, he notes, are three: some say that air does not gravitate but rather levitates; others hold that air and water are heavy in a qualified way, and so all elements, fire alone excepted, gravitate in their own spheres; and finally there are those (including Ptolemy and Themistius) who answer completely in the negative, saying that no element either gravitates or levitates in its own sphere. To understand the question, he goes on, one must note that there are different ways of speaking of gravity and levity and also of the motions associated with them, frequently designated by expressions such as "to gravitate" and "to levitate." One could use both sets as meaning roughly

[59] Cod. SB Bamberg 62-4, fols. 94r-101v.

[60] Sec. 5.2.b; see also *Opere* 1:285-289.

the same and say, for example, that water gravitates because it is endowed with both gravity and levity but one of these, gravity, predominates; or one could say, alternatively, that water has only one motive quality but that it is heavy and thus gravitates in relation to air, because of the way these two elements are ordered. Again, one could separate the meanings and maintain that all elements simply remain in their own spheres without any tendency or inclination to another place, or that, even within their own spheres, they gravitate or levitate in second act, that is, they actually go into motion because of the motive qualities with which they are endowed. Valla's principal conclusions are based on his separation of the meanings, as follows: (1) all elements, regardless of where they might be located, retain their habitual motive powers even though they remain at rest; and (2) no element, when in its proper place, actually goes into motion as a consequence of its motive power(s), and thus, properly speaking, the elements do not gravitate when in their proper spheres.[61]

His second conclusion, Valla goes on, is the teaching of Archimedes in his *De ponderibus*, where he asserts that no body is heavy or light in itself and thus that water does not gravitate in water nor oil in oil. It can be proved by a variety of arguments, including that of Ptolemy, who pointed out that men under water feel no such gravity, for underwater swimmers sometimes descend to the bottom of the sea and have a great quantity of water above them, and nonetheless they do not sense any weight. Similarly, we live under air and do not sense its weight, despite the huge quantity of air over us. Another proof is that of Themistius, who argued that no element in its proper place is inclined to be outside of it, but that if the elements were to gravitate and levitate naturally in their proper places they would incline to be outside of them. As a confirmation of this, one should consider that the parts of the elements do not have place by themselves but only by reason of the whole as long as they are in the whole; but when the elements are in their proper places the whole element is in place; therefore there is no reason why the parts should levitate or gravitate. On the other hand, if parts become separated from the whole by some element of a different kind, then they do gravitate or levitate in order to be joined to the whole and to be in place by reason of it. Thus, if a part of earth is placed above water, the earth will descend and the water will ascend even though it was previously at rest; similarly, if a hole is made in the earth, a part of earth can descend in it, even though previously it did not descend, because it is now separated from the earth and a lighter element has gotten below it.[62]

[61] Cod. APUG-FC 1710, Tract, 5, Disp. 2, De elementis in particulari, Quaes. ult., An aer, aqua, et terra gravitent in suis spheris.

[62] Ibid., Secunda conclusio.

The contrary opinion, and especially the one holding that all elements, fire alone excepted, gravitate in their proper places, is attributed by Valla to Aristotle and Averroes, and in its support he offers five arguments, each of which he proceeds to refute. All of these are taken up by Vitelleschi in more detail, and thus their enumeration may be better considered in the context of Vitelleschi's discussion of the problem.

In this matter Vitelleschi takes essentially the same position as Valla, although he expresses it in different terms and relates it more fully to authors mentioned in Galileo's logical and physical questions—thereby taking on greater interest for the purposes of this study. There is no doubt, Vitelleschi begins, that the elements, while at rest in their proper places, retain their motive qualities and are truly heavy and light in first act. This is obvious because such qualities, flowing as they do from the substantial forms and natural dispositions of the elements, cannot be separated from the elements when these are in their natural states. Again, experience shows that the elements do not acquire their motive powers at the moment they start to move, but rather that the motive force they already possess goes from first to second act. Yet again, motive powers are in the elements not only that they may move to their proper places but also that they may come to rest in them. Therefore, while the elements are at rest they retain their motive qualities, and this is beyond question. The problem is whether the elements, when in their proper places, have *gravitatio* and *levitatio*, that is, some second act or effect of these qualities, so that they can be said to gravitate and levitate and not merely to be heavy and light. For there is a threefold effect that can be attributed to motive qualities: the first is the motion they cause to their proper place when nothing impedes this; the second is a tendency to motion when the latter is impeded, such as one senses when carrying weights on the shoulder, for although they do not move downward they gravitate and press down on the shoulder; and the third effect is to remain immobile and at rest at the terminus of a natural motion, for when the proper place has been reached the motive quality connaturally resists being drawn from it. It is these effects that must be examined carefully if one is to answer whether and how the elements gravitate and levitate in their own spheres.[63]

It seems, Vitelleschi goes on, that the elements gravitate and levitate in the second way just noted. In support of this he gives nine arguments, most of them drawn from Aristotle's *De caelo* or from the *Questiones mechanicae* then commonly attributed to Aristotle. (1) Wood of a hundred pounds is heavier in air than lead of one pound, whereas in water such wood floats whereas such lead sinks, and therefore the former is lighter

[63] Cod. SB Bamberg 70, fol. 369r-v.

than the latter. The cause of this is that while the wood is in air it is impelled downward by the weight of the earth, water, and air itself, whereas while it is in water it is pushed down by the weight of the earth and water but not of air; and because lead of one pound has more weight from water and earth than wood of a hundred pounds; therefore it gravitates more in water. (2) A bladder inflated with air has greater weight than when empty; therefore air adds gravity to it. (3) If air is taken away water descends naturally, and similarly air if water or earth is taken away; on the other hand, if fire is taken away air does not ascend naturally but by force, and similarly water if air is taken away; therefore air and water gravitate in their own spheres. (4) Shouts are better heard from a lower place than from a higher, and the reason is that air of the voice contains moisture which makes it tend downward and not rise up. (5) Some try to explain why the entire mass [of dough] is lighter than the water and wheat from which it is made by saying that air is entrapped during the mixing and this makes the mass lighter; but Aristotle rejects this explanation, saying that even though the air be mixed with the water, this does not make the mass lighter but heavier; therefore air gravitates in air. (6) Animal spirits are of air and yet they do not rise from the body of the animal; therefore they have gravity. (7) If someone tries to move earth from its proper place it resists, and so do fire, etc.; therefore these elements gravitate and levitate. (8) The parts of earth that are closer to the center are denser, for they are compressed by the parts above them that push downward. And (9) experience shows that an empty vase floats in water whereas a full one sinks; the cause of this cannot be other than the gravity of the water, which moves downward even in its proper region.[64]

Before proceeding to his evaluation of these and similar arguments, Vitelleschi first notes that there are only two main opinions on this. The first is that of Aristotle, who holds that all elements except fire gravitate in their proper regions, and in this he is followed by Averroes, Nifo, the Conciliator of Abano, Gentile, and Borro in his treatise *De motu gravium et levium*. The second opinion denies that the elements gravitate and levitate in their proper places, and here Vitelleschi names Themistius, as cited by Averroes; Simplicius, who also cites Ptolemy; and Syrianus Magnus; then he names additionally Archimedes in his *De ponderibus*, Philaltheus, Hugo, the *Parisienses*, and Vallesius; and he concludes with the cryptic observation that this is the common opinion of the peripatetics, thus (perhaps unintentionally) excluding the Averroists *en masse* from the peripatetic tradition.[65]

Vitelleschi's own conclusions follow directly from the threefold way

[64] Ibid., fols. 369v-370r.

[65] Ibid., fol. 370r.

in which he has said the terms *gravitatio* and *levitatio* can be understood and thus can be expressed in three different propositions.

The first conclusion is that, if *gravitatio* and *levitatio* refer to the elements' remaining in their proper places and resisting attempts to remove them forcibly therefrom, then fire can be said to levitate, earth to gravitate, and the intermediate elements (air and water) to both levitate and gravitate. Vitelleschi's argument here is that it is natural for the elements to conserve themselves and resist efforts to dislodge them, since this is one of the reasons why they have motive qualities; thus it is natural for fire to resist by levitating, earth by gravitating, and the intermediates by gravitating if urged toward a higher region and by levitating if urged toward a lower. Another argument is that *gravitatio* and *levitatio* are effects of gravity and levity, and nonmotions may be included in these effects; thus, if "to gravitate" and "to levitate" can be used to designate motions, they may also be used to designate resistances that arise from the same source.[66]

His second conclusion is that, if *gravitatio* and *levitatio* are taken to refer to motions up and down, then the elements neither gravitate nor levitate when in their proper regions. The proof is that if the elements were to do this, it could only be so that they might end up under the lighter and above the heavier, but this is what they have already accomplished by being heavy and light. Thus, when in their proper places they neither gravitate nor levitate *per se* in this sense, though they might do so *per accidens*, for example, if impelled in one direction or another by an external agent, or should they become heavier or lighter, the way in which vapor ascends from water.[67]

The most difficult problem to resolve, of course, comes when *gravitatio* and *levitatio* are taken for neither rest nor motion but for the tendency to motion explained above as the second effect that can be attributed to the motive qualities. This Vitelleschi replies to negatively in his third conclusion: no element gravitates or levitates in its proper place when these expressions are taken to designate tendencies to motion. He offers various proofs of this, the first being that a *gravitatio* and *levitatio* of this type is a tendency or propensity toward a natural place, and this does not exist when the body is in its natural place. If it had such a propensity while in its proper region, it would tend to another place and then would not rest naturally where it was, namely, in its natural place. Again, the same nature that is the principle of gravitational motion when the body is outside its natural place is also the principle of rest when it is in that place. Another proof is that if everything but fire were to gravitate,

[66] Ibid., fol. 370r–v.

[67] Ibid., fol. 370v.

nothing would go upward by its nature but only when propelled there from below, and this would be contrary to the teaching already sketched when discussing gravity and levity. Yet another is that if air were to gravitate, the same thing would feel heavier if put in a large vase than in a smaller, because the former contains more air. Then there is the argument from our not feeling the weight of the air surrounding us and from underwater swimmers not feeling the weight of the water pressing on them; this gains greater force from the fact that we feel the weight of a body when supporting it by hand in air, whereas we might not feel it in water. Why is it, Vitelleschi asks, that when we draw water from a well we do not feel its weight as long as the container is in the water, yet we do feel it as soon as the container leaves the water? Such would not be experienced if water gravitated in its proper place.[68]

After proposing these and other proofs Vitelleschi returns to the nine arguments drawn mainly from Aristotle's texts and gives lengthy replies to each in ways consistent with the three foregoing conclusions. It would be tedious to consider all of these in detail here, but the first two and the ninth are of interest because of their relevance to Galileo's early drafts on motion.[69] To the first Vitelleschi disagrees with the causes assigned for the different motions of the wood and the lead in air and in water, and for the proper explanation refers the student to Scaliger (Exercitatio 143, n. 1) and to Archimedes in his book *De iis quae aqua invehuntur*. To the second he questions the facts as they are proposed: Themistius, he says, denies that this takes place; Ptolemy found that the inflated bladder is lighter, whereas the Conciliator found the opposite; and Simplicius discovered the same weight whether the bladder was inflated or not. Averroes contended that it is difficult to perform the experiment properly, and St. Thomas assigned the reason for this: sometimes air is heavier, sometimes it has light exhalations mixed with it; and sometimes it stands midway between the two. Moreover, it can happen that the bladder seems heavier when inflated because it is inflated with the breath of the mouth, as Simplicius notes, and since this is denser, more watery, and perhaps causes water to condense in the bladder, it actually is heavier. As for the ninth argument, Vitelleschi replies that the sinking of the vase does not come about because the water that fills it is gravitating, but because the air that was in it when empty no longer supports it when full; and still, not every vase that is full of water will sink, but only one made of a material that does not float in water. One should not be concerned, he concludes, with what Aristotle has held on this, since practically the entire peripatetic tradition has deserted him here, and it

[68] Ibid., fols. 370v-371r.

[69] Sec. 5.2.

is safer to abandon him than try to interpret him, since his authority should be used to confirm the truth, not abandon it, considering that truth is the goal of all philosophers.[70]

d. VELOCITY OF FALLING MOTION

A final problem relating to gravitational motion is that concerning the velocities with which things fall, and particularly why falling motion is faster at the end than at the beginning. Both Vitelleschi and Rugerius devote questions to this topic, which is discussed at length by Galileo in his early *De motu*.[71] Vitelleschi also has a preliminary treatment of velocity when analyzing the motive qualities of the intermediate elements, and since this supplies a good background for the discussion of acceleration, it offers a convenient starting point for what follows.

The context of Vitelleschi's first remarks on the velocity of motion of the elements occurs in a question where he is attempting to determine whether the motive qualities of the intermediate elements, air and water, are composites of those of the extreme elements or whether they are simple and specifically distinct in their own right.[72] The differentiation of motive qualities, he states, will be based on the differentiation of the motions to which these give rise, and such motions in turn will be different according as they are directed to different termini or effected with different velocities. Thus fire and earth go upward and downward respectively with very great velocity, while air and water do so more slowly. To understand how gravity and levity effect this velocity difference in motions, he goes on, one must understand that these motive qualities can themselves be compared in two ways: intensively, because one has more degrees and produces a greater effect than the other; and extensively, because the body in which they are has more parts of the same kind and with the same degree of intensity. Some think that greater or less velocity in downward motion arises from a greater or lesser gravity understood intensively alone, and that extensive differences have no effect. But Vitelleschi has serious doubts about this, particularly because of what Aristotle says about greater extensive gravity producing a greater velocity. Experience also teaches that in other actions the union of more parts makes for greater velocity of action, so perhaps this is true in local motion as well; for example, a thousand men can pull a ship a certain distance, whereas one man alone could not move it a thousandth of the distance. And yet others say that the contrary is shown by experiments,

[70] Cod. SB Bamberg 70, fols. 371r-373r.

[71] Sec. 5.2.b-c.

[72] Cod. SB Bamberg 70, fols. 363v-369r.

to wit that the smallest stone falls faster than an enormous piece of wood, provided that the stone is not so small as to be moved by the air and that the shape is not notably different, for shape makes a considerable difference in velocity. Here Vitelleschi refers to Thomas Bradwardine in his treatise *De proportione motuum* and to Joannes Thaisnerus in his treatise on the same subject; the latter, he says, claims to have demonstrated it. This is an interesting observation on Vitelleschi's part, for it shows that he was acquainted with Benedetti's work on falling bodies, with which Galileo's early *De motu* is thought to have some affinity, and which was plagiarized by Thaisnerus and presented under this title.[73] Vitelleschi goes on to remark that in this matter he does not have any certain experiments, and those that he has seem to go contrary to Aristotle. In such a matter, he adds, the data of experience are of crucial importance, and yet there are many things that make him suspect all experiments of this kind. For one, it seems hardly possible that the shapes of the bodies that move, the resistances of the media, and all other things that influence their motions should be equal. For another, one cannot perceive a difference in velocity of this kind unless the motion takes place over a very great distance, but over such a distance many things can happen that detract from the certitude of the experiment.[74]

However this may be, Vitelleschi maintains that it seems certain that if one of two bodies is heavier intensively, and if the two are of the same size, shape, etc., and other things are equal, the heavier body will fall faster. Similarly, if one body appears heavier than another, not only in air and light media but also in water and heavier media, one can be certain that it is heavier intensively, unless something else nullifies the experiment. For example, despite the fact that a large piece of wood seems in air to be heavier than a small piece of lead, if the lead sinks in water and the wood floats one must conclude that the lead is intensively heavier than the wood. This, therefore, is Vitelleschi's rule for recognizing intensive gravity: any body that approaches the center to a greater degree, or ends up under all others, and this is shown by certain experiments, must be regarded as having greater intensive gravity. Bodies that are extensively heavier, on the other hand, being generally larger, do not move through the medium as easily insofar as the latter offers more resistance to their parts. Perhaps this explains why some bodies are held

[73] For details of Benedetti's work and the plagiarized version of Thaisnerus, see Carlo Maccagni, *Le speculazioni giovannili "de motu" di Giovanni Battista Benedetti*. Testimonianze di Storia della Scienza, 5 (Pisa: Domus Galilaeana, 1967). See also secs. 5.1.c and 5.2.b.

[74] Cod. SB Bamberg 70, fols. 363v-365v; for a summary account of Vitelleschi's teachings as referenced in this and the following notes, see Wallace, *Prelude to Galileo*, pp. 114-120, 312-315.

up by the medium: for example, lead in the form of a ball will sink in water, whereas when stretched into a long filament it can be made to float. Here Vitelleschi references Cardanus and Scaliger for fuller details, and then speculates as to why no metal except gold sinks in mercury, drawing on Agricola and Fallopius for possible answers to this question.[75]

This entire discussion is apparently a digression from the problem at hand, namely, the specification of the motive qualities of the elements. Vitelleschi therefore returns to this and observes that motive qualities can be more or less intense in two ways: within the same species, because there are more or less degrees of quality of the same kind, for example, a greater gravity of water; and combining different species, when degrees of qualities of different kinds are conjoined in the same subject, for example, a particular levity of fire with a similar levity of air. In the latter way, although in a particular compound the gravity of one element might be very weak, the overall gravity would be intense if this were added to a stronger gravity of a different kind, say the gravity of earth, which is capable of producing a greater effect than the gravity of water. And in this second way, particularly when speaking of compounds, one can say that their motive qualities are more or less intense. With regard to the elements, and Vitelleschi concludes with this after yet further discussion, the motive qualities of the intermediate elements are specifically distinct from those of the extreme elements, since they move the intermediate elements to distinctive termini, even though they need not always do this with the same velocity because of the many other factors that affect the resulting motion.[76]

Among such factors, resistive forces are obviously of considerable importance, and Vitelleschi discusses these in several different contexts. The first is when analyzing the regularity to be found in the motions of the non-living, for the movements of the elements, he observes, are not completely regular—seeing that projectiles slow down and falling bodies speed up as they move. Vitelleschi thereupon enumerates five different causes that can produce such irregularities in the velocities of moving bodies. These are: (1) the mover, as this has greater or less force to effect motion; (2) the thing moved, as this more or less resists the mover; (3) the thing moved again, as this is also endowed with more or less motive power; (4) the medium through which it moves, as this is thicker or thinner and so variously impedes the motion; and (5) other factors, particularly the shape of the moving body, which makes it more or less suited to cut through the medium.[77] All of these factors can be brought

[75] Cod. SB Bamberg 70, fols. 365v-366r.

[76] Ibid., fols. 366r-367v.

[77] Ibid., fols. 257r-258r.

to bear on the analysis of the velocities involved in gravitational and projectile motion, as will be seen in the development to follow.

Vitelleschi's treatment of the composition of motions is another context wherein resistance enters incidentally, though with a slightly different nuance. He begins this topic by noting that, according to Aristotle, some motions are simple whereas others are composite. To explain how simplicity and composition can be applied to motion, he then proposes to consider three different factors related to these: the motion itself; the power that causes it; and the distance over which or around which it takes place. With regard to the third, the distance traversed, Vitelleschi takes the position that straight and circular distances are simple, whereas others made up of these—that is, those partly straight and partly circular—are composite. Similarly, with regard to the first point, the motion itself, the downward motion of an element is simple, whereas the progressive motion of an animal is composite. There remains then the second point of comparison, the power causing the motion, and from this viewpoint a motion is simple if it comes from a simple power, such as gravity or levity, composite if it comes from several powers, say, from gravity and levity at the same time. The first kind of motion obviously characterizes the elements, for these have only one motive power; compounds, on the other hand, can have two powers—for example, different degrees of gravity and levity—as he has already explained. In the case of compounds, moreover, a particular element always predominates, with the result that the motive power of the predominant element exceeds any power opposed to it, and an effect is produced that is really the result of interacting forces. This opposition of motive powers within compounds can therefore be seen as generating a type of internal resistance that affects the resulting motion and renders it composite.[78] In the case of a compound, Vitelleschi notes elsewhere, its substantial form is the primary internal cause of its motion, though the motive quality of the predominant element serves as its instrument. So he maintains that the form of the compound, much like the elemental substantial form, has the force of effecting its motion through a motive quality it derives from the form of its predominant element.[79]

All of these matters are brought together by Vitelleschi in his discussion of what makes falling bodies move faster as they fall. He takes it as a fact of experience that when heavy and light bodies are moved by nature they move more rapidly at the end of their motion than at the beginning, whereas when they are moved from without by force they move more rapidly at the beginning and more slowly at the end. With regard to the

[78] Ibid., fol. 259r-v.

[79] Ibid., fol. 131r-v.

natural motions, the following seem to him to be the possible explanations of the change in speed: since two things are involved, the body moved and the medium through which it moves, either the cause of the velocity increase is extrinsic to the thing moved and is to be sought in the medium, or it is in the body itself, and then it arises either from the fact that proximity to its proper place increases the motive power or because some power that opposes the motion is gradually diminished. Which of these possibilities Vitelleschi regards as correct can be seen from the way in which he discards the various alternatives. The first to be eliminated is the explanation favored by Galileo in his early writings: contrary to Hipparchus and others, the velocity increase is not caused by a gradual overcoming of the residual opposing force left in the body.[80] Again, the greater velocity does not arise from the motive power's being increased or strengthened as the body gets closer to its natural place. Similarly, contrary to Iamblichus and Syrianus, the velocity increase is not traceable to a decrease of resistance on the part of the medium, at least not in the sense that the medium becomes more easily separable or because the air closer to the earth is not as light and so offers less resistance to the descending body, as Durandus thought. Yet Vitelleschi admits that the medium plays a significant role in explaining the acceleration, provided it is taken to be in interaction with the falling body itself.[81]

The explanation he finally prefers, following Zabarella, is that the earlier part of a body's fall causes a greater velocity in the later part of its fall, because it then causes the medium to resist less. The basic mechanism is easily understood on the analogy of a body moving against a flow of water; obviously the body will move more quickly if the water is at rest, and more quickly still if it moves in the same direction as the flow. In a similar way, the falling body propels the first part of the medium, and then when it reaches the second part it propels that more quickly in the same direction, and the third part more quickly still; as a result the medium comes gradually to impede the movement less and less. And since the velocity of any motion results from an excess of the motive force over the resistance encountered, the velocity of the motion increases as the resistance grows less. Stated otherwise, since the resistance of the medium continually decreases while the motive force remains constant, the difference between them increases and with it the velocity of fall. In the case of projectile motion, on the other hand, the motive force is always being weakened and lessened, and so it cannot effect any decrease in the resistance of the medium; that is why its motion

[80] Sec. 5.2.b.

[81] Cod. SB Bamberg 70, fols. 380v-382v.

is more rapid not at the end but at the beginning, for then its motive power is the stronger.[82]

Vitelleschi admits a difficulty with this explanation in that Aristotle seems to claim that the greater velocity at the end of falling motion results from an increase of gravity. His reply to this is based on a distinction already explained. Gravity can be taken in two ways: in first act, usually referred to as *gravitas* and designating the motive power that produces the motion; and in second act, generally referred to as *gravitatio* and designating the motion that results from it. *Gravitatio*, as Vitelleschi sees it, is nothing more than the excess of the motive force over the resistance that opposes it. Therefore, even though the *gravitas* remains the same throughout the fall, the *gravitatio* increases, and this is what Aristotle means. In other words, the velocity increase comes from the excess of the motive power over the resistance, and this excess derives not from an increase of gravity in first act, but rather from a decrease in the resistance encountered as the body falls.[83]

Rugerius, as already noted, takes up this same problem when inquiring how the motions of heavy bodies are related in terms of velocity at the beginning, the middle, and the end of their motions. The most significant difference between his and Vitelleschi's treatment is that Rugerius's is more eclectic, while incorporating most of the arguments already found in his predecessor's account. Yet Rugerius does introduce an element not found in Vitelleschi, namely, he ties this particular question with Aristotle's discussion of the rules or theorems relating to the comparison of motions with respect to speed and slowness in the seventh book of the *Physics*. Rugerius points out that some of these theorems labor under severe difficulties, and for a fuller discussion of the ways in which they might be revised one should read Toletus and Domingo de Soto, among others.[84] The reference to Soto is, of course, especially felicitous, for of all the scholastics Soto is the only one who held explicitly that falling bodies accelerate uniformly as they fall, that is, that their motion is *uniformiter difformis* with respect to time, thereby adumbrating the law of free fall as this was to be adopted by Galileo years later in his *Two New Sciences*.[85] Another difference between Vitelleschi's and Rugerius's accounts is that the latter makes much more use of the *Quaestiones mechanicae* in analyzing the various factors than does the former, and so ties

[82] Ibid., fols. 383r-384r.

[83] Ibid., fols. 384v-385r.

[84] Cod. SB Bamberg 62-4, fol. 101r-v.

[85] Secs. 5.4.c-d and 6.3.d; for fuller details, see the essays in part 3 of Wallace, *Prelude to Galileo*, especially pp. 99-109.

this question, usually only argued in the context of materials found in the *Physics* and the *De caelo*, with the science of mechanics as this was being developed among contemporary mathematicians.[86]

The natural motion of heavy bodies, Rugerius states, is fastest at the end; this is in opposition to their violent motion, which is slowest at the end, and to the motions of projectiles and of animals, which are faster at the middle than at the beginning. Two experiments show this, namely, water falling from downspouts or dripping off tiles, and one stone falling from a greater height than another. Regarding why this is so, Rugerius observes that as many causes have been assigned as there are philosophers. In his view the argument of Hipparchus is not satisfactory because it is only valid in cases where natural motion follows immediately after violent motion, and not for the general phenomenon that is to be explained.[87] Again, it depends on the *virtus impressa*, about whose existence Rugerius expresses some doubts, as will be seen presently. Nor is the argument that the motive quality increases as the moving body approaches its natural place satisfactory: a body falling from a high place should fall faster than one from a low place, but according to this explanation they would be equally close to their natural place and thus fall at the same speed. Nor is Iamblichus's argument that the medium resists less at the end of the fall, for the same reason; similarly to be rejected is Durandus's explanation that air closer to the earth is not so light and thus works less against the motion. Rugerius's own theory is more Averroistic than Vitelleschi's: the basic cause of the phenomenon is the medium, which propels the falling body from behind when it itself is impelled—similar to the traditional peripatetic explanation of projectile motion. And, since the medium is moved by the falling body, the medium's motion gives more and more assistance to the body's fall as it progresses, rather than steadily impeding its downward motion. There are many other factors that promote the velocity increase, Rugerius concludes, and among these he enumerates five: (1) the medium is cut through more easily toward the end of the motion; (2) the medium is broken up more throughout the fall and does not resist as it did at the beginning; (3) there is nothing that would prohibit gravity and levity being reinforced by proximity to natural place; (4) should there be such a thing as the *virtus impressa*, this would probably be generated by the fall itself and thus increase throughout the fall; and (5) if an extrinsic *virtus* were work-

[86] See sec. 4.4.

[87] Cod. SB Bamberg 62-4, fols. 112v-113v. Galileo's explanation of the increase of speed of fall in his *De motu antiquiora* offers a similar critique of Hipparchus's theory; see sec. 5.2.b *infra* and note 57 to chap. 5.

ing against the downward motion of the body, this would diminish the longer the action continued against it, along lines suggested by Hipparchus.[88] Thus Rugerius has effectively made a synthesis of all the explanations that have been proposed, but without making any substantive advance as to the mechanism that causes the velocity increase or how this might be described in quantitative terms.

3. Projectile Motion

The final problem relating to local motion that was treated at some length by professors at the Collegio Romano was that of the projectile, namely, what is the cause of the projectile's continued motion after it leaves the projector, seeing that this is a violent motion and cannot be explained in terms of natural causes. Menu, Valla, and Rugerius all address this question at some length, the first two proposing solutions that are quite similar to that advanced by Galileo in his early *De motu*,[89] and the last returning to a more traditional peripatetic explanation. Vitelleschi's question devoted to this topic is no longer extant, though from the way Rugerius treats the subject one can surmise that Vitelleschi took an intermediate stance between him and Valla on it. In what follows, therefore, major attention will be given to the teachings of Menu and Valla, after which an attempt will be made to situate Vitelleschi and Rugerius against the background the former provide.

a. MENU

Menu's discussion occurs in his question on the violent motion of heavy and light projectiles, at the outset of which he refers the student to the works of Themistius, Simplicius, Philoponus, Albertus Magnus, Buridan, Albert of Saxony, Gratiadei, Paul of Venice, Scaliger, Domingo de Soto, and Pererius for fuller treatments of the subject. The two competing explanations he proceeds to juxtapose are that the projectile is moved by the medium, that is, by the air through which it is thrown, which is the traditional peripatetic view, and that the projectile is moved by a *virtus impressa* or impetus, the view of the later scholastics. Menu decides that both are correct, and so comes to two conclusions: it is probable that projectiles are moved by the media through which they

[88] Cod. SB Bamberg 62-4, fols. 113v-114v.
[89] Sec. 5.2.b.

pass; but it is also probable, and indeed more probable, that they are moved not only by the medium but also by some quality such as a *virtus impressa* that inheres in them.[90]

That the projectile is moved by the air, says Menu, can be seen to be true provided one understands the mechanism that brings this about. The thrower impels the air by his hand and communicates to it a *vis impellendi* that spreads through its various parts; this results, as explained by Giles of Rome and Walter Burley, from alternate compressions and expansions, and in the process the *vis* loses its force because it spreads out like a wave, which serves to explain why the violent motion ultimately ceases.[91]

Arguments that are brought against this type of explanation, as Menu sees it, are eight in number: (1) effectively the problem has been transferred from the projectile to the air, and then the question remains as to what moves the air; (2) violent motion in a void would be ruled out because of the absence there of air or another medium; (3) on this explanation, a strong contrary wind should stop the projectile; (4) the nature of air works against the explanation, since air diffuses and does not push directly; (5) again, air gives way and does not impel; (6) a feather would be projected farther than a heavy object, and this is not the case; (7) many experiments cannot be explained through the motion of the medium, for example, (i) why a ball thrown to the ground bounces back; (ii) why a sling throws a stone faster and farther than the hand; (iii) why a thrown ball hurts less when it impacts closer rather than farther away; (iv) why a taut string continues to vibrate after being struck; (v) why a top, wheel, or hoop continues to rotate so rapidly; and (vi) why air close to a stone, when moved, does not move the stone; and finally, (8) the proposed explanation does not say what air moves the projectile, namely, that preceding it, or following it, or traveling along by its side.[92]

For all of these, Menu attempts an answer, usually brief and yet consistent, more or less, with the two conclusions already noted. In particular: (1) it is true that the problem has been transferred from the projectile to the air, and thus a *vis* must be impressed on the air; (2) this is also true, since according to this explanation violent motion would be im-

[90] Cod. Ueberlingen 138, In libros De generatione, Tract. 4, Sect. 1, Disp. 5, Cap. 6. De motu violento proiectorum gravium et levium. The entire teaching of this *capitulum* is summarized in Wallace, *Prelude to Galileo*, pp. 325-330; the Latin of Menu's references and his two conclusions is given on p. 338, notes 8-9.

[91] Cod. Ueberlingen 138, De generatione, Tract. 4, Sect. 1, Disp. 5, Cap 6, Notandum secundo. . . .

[92] Ibid., Quod non moveantur ab aere. . . . These arguments are given at the beginning of the *capitulum*, immediately following the citation of authorities. They are numbered here in a distinctive way so that they may be correlated with the responses given to them in

possible in a void; (3) a strong contrary wind would not necessarily stop the projectile, since it can receive a greater *vis* from the impelling air than from a contrary wind; (4) far from working against the explanation, it is the nature of air that permits the impelling mechanism to work; (5) here one must understand that the swiftness of the motion makes the air press harder and so impel the projectile; (6) the problem here is not with the mover but with the feathers, which are hard to move anyway; (7) the experiments adduced are all capable of different interpretation, to wit: (i) the ball bounces back because the air, too, is reflected back and pushes upward; (ii) the sling moves the stone faster because it has more leverage, according to the principles explained in the *Quaestiones mechanicae*; (iii) the ball's motion is faster at the beginning, but later the motion is more level and uniform and thus produces the greater impact; (iv) the taut string continues to vibrate because it moves the air, which continues to move it; (v) air can assist the motion of the wheel, and it can also explain why any rotating object comes to rest; and (vi) air pushed from behind cannot be impelled as well as a stone pushed directly; and (8) the air impels the projectile from behind, according to the mechanism adopted when explaining the conclusion.[93]

Menu's second thesis, as already noted, is that it is also probable, and indeed more probable, that the projectile is moved not by the air alone but by some quality such as a *virtus impressa*. For this he offers eight proofs, some of which are obviously based on the arguments he has listed against his first conclusion. These are all brief, and may be summarized as follows: a lance with a sharp cone at its rear cannot be impelled by air alone; if a cloth or fan of some kind is used to shield the air from a rotating wheel, the wheel still continues to rotate; air is easily divisible and gives way without effort, and so it is incapable of supporting weights of three pounds or more; a stone suspended by a thread is not moved by air agitated in front of it; the experiments already discussed are more readily explained by a *virtus impressa* than by the motion of the medium; a man whom one hits with the fist is not moved by the force of the air; even air cannot be compressed without some type of quality being impressed upon it; and if a circular segment is removed from a tablet and then inserted back into the hole it previously occupied, it can be made to rotate and will continue to do so, even though there is no space for the air to act upon its circumference. The last proof, it may be noted, is taken from Scaliger, for it uses the same words and the same Latin

the following paragraph. A similar numbering technique is employed for the sets of objections and replies referenced at notes 95 and 96, *infra*.

[93] Ibid., Ad argumenta. . . .

constructions as are found in the latter's text, which are quite different from those normally employed in the Collegio lectures.[94]

Against this conclusion Menu has also listed seven different objections, which may be abbreviated in the following terms: (a) such a doctrine was not taught by Aristotle and the peripatetics; (b) a difficulty presents itself on the part of the thing moved, for, when a *virtus impressa* is invoked, either the projectile would be moving from an internal principle and thus naturally, or it would be moved violently and yet by itself, which implies a contradiction; (c) more serious difficulties arise from the nature of qualities, to wit: there is no species of quality in which such a *virtus* can be placed, for its existence would posit two contraries opposed to only one natural quality (e.g., *levitas* and the upward *virtus*, both opposed to *gravitas*); or the *virtus* would be natural and thus a property of the body; or it would offer resistance and thus be produced in time with greater or less intensity, and so could not come to be instantaneously; or it would actually be *gravitas* or *levitas*, and there would be no way of distinguishing it, for example, from the *levitas* of fire, since it would exhibit the same properties; (d) there are other problems associated with the way in which such a *virtus* would be generated, for example: no quality can be produced by local motion alone; or the quality would be acquired by squeezing a stone in one's hand, which in fact does not happen; or an infinite number of entities would be continuously generated, and this is absurd; (e) the manner of its generation is also difficult to understand; for such a quality would have to be produced throughout the entire projectile, and this is opposed to its being indivisible; or it would exist only in part of the projectile, and this is opposed to the *virtus motiva*'s moving the whole; (f) there are difficulties arising from the motion's being violent: for on this explanation the projectile's motion would be one and continuous, contrary to Aristotle; or there would be no reason why its motion should be faster [in the middle] than at the beginning, again contrary to Aristotle;

[94] Ibid., Dicendum secundo. . . . Menu's text reads as follows: "Octavo, si levissima tabula ex qua eximatur orbis trochus ita ut sine ullo attritu mutuo intra illud cavum circumduci queat, deinde tabula alicui affixa vectem cum manubrio illi orbi infigito quod manubrium singulis utrinque furcillis sustineatur, tunc manifeste videbis circumactum iri intra illud spatium orbiculatum nullo aere impellente." In his *De subtilitate ad Hieronymum Cardanum*, Exercitatione 28, Scaliger writes: "Sit levissima tabula, ex qua eximatur orbis torno, aut circino incidente: ita ut sine mutuo attribu orbis ille intra illud cavum circumagi queat. Tabula igitur alicubi defixa, vectem cum manubrio illi orbi infigito. Quod manubrium singulis utrinque furcillis sustineatur. Tunc manifeste videbis, circumactum orbem intra illud spatium tabulae orbiculatum, moveri a moto motore, nullo aere impellente."—Lyons: A. de Harsy, 1615, p. 103. The *De subtilitate*, it may be noted, was first published in 1557.

and finally, (g) motion in a void would be possible, yet again contrary to Aristotle.[95]

For each of these arguments Menu has a reply, and the responses are most helpful for clarifying his understanding of impetus and how it is produced. (a) With regard to the Aristotelian tradition in this matter, for example, he passes over the complex problem of ascertaining what Aristotle himself actually held, noting that both Themistius and Simplicius placed a *virtus impressa* in the air, and that since putting it in the air is little different from putting it in the stone, Philoponus expressly located it in the projectile—also to allow the possibility of motion in a void. Menu here attributes Philoponus's teaching to St. Thomas, the *Doctores Parisienses*, Albert of Saxony, Paul of Venice, Buridan, Scaliger, and others. (b) With regard to the naturalness of the motion, he explains that local motion is not always natural, unless it comes from a form that is natural to the object moved and exists in it; also the projectile does not move itself, since the projector moves it by means of the *virtus impressa*; again, the *vis* is not a property of the stone but rather is the instrument of the projector, much as is the magnetism impressed on iron by a magnet. (c) On the nature of the quality involved, Menu claims that the *virtus impressa* is reducible to a disposition of the first species of quality, since it is *facile mobilis*; also, there is no incompatibility in the *gravitas* of the stone having two contraries, provided one is natural and the other violent; again, the *virtus* is an imperfect being and as such does not have a perfect *esse*, thus does not have to be natural to the stone; moreover, this particular *virtus* is like a spiritual quality, and so can be induced instantaneously; yet again, such a quality, properly speaking, is not contrary to the *gravitas* of the stone, especially in act, for two reasons: contraries have to be of the same kind, but the impressed quality is an intentional quality whereas the *gravitas* of the stone is a material quality; and if it were a contrary, it could not be impressed on the stone without diminishing its *gravitas*, and this is contrary to experience; finally, such a quality is itself neither heavy nor light, but rather a type of force that is able to move up or down or back or forth, just as the magnetic quality induced in iron or the force that moves an element to fill a vacuum. (d) Respecting the generation of the quality, Menu answers that it is possible for local motion to produce an imperfect quality, such as occurs when sound results from rubbing or friction; the impressed force, moreover, is produced by the *virtus motiva* of the projector, which is lacking in the

[95] Cod. Ueberlingen 138, De generatione, Tract. 4, Sect. 1, Disp. 5, Cap. 6, Ex alia parte non videtur fieri a virtute impressa. These arguments follow directly after those already referenced in note 92, *supra*.

pressure of squeezing; and there is nothing to prevent imperfect entities from being generated a great number of times. The remaining difficulties are answered similarly by him: (e) the quality is produced throughout the entire projectile, and this is possible for a quality that has a spiritual and intentional character, similar to magnetism; (f) the inference is correct, for the projectile's motion is continuous and uniform, and it is rather the contrary that is absurd; again, the greater velocity comes from the motion's being more level and more uniform, as already explained, and not from the force that moves the projectile; and (g) the conclusion is correct and is to be conceded, for motion in a void is truly possible.[96]

Menu concludes this exposition of the projectile problem by appending two remarks. The first clarifies how the *virtus impressa* can explain the bounce of a ball—not by having it corrupt at the point of impact but by having it reflected, a phenomenon not uncommon with intentional species. The other explains that the *virtus* does not corrupt instantaneously, even though it comes into being in this way, since it is an imperfect quality that gradually weakens once it has come into existence.[97] All of these observations, it may be noted, are quite consistent with the view of impetus advanced by Galileo in his *De motu antiquiora*, as will be explained in the next chapter.

b. VALLA

Valla, as already noted, has the fullest treatment of impetus of any of the four Jesuits whose views on local motion are being examined. There are noticeable similarities between his lecture notes on this subject and those of Menu, which gives good indication that the latter's materials were probably available to him when preparing his own presentation. Although Menu's ideas are generally endorsed by Valla, however, there is also an independent development, for Valla's explanations are fuller and better thought out, and he is even more sympathetic to the acceptance of the *virtus impressa* than is Menu. There are also similarities between his terminology and that of Vitelleschi, which suggests that perhaps there are lines of influence between them, to be indicated shortly.

Valla begins his question on projectiles with the notation that the term "projectile" means anything moved from without in such a way that the mover does not accompany the object moved, but first pushes it and afterwards no longer touches it. The problem then is this: granted that the projector is the mover of the projectile, what is the instrument through which he moves it after he no longer touches it? Is it some *virtus impressa*

[96] Ibid., Ad argumenta. . . . [*bis*]

[97] Ibid., At dicas . . . Dicas secundo. . . .

inhering in the projectile, or the action of the medium through which the projectile moves, or something else entirely? In order to appreciate the problem, Valla notes that there are a number of difficulties that have to be clarified relating to the concept of *virtus impressa* and to the way in which the medium might move the projectile, and he discourses on these before giving his conclusions and the arguments in their support.[98]

The observations on the nature of *impetus* and its mode of generation and corruption are particularly significant. The *virtus impressa*, for Valla, is an imperfect quality after the fashion of an intensional or spiritual form, somewhat like light and color, and as such has no contrary in first act although it has a contrary in second act. Because it has no contrary in first act it can be introduced in a stone instantaneously. When the stone is thrown upward, the *virtus* is not contrary to the stone's *gravitas* in first act, though it is contrary to it in second act; as a consequence the stone's *gravitas* remains in the stone, even though it is impeded by the *virtus* in second act, that is, in its motive effects. That the two are not contraries in first act is obvious from the fact that *gravitas* is a natural quality whereas the *virtus impressa* exists after the fashion of an intensional or spiritual quality. One need not fear, therefore, that if a massive *impetus* were imparted to the stone sufficient to impel it all the way to the inner concavity of the orb of the moon, the stone would lose its natural gravity; it would still retain the latter in first act, though its second act would be impeded by the action of the *virtus*.[99]

The problem of locating the *virtus impressa* in a species of quality is solved by Valla in two ways. According to Marsilius of Inghen, he says, it is reducible to the first species of quality, to a disposition that is easily impressed and just as easily removed. According to others, and Valla prefers this view, it is a *qualitas passibilis* (therefore in the third species of quality) that is impressed on an object the way in which magnetism is impressed on iron. One could call the magnetic quality *levitas* when it moves the iron upward and *gravitas* when it moves the iron downward, but properly speaking it is neither *levitas* nor *gravitas*; rather it is a kind of quality that is capable of moving the object moved in any way the one impressing it desires. Similarly the *virtus impressa* can move the projectile in any direction whatever, and, being like a *species intensionalis*, it can do so without being the contrary of any of the projectile's natural qualities.[100]

[98] Cod. APUG-FC 1710, Tract. 5, Disp 1, Pars 5, Quaes. 6. A quo moveantur proiecta. Notandum est. . . . See Wallace, *Prelude to Galileo*, pp. 330-333 for a summary of the content of this *quaestio*, and pp. 338-339 for transcriptions of the significant Latin texts.

[99] Cod. APUG-FC 1710, Tract. 5, Disp. 1, Pars 5, Quaes. 6, Prima dubitatio. . . .

[100] Ibid., Dubitatio secunda. . . .

Regarding the *virtus*'s mode of production and corruption, Valla holds that it is produced by local motion, just as is heat, from the motive power of the thrower; thus it is generated either in an instant or during the time the thrower is touching it. Since the *virtus* is, like light, a diminished and imperfect entity (*diminuta et imperfecta entitas*), it is easily corrupted. It is distributed throughout the entire volume of the projectile, just as the magnetic quality is distributed throughout the magnetized iron, and this presents no difficulty for intensional species of this type. Though produced instantaneously, it corrupts successively and in time, for it has degree-like parts (*partes graduales*) and so loses existence gradually, becoming weaker the more it is removed from its source. The quality is corrupted, Valla says, from the fact that it is resisted in second act; its second act is contrary to the second act of the *virtus motiva* of the stone, and therefore there can be succession in this corruption. Nor is there any absurdity in something being produced in an instant and corrupted in time, because in its production it has no contrary and is produced by the motion of the projector, whereas in corruption it does have a contrary in some way, and it corrupts because it gradually loses existence through parts.[101]

These preliminaries aside, Valla lists two opinions and their proofs before proceeding to his own conclusions. The first opinion is that the projectile is moved by the medium in the way he has already explained, and he attributes this to Burley, Aristotle, Themistius, Philoponus, Averroes, St. Thomas, and others, adding that all of these admit that some *virtus impressa* is also present in the medium. The proofs he offers for this include various texts of Aristotle that speak of the air alone moving the projectile, and then the various arguments already given by Menu against the existence of such a *virtus* in the projectile itself. The second opinion, he says, holds that the projectile is moved by a *virtus impressa*, and this is the teaching of Scaliger, Albert of Saxony, and Buridan. Paul of Venice and Simplicius, he adds, seem to put this *virtus* in the medium; if it is there, it might just as well be put in the projectile, and then violent motion would be possible in a vacuum. St. Thomas and the *Doctores Parisienses*, he continues, admitted such a *virtus* in the medium, and so did practically all of the ancients, including Aristotle himself. The arguments he then offers in support of this opinion are basically those given by Menu against the projectile's being moved by the medium alone. To these he adds only one new argument: if this were so, then a projectile would be projected more easily in water than in air, since water is better able to propel an object than is air, but such is not found to be the case.[102]

[101] Ibid., Tertia dubitatio. . . .

[102] Ibid., Prima opinio . . . Secunda sententia. . . .

Valla's conclusions, finally, are stated somewhat more positively than the corresponding conclusions in Menu. The first is that the medium alone is not sufficient to explain the motion of the projectile, and that some other *virtus* must also be admitted. The proofs offered in support of this conclusion are essentially those given by Menu to justify his second conclusion, including the experiment taken from Scaliger, whose source is explicitly acknowledged by Valla. After this the second conclusion follows, without any further proofs, stating that, although the *virtus impressa* is sufficient to cause the motion of projectiles, almost always such a motion is aided by the motion of the medium. In this way, Valla concludes, one can "save" Aristotle and the peripatetics and all the experiments that have been discussed: the *virtus* is sufficient to move the projectile without the medium, but without the *virtus* the medium can move the projectile either not at all or merely for a very short distance.[103]

There remains only Valla's resolution of the various difficulties that have been raised against his position. Of these one is noteworthy, namely, the way in which he replies to the objection that the motion of the projectile would then be one and continuous, which is against the text of Aristotle. Valla's answer is that continuous can be taken in two ways: first, for a motion that is not interrupted by any intermediate point of rest; second, for a uniform and regular motion. If taken in the first way, he says, he concedes the inference; nor did Aristotle deny this, nor could he deny it, since experience shows that the stone, when thrown upward, does not come to rest in any way. If taken in the second way, he denies the inference and claims that this is what Aristotle meant in that text. For Aristotle concluded from the fact that the motion of the projectile is not regular and uniform that it could not be perpetual, and it is certainly true that the motion is not regular, for the *virtus impressa* gradually weakens and the medium that assists the motion is not always uniform.[104] Of special interest is Valla's admission that the projectile thrown upward does not come to rest "in any way," which was also the position taken by the young Galileo against the traditional peripatetic teaching of his day.[105]

c. RUGERIUS AND VITELLESCHI

Rugerius takes up the motion of projectiles at the end of his questions on the *De caelo* in a tractate devoted entirely to the local motion of heavy and light objects. He incorporates much of the material that is found in

[103] Ibid., Prima conclusio . . . Secunda conclusio. . . .

[104] Ibid., Ad sextum respondeo. . . .

[105] Sec. 5.2.a-b; cf. Wallace, *Prelude to Galileo*, p. 333.

Menu and Valla, but organizes it differently and comes to a contrary conclusion. Rather than favoring the *virtus impressa* explanation over the media explanation while admitting the possibility of both, he accepts the concept of impetus in a qualified way but expresses his preference for the more traditional Aristotelian view. Although it is not yet too improbable to posit this *virtus impressa*, he concludes, it is more philosophical and more in conformity with peripatetic principles to attribute the motion of projectiles to the medium alone, and not to any *virtus* newly impressed.[106]

The explanation Rugerius offers for the qualifying clause with which he begins his conclusion is of value for its description of the way in which he conceives impetus, even though he is not fully prepared to accept the concept. The reason why he does not reject the *virtus impressa* entirely, he says, is because most of the difficulties that have been brought against it can be solved by maintaining that this *virtus* is different in kind from the natural motive qualities, that it is of only one species, and that it moves up and down and in different directions according as the impeller begins the movement. Again, it is corrupted in two ways: one by the natural motive quality of the body on which it is impressed, since this opposes it; and the other by rest. Likewise, it undergoes intension and remission, it is produced in time, and it is received in the entire object that is moved. Nevertheless, adds Rugerius, the motion that results from this *virtus* is not natural, because of the existence in the body of another *virtus motiva*; it is this that makes the motion arising from the *virtus impressa* violent.[107]

In another way of looking at it, Rugerius goes on, one could maintain that the *virtus impressa* is not a really distinct quality that is newly produced, but is a kind of modification of *gravitas* in the heavy element, or in some way mixed with its *gravitas*, and that the natural quality is modified by the one impelling the object so that the *gravitas* inclines the object in the direction intended by the projector. The modification could be more or less according as the object is impelled with greater or lesser force, within a certain latitude of *gravitas*. Even on this explanation, however, the motion would not be natural, for the object in this case is not carried along by its *gravitas* with the modification the motive quality has by nature, but rather with the modification it receives from the projector.[108] This view of the *virtus impressa* is indeed quite original, and seems to be a development of ideas contained in Menu and Valla, while

[106] Cod. SB Bamberg 62-4, fols. 101v-106r; this teaching is summarized in Wallace, *Prelude to Galileo*, pp. 335-336, and significant Latin texts are given on pp. 339-340.

[107] Cod. SB Bamberg 62-4, fol. 104r, Propositio. . . .

[108] Ibid.

differing from them in significant respects. The question naturally arises whether the alternate way of conceiving impetus was Rugerius's own, and if so, why he was not prepared to give a wholehearted endorsement of the concept. It seems more probable that the proposal came from another source, possibly Vitelleschi, and that this was Rugerius's way of acknowledging the cogency of the explanation while himself preferring the more peripatetic view.

With regard to Vitelleschi, as has been noted,[109] the surviving *reportationes* do not contain his treatise on impetus and thus his evaluation of the *virtus impressa* can only be inferred indirectly from his treatment of related subjects. A number of his statements seem favorable to the impetus concept, but in contexts that would not entail the rejection of the idea that the projectile is also moved by the medium, and thus it seems likely that he held a position similar to those of Menu and Valla, as already explained. His willingness to acknowledge the possibility of a motion intermediate between the natural and the violent that would go on perpetually, for example, would involve a recourse to some internal active principle in order to safeguard the principle, whatever is moved is moved by another. Again, in discussing the ways in which natural motions are differentiated from non-natural ones, Vitelleschi admits that the fact that a motion proceeds from an internal principle does not qualify it automatically to be regarded as natural: a violent motion could also proceed from an internal principle if this were not connatural to the object moved—which is precisely the way in which one would view the *virtus impressa*. Likewise, in discussing how *gravitas* is an instrumental cause of the generator and the substantial form in effecting the natural motion of the elements, Vitelleschi clearly leaves open the possibility of the *virtus impressa* being the instrumental cause of the projector in effecting the violent motion of these and other heavy bodies, for everything he says about the causality of gravity would apply, *mutatis mutandis*, to impetus or other impressed quality.

But it is in his discussion of the point of rest in reflex motion that Vitelleschi implicitly invokes the *virtus impressa*, for without this or a similar concept it seems impossible to make sense of the explanation he offers for that phenomenon. It is here that the marginal note referring to this *virtus* occurs in his manuscript.[110] Similarly, when contrasting the different velocity variations at the end of natural and projectile motions, respectively, Vitelleschi speaks of a *virtus motiva* being involved in both types of motion, and attributes the slowing down of the projectile to the weakening of its motive force, which is the way one would normally think of the *virtus impressa*. The only teaching of Vitelleschi that seems

[109] Sec. 4.1.d.

[110] See *supra*, p. 164.

to go contrary to this concept is his insistence that local motion would be impossible in a vacuum, on the grounds that the medium is necessary for the resistance it offers to the motion. This in itself would not be a reason for rejecting impetus, although it might support the view that the medium was essential to every local motion and thus would have to be taken into account in projectile motion also. But, as has been seen, none of the Jesuit professors ruled out the role of the medium in violent motion even while invoking the *virtus impressa*, and thus Vitelleschi could have seen the two explanations as complementary, just as did Menu and Valla. Whether he inclined more to the position of Valla or to that of Rugerius is difficult to say, but fortunately that question need not be resolved for purposes of this study.[111]

4. The Science of Mechanics

From the account thus far, we can see that the Jesuits at the Collegio Romano regarded the study of the motion of heavy and light bodies as an integral part of the science of nature, but not in such a way as to preclude motions that came from extrinsic principles and thus were in some way violent or preternatural. Again, when analyzing *gravitas* and *levitas*, they took a view of motive qualities that saw them as accounting not only for the motion of such bodies but also for their rest. Similarly, they regarded the pseudo-Aristotelian *Quaestiones mechanicae* and Archimedes' works on weights and on bodies that float in water as equally valid sources, together with Aristotle's *naturalia*, for resolving problems relating to local motion. All of this suggests that they operated under a conviction that there were general principles that grounded the possibility of a science of mechanics, as the study of local motion is known in the present day, and that such principles were equally applicable to the two branches of that science now known as statics and dynamics. This attitude of mind was not necessarily shared by other authors in the late sixteenth century, especially those who had been trained as mathematicians and were working in the tradition of mechanical treatises deriving from Greek antiquity, and thus a few words devoted to the science of mechanics as this was then conceived may not be out of place.

Although there were many differences on points of detail, mechanics was generally regarded in the sixteenth century as a mixed science in the sense described in the previous chapter, wherein mathematical principles

[111] See the discussion in Wallace, *Prelude to Galileo*, pp. 334-335, 116-117.

were applied to the study of physical objects.[112] As a science it would have to satisfy the requirements of the *Posterior Analytics*, and usually these were understood to be those specified for a subalternated science whose principles derived, at least to some extent, from the subalternating science of mathematics. The model set out by Clavius in the introduction to the second edition of Euclid's *Elements* was generally adopted,[113] and those who attempted to be more rigorous in developing the science usually departed from the discursive mode of treatment found in the Aristotelian *Quaestiones mechanicae* and simulated instead the axiomatic mode of the *Elements*. Thus they would begin their treatises with lists of common axioms, definitions, and postulates, and from these proceed to deduce propositions or theorems that would express the main conclusions of their discipline. In such a context the term *suppositio* occurs with some frequency in sixteenth-century expositions of mechanics as a science, with this term being taken in the sense described in the previous chapter.[114] Apparently no one denied that *suppositiones* would be necessary to put the science on a sound foundation: the point of contention that soon arose, however, was precisely what *suppositiones* would be allowed, or would have to be accepted as true, for the subsequent development to be strictly scientific. Galileo's position on this point was quite distinctive, as will be seen in the next chapter, and strongly influenced the way in which he sought to develop his *nuova scienza* of motion.[115]

a. TARTAGLIA AND GUIDOBALDO

Niccolò Tartaglia is a convenient figure with which to begin, since his writings on mechanics initiated the revival of that science in the sixteenth century and continued to be influential to the time of Galileo and beyond. Tartaglia begins the first book of his *Nova Scientia* of 1537 with fourteen definitions, follows these by five suppositions, then lists four axioms, and from all of these as principles proceeds to deduce six conclusions or propositions relating to the science of motion. Among the definitions are those for time and the instant, natural and violent movements, motive powers and resistances; among the suppositions are statements relating to the equality and inequality of effects deriving from similar and dissimilar motive powers and resistances; and among the axioms are statements relating altitudes and resistances on a similar basis of equality and inequality. The propositions derived from these then state, in a general

[112] Sec. 3.4.a.

[113] Sec. 3.4.b.

[114] Secs. 3.1.a, 3.2.a, and 3.4.c.

[115] Secs. 5.1.a, 5.2.a-b, 5.3.b-c, 5.4.a, 5.4.e, 6.1.d-e, 6.3.d-f, and 6.4.a.

quantitative way, how the speed of bodies varies in natural and violent movement, for example, that heavy bodies fall more rapidly the more they depart from the beginning of their movement and the more they approach its end; that those falling from higher altitudes fall faster toward the end of their fall; and that heavy bodies in violent motion move more slowly at the beginning and at the end of their movement.[116] These statements are similar to those already seen in the Jesuit *reportationes*,[117] the main difference being that they are framed more explicitly in terms of equalities and inequalities after the fashion of a mathematical treatise.

It is in the second book that Tartaglia breaks new ground, as it were, and sets the stage for mixed reactions to his teachings. Here he begins to treat of motions that are curved as well as straight, and how such motions stand in relation to the horizon and to the earth's spherical surface. Again he lays out fourteen definitions, then four suppositions, and from these he deduces nine propositions with a few accompanying corollaries. His first supposition is of special interest because it states that all natural movements of heavy bodies are parallel both to each other and to a perpendicular erected to the plane of the horizon. In explaining this, Tartaglia points out that he is aware that such motions can never be perfectly parallel to one another or to the perpendicular, since ultimately they must converge at the center of the earth. But, he goes on, since this error is undetectable in a short space, one may *suppose* them all parallel to one another and to the perpendicular from the horizon as well.[118] This *suppositio*, unfortunately, did not gain uniform acceptance among those who read his treatise, as will be seen, and indeed cast doubt on whether it would ever be possible to construct a strict science of local motion according to Euclidean canons.

In his *Quaesiti et inventioni diverse* of 1546, Tartaglia again mentions the need for suppositions, this time for erecting a science of weights (*scientia de ponderibus*), which he is discussing with his student, Diego Hurtado de Mendoza, after having reviewed with him in a general way the principles behind Aristotle's *Quaestiones mechanicae*. The science of weights requires definitions of its terms, and, over and above this, he says, it also requires principles that are first and indemonstrable, which, being conceded or assumed, afford the means for deducing the propositions that make up the content of the science. There are various terms

[116] These passages are translated by Stillman Drake and appear in *Mechanics in Sixteenth-Century Italy*, ed. Stillman Drake and I. E. Drabkin (Madison: The University of Wisconsin Press), 1969, pp. 70-81.

[117] Secs. 3.1.a, 3.2.a, and 3.4.c.

[118] Drake and Drabkin, *Mechanics in Sixteenth-Century Italy*, pp. 81-97; the first supposition is stated on p. 84.

used to designate such principles, he goes on, some calling them axioms, since they prove other propositions but cannot be proved from others; others calling them suppositions, since they are supposed to be true in the given science; and yet others calling them petitions, since, if one wishes to debate such a science and sustain it with demonstrations, one must first request the adversary to concede them.[119] If he does not, of course, the entire science would be denied and one could not debate further over it. This last terminological usage Tartaglia finds somewhat more pleasing than the others, and so he proceeds to list a number of *petitiones* on which he erects the subsequent science, concluding with the sixth petition, namely, that water has no weight in water, nor has air any weight in air, etc.—which may readily be conceded, he says, because experience makes this manifest.[120]

It should also be noted that Tartaglia was well acquainted with medieval formulations of the science of weights, and edited a version of Jordanus de Nemore's treatise *De ratione ponderis* (which Tartaglia entitled *De ponderositate*) that was published at Venice in 1565.[121] The medieval *opuscula* on this subject invariably began with a number of *suppositiones* that served as principles for the propositions or theorems that would be deduced from them, usually without definitions—as these were formulated in the sixteenth-century treatises on mechanics.[122] The medieval version of Archimedes' *De insidentibus in humidum*, however, employed *petitiones* instead of *suppositiones*, and it is perhaps noteworthy that Federico Commandino's *De centro gravitatis solidorum* of 1565 likewise lists two *petitiones* after its *definitiones*, following Archimedean practice—since he was the editor and translator who made many of Archimedes' works available in Latin to Renaissance readers.[123] Jordanus's *Liber de ponderibus* had earlier been published in 1533 by Peter Apianus, and this employed a *suppositio* relating to positional gravity that was similar to Tartaglia's on the parallel paths of falling bodies, and which was to be rejected along with Tartaglia's by later sixteenth-century mechanicians.[124]

[119] Cf. secs. 3.1.a and 3.2.a.

[120] Drake and Drabkin, *Mechanics in Sixteenth-Century Italy*, pp. 98-143; the passage relating to principles, axioms, suppositions, and petitions occurs on pp. 115-116.

[121] *Iordani opusculum de ponderositate, Nicolai Tartaleae studio correctum, novisque figuris auctum* (Venetiis: Apud Curtium Troianum, 1565). Various *suppositiones* are noted on fols. 3r, 17r, and 17v of this edition.

[122] For an account of these *opuscula* and some of the *suppositiones* they employ, see J. E. Brown, "The Sciences of Weights," in *Science in the Middle Ages*, ed. D. C. Lindberg (Chicago: The University of Chicago Press, 1978), pp. 179-205.

[123] These *petitiones* appear on fol. 1v of the Bologna (Benacii) edition of 1565 of Commandino's work.

[124] For suppositions relating to positional gravity, see Brown, "The Science of Weights,"

Guidobaldo del Monte studied with Commandino and in 1577 published his *Liber mechanicorum* wherein he attempted to restore mechanics to the status of a rigorous science, and in so doing to point out the defects in medieval proofs deriving from Jordanus and revived by Tartaglia some years earlier. One principle he explicitly attacked was the supposition already noted, which he says was employed by both earlier writers, viz., that the motions of bodies separated by some lateral interval on the earth's surface could be regarded as tending in parallel lines toward the earth's center.[125] Another principle on which Guidobaldo insisted was that the power to sustain a weight must be less than the power to move it—a principle we have seen invoked by Vitelleschi to argue that a point of rest must always intervene between the stone's upward motion and its downward fall.[126] Unfortunately such a principle, if consistently applied, eliminates the possibility of applying the principle of virtual work to the solution of problems in statics, as this was adumbrated by Jordanus, and thus of constructing a unified science of mechanics that would embrace both dynamics and statics under a common set of principles. The appeal to rigor was so great among sixteenth-century mathematicians, however, that Guidobaldo was soon followed by Giovanni Battista Benedetti—who, as we have earlier noted, was admired by Clavius[127]—in his rejection of the *suppositiones* employed by Jordanus and Tartaglia. The basic error into which these men fell, writes Benedetti in the same vein as Guidobaldo, is attributable to the fact that they took lines of inclination, that is, the respective lines along which weights tend to fall in their path to the earth's center, as parallel to each other, whereas in strict mathematical reasoning they can never be such.[128]

b. BLANCANUS AND GUEVARA

From these brief indications it can be seen that the attitude implicit in the *reportationes* from the Collegio Romano—namely, that a single science of mechanics that includes the study of bodies in local motion as well as

---

pp. 191-192; the relevant treatises are translated in E. A. Moody and Marshall Clagett, eds., *The Medieval Science of Weights* (Madison: The University of Wisconsin Press, 1952).

[125] The relevant passages are translated by Drake, *Mechanics in Sixteenth-Century Italy*, pp. 262 (where Jordanus and Tartaglia are explicitly named by Guidobaldo), 275, 278-281.

[126] Ibid., p. 300; on Vitelleschi, see the text cited *supra* at note 28.

[127] Sec. 3.4.b.

[128] I. E. Drabkin has translated the passages from Benedetti's *Diversarum speculationum mathematicarum et physicarum liber* (Turin 1585), where the *suppositiones* of Jordanus and Tartaglia are rejected, in Drake and Drabkin, *Mechanics in Sixteenth-Century Italy*, especially pp. 177-178, where reference is made to passages in Tartaglia given earlier on pp. 141-142 in the translation of Stillman Drake.

those at rest can be formulated, employing the general principles of an Aristotelian science of nature and yet being able to account for all known mechanical phenomena—was not uniformly accepted in the latter part of the sixteenth century. Precisely how the Jesuit mathematicians of that period would go about formulating such a science is not easy to ascertain, since, to the author's knowledge, there are no extant mechanical treatises in the archives of the Collegio. Thus it is difficult to know how they would have applied the teachings of the *Posterior Analytics, Physics, De caelo*, etc., to answering the questions raised by Tartaglia, Guidobaldo, and Benedetti in their differences over the principles of mechanics. Clavius, as was noted at the end of the previous chapter, included under the subject matter of such a science everything that has the power to move matter, and particularly the study of weights, whose disequilibrium he regarded as the cause of motion and whose equilibrium the cause of rest.[129] Apparently he did not go beyond this in any of his extant writings, but his student Blancanus gives a few indications that prove helpful for reconstructing his general line of thought.

In his *Dissertatio* on the nature of mathematics of 1615, Blancanus, as already observed, maintains that the adequate total object of the mathematical sciences is terminated quantity, which is referred to as intelligible matter to differentiate it from the sensible matter studied in the physical sciences.[130] The type of abstraction whereby intelligible matter is attained, he says, confers on it a kind of perfection not found in nature, making it easier for the mathematician to detect principles and causes with exactitude, rather than having to search for effects and the more occult causes associated with sensible matter after the fashion of the physicist.[131] Those who complain, therefore, that the subject of mathematics is ignoble because it is only an accident overlook the fact that it is a special type of accident, viz., one that is both immaterial and abstract, for which reason it is located between the objects of physics and metaphysics. On this account it is much better to arrive at numerous and marvelous truths about such an accident than it is to be concerned with a thousand differences of opinion about material substance, a true knowledge of which will never be attained. Yet again, the intermediate branches of mathematics—and among these Blancanus has already enumerated optics, mechanics, music, and astronomy—enjoy a special status altogether, for their practitioners study not nude quantity (*non nudam quantitatem*) but rather the matter of the heavenly bodies, musical sounds, visual rays, and the causes of forces in machines (*causas virium machinarum*) to the same end as other philosophers study their respective subject matters.

[129] Sec. 3.4.b.

[130] *Dissertatio*, p. 6; see also sec. 3.4.c.

[131] *Dissertatio*, p. 25.

Finally, when people object that mathematicians are not able to formulate perfect demonstrations, they never bring up the *scientiae mediae* (and here Blancanus again mentions *mechanica*), for in these all are willing to concede that the true character of a demonstrative science is to be found.[132] Later Blancanus gives a few examples of the remarkable results attained in the mixed sciences: for mechanics he merely inquires why it is that the wedge and the screw generate such forces, and asks if there is anything more wonderful than that the power of an ant can move any weight whatever. He points, too, to the subtlety of studies into the center of gravity, such as those pioneered by Archimedes and brought to perfection in recent times by Fredericus Commandinus and Lucas Valerius.[133]

Blancanus has only a few references to mechanics in his *Apparatus ad mathematicas addiscendas et promovendas* of 1620, but these also are worthy of note. In describing the *scientiae mediae* he characterizes mechanics as concerned with machines or, as Aristotle says, as concerned with artifacts in the same way as the natural sciences are concerned with natural entities. It demonstrates properties of the six principal machines: the balance, the lever, the pulley, the wheel and axle, the wedge, and the screw. Because it considers in these artifacts the quantities of moving forces, weights, movements, and times in which these occur, and regards such machines as lines revolving around centers, it is subalternated to geometry and thus it demonstrates geometrically. Its most profound part is that dealing with centers of gravity of planes and solids.[134] For selected works from which the principles of the science may be drawn Blancanus cites the Aristotelian *Quaestiones mechanicae* and the basic works of Archimedes and of Hero of Alexandria, the *De ponderibus* of Jordanus Nemorarius, and then recent authors such as Guidobaldo del Monte, Lucas Valerius, Marinus Ghettaldus, Benedetti, and Galileo—the latter for his discourse on bodies in water.[135] Methodologically, he goes on, Euclid is most important, and thus the first six books of the *Elements* must be known before one begins the study of the works already enumerated.[136] Advanced work in the field focuses on finding the centers of gravity of various plane and solid figures, for not all of these are known, especially those bounded by circles, ellipses, and hyperbolas.[137]

Apart from these emendations to Clavius's general ideas, as already noted, there is no full-blown development of the science of mechanics by any Jesuit professor of the period. Fortunately for purposes of this account, however, the writings of another mathematician who addressed

[132] *Dissertatio*, p. 26.

[133] *Dissertatio*, pp. 30-31.

[134] *Apparatus*, p. 389; see chap. 3, note 146, *supra*.

[135] *Apparatus*, p. 395.

[136] *Apparatus*, p. 402.

[137] *Apparatus*, p. 413.

himself specifically to this development, and who was a friend and correspondent of Galileo, can serve to fill out the needed background. This was Giovanni di Guevara, a priest of the order of Clerks Regular Minor, who was only a few years older than Galileo and whose terminology suggests that he had either studied with the Jesuits or was acquainted with the matter they taught, particularly with regard to the subjects and principles of a science. Guevara published a commentary on the Aristotelian *Quaestiones mechanicae* under date of 1627, though it did not appear until 1629, by which time he was General of his order and then bishop of Teano.[138] In his introduction to that treatise he fortunately has a lengthy disquisition on the science of mechanics as viewed from an Aristotelian perspective. A review of its contents will be of value for complementing the study of local motion presented in the earlier sections of this chapter and for preparing the ground for an examination of Galileo's attempts at constructing a similar science in subsequent chapters.

Guevara proposes his work as a commentary on Aristotle's *Mechanica*, together with some additions, wherein he will first lay out a general teaching on the nature and object of mechanics and on its causes and principles, and then will work through the thirty-five questions contained in Aristotle's work and show how these can be answered in terms of the general teaching. The principles that must be foreknown (*praecognita*) for one to demonstrate in this science, apart from the elements and theorems of geometry, he notes as being concerned with the nature and properties of the circle, as these were used by Aristotle, to which he would add others relating to the center of gravity, as these were developed by Archimedes, Hero, and Pappus.[139] Thus he apparently sought to make a synthesis of the materials contained in the Aristotelian and the Archimedean traditions, sometimes referred to as the dynamic and static traditions respectively, and from these construct a unified science of mechanics.

The general subject matter of the mechanics, following Aristotle's introduction, is noted by Guevara as being remarkable things that are not done by nature but are beyond nature (*praeter naturam*) and can be effected by human ingenuity. The problems considered in the discipline are not completely the same as those of the natural sciences, nor are they completely different from them, and at the same time they have elements in common with both physics and mathematics. Guevara elaborates on this by saying that the principal concerns of the mechanician, namely,

[138] Ioannis de Guevara, *In Aristotelis Mechanicas Commentarii, una cum additionibus quibusdam ad eandem materiam pertinentibus* (Romae: Apud Iacobum Mascardum, 1627); henceforth cited as *Mechanica*.

[139] *Mechanica*, pp. 1-2.

weight and the force that moves it, are materially contained under the adequate object of physics and that demonstrations concerning them use not only geometrical but also natural arguments; therefore, by reason of its material subject and a number of its conclusions that flow from the principles of either science, it has elements in common with both.[140] In itself it is both an art and a science, the first as opposed to the liberal arts and embracing everything that can be done by mechanical means, and the second insofar as it is concerned with the principles behind the motion and rest of heavy and light bodies as these are effected by artifice.[141]

The goal or end of mechanics, Guevara adds, is especially to gain knowledge of motions that are effected by small forces and that are remarkable in the way they are produced, whether such motions are in accord with nature, contrary to it, or beyond it. To gain this type of knowledge, two things are particularly important: the quantity of weight in the body to be moved, and the quantity of motive force in the mover, whether this comes immediately from some intrinsic power, from an impetus impressed from without, or from some type of instrument. The mechanical art is based on knowledge of such factors, for it is these that permit one to calculate how to supplement natural forces to achieve a desired result. Such calculations, moreover, do not require that the mechanician understand *gravitas* and *levitas* in an absolute way, the way these are taken by the physicist; rather, a relative understanding is sufficient, since this is adequate to achieve the goals sought in the discipline.[142]

This understood, Guevara goes on, the adequate material subject of mechanics is the heavy and the light, or the quantity of their weights, and the force required to move or detain them. The reason for this is that the adequate material subject of a science should include everything that is treated in the science, and everything treated in mechanics is reducible to the heavy and light body that is to be moved or detained, and the quantity of the force required to move or detain it. The formal object, on the other hand, is the mobility and rest of such bodies as this is remarkable and effected by artificial means, whether the motion and rest be natural or violent, such as that brought about by impressed forces. The total and adequate object of the discipline as a whole then combines both the material and formal considerations, and may be stated as the heavy and the light as these are capable of being moved and brought to rest through the use of artifacts of various kinds.[143]

It is obvious, says Guevara, that mechanics is an intellectual habit or discipline, and as such it qualifies as both an art and a science. An art is a habit that is productive of things that can be otherwise, and so it is

[140] *Mechanica*, pp. 5-6.

[141] *Mechanica*, pp. 6-7.

[142] *Mechanica*, pp. 9-10.

[143] *Mechanica*, pp. 10-12.

usually differentiated from a science, which is concerned with things that are necessary and so cannot be otherwise; yet Aristotle admits that practical sciences such as medicine and architecture can be referred to as arts, and he also speaks of mathematical disciplines as being both arts and sciences. Mechanics clearly qualifies as an art because it is concerned with doing things, namely, effecting the motion and rest of heavy and light objects that can be acted on by artificial means in a great variety of ways and so can always be otherwise. It is also a science, moreover, because scientific knowledge is knowledge through causes that make things be the way they are, and in virtue of which they cannot be otherwise. Now mechanics is such a type of knowledge, for it fulfills the two conditions requisite for scientific knowing. First, it is an intellectual knowledge of everything associated with the motion and rest of heavy and light bodies that is attained through the foreknowledge of principles, whether these are *per se nota* or demonstrated in a superior science, which principles are the causes of the properties attributed in the conclusions. In other words, mechanics is a science because it yields knowledge of effects from causes that are foreknown. The second condition is that these be proper causes from which effects necessarily follow, and this is verified in mechanics also: the motions of levers and balances, for example, do not happen in a fortuitous way, but in ways that are necessary and predetermined by the forces that are acting on them. What Aristotle says about arts being concerned with things that can be otherwise does not preclude this knowledge, for he is thinking of singular and individual things that are greatly variable; scientific knowing of this type, on the other hand, is concerned with *universalia factibilia*, that is, things to be done in a general or universal way, which require analysis through the causes that produce them. From this it follows that mechanics is a practical science, that is, a science that does not terminate in the contemplation of truth alone but rather passes on to operations or works to be done. Again, this does not rule out that some of its propositions, considered in themselves, might be speculative or theoretical; the point is that they are considered in the science precisely as they have an ordination to practice, and this is what makes the discipline of which they are a part essentially a practical science.[144]

The next question Guevara addresses is whether mechanics pertains to physical science, or to mathematical science, or partially to both—a question, he notes, that is not easy to answer. It would seem that it is part of physical science, he begins, because the subject with which it is concerned is physical in nature, namely, the heavy and the light and the

[144] *Mechanica*, pp. 12-16.

forces that can move these and bring them to rest; again, because it treats this subject in a physical way, inquiring if motions come from nature or are impressed from without; and yet again, because its foundations are to be sought in principles that are purely physical and universal, even though it employs mathematical principles to make its reasoning clearer and more evident. Mechanicians also use mathematics to make detailed applications to particular cases, where diagrams can be drawn and estimates thus made of magnitudes and distances, but this in itself does not make mechanics a mathematical discipline. And finally, even though the mechanician takes much help from mathematics, and even appropriates principles from geometry and arithmetic, this does not mean that his science is subalternated to mathematics, for neither physics nor natural theology are subalternated to metaphysics even though they draw heavily on that discipline.[145]

As against this way of arguing, Guevara notes that all philosophers and mathematicians who have written about this science, such as Hero, Pappus, and their followers, hold that it takes its origin from physics and geometry; on this account mechanics should be regarded as a mixture of the two and thus a third type of science. It is this position that has been taken by Guidobaldo del Monte in the preface of his *Liber mechanicorum*, says Guevara, and it is supported by Aristotle's own statements in his introductory remarks to the *Quaestiones mechanicae*.[146]

Neither of these positions, however, is subscribed to by Guevara himself, who holds that mechanics is subalternated to mathematics and as such should be regarded as essentially a mathematical discipline. To make his stand clear, he first explains what it means for one science to be subalternated to another. The first requirement is that the principles that are dealt with in the subalternated science cannot be seen as evident on their own merits, but must be demonstrated in the subalternating science, on which they therefore depend intrinsically and essentially. The second is that the objects of the two sciences must be the same when considered under a particular formality. This formality will be verified in a more universal and simpler mode in the subalternating science, and it will be restricted and contracted by some accidental difference in the subalternated science; thus music is subalternated to arithmetic, since the latter considers number in itself whereas the former considers number in sounds, and similarly optics is subalternated to geometry, since the latter considers lines in themselves whereas the former considers lines as they can be seen in visual rays.[147]

[145] *Mechanica*, pp. 17-18.

[146] *Mechanica*, p. 18.

[147] *Mechanica*, pp. 18-19.

These points understood, Guevara maintains that mechanics, considered absolutely and completely, is subalternated not to natural philosophy but to mathematics, and that this is what Aristotle meant when he said that its subject is physical but that its mode of consideration is mathematical; practically all philosophers and mathematicians follow him in this, for they list mechanics under the mathematical disciplines and regard it as subalternated to geometry. The argument in support of this view is based on the notion of subalternation: mechanics considers a subject that shares in the same formality as that of geometry, and it takes the conclusions of the latter as principles in its own reasoning. Thus the subject matter of mechanics is the heavy and light movable body, but it considers this under the formality of the quantity of weight the body has or the quantity of force necessary to move or support it. Such quantities are considered according to the ratios they have to one another, and as they can be abstracted from matter; they are related, moreover, according to distances and figures associated with the bodies that serve as movers and moveds, that is, according to the continuous quantity of such bodies as this is studied in geometry. Again, mechanics makes use of principles taken from geometry, and in fact resolves its conclusions back to such principles, mainly as these are proved in the third and sixth books of Euclid's *Elements*. To verify this, Guevara says, all one need do is consult the writings of Archimedes, Pappus, and other mechanicians.[148]

With regard to the arguments offered in support of the first position, Guevara agrees that the subject of mechanics is physical in nature when considered as a kind of being, but not when considered under the aspect of its being known in that discipline: materially the subject pertains to physics, but formally the consideration is quantitative and mathematical. Again, it is true that the mechanician considers heavy and light bodies in a physical way, for example, as these are movable by nature or from extrinsic principles, but he does not focus on them from the viewpoint of the ends intended by nature, but rather considering how they can be moved or brought to rest by machines or how they can achieve goals intended by the mechanician. Yet again, physical principles are not sufficient of and by themselves for the demonstrations sought in mechanics; it is true that some properties of heavy and light bodies can be proved in physics, and yet others will be based on suppositions of that discipline, but over and above these are also required principles or conclusions that can only be demonstrated in geometry. The properties of the circle as these are employed by Aristotle in his *Quaestiones mechanicae*, for example,

[148] *Mechanica*, pp. 19-21.

are essentially mathematical and can be known only in geometry. Thus the use of mathematics in mechanics is not confined to conferring clarity on its demonstrations, or on making applications to particular figures, but rather enters intrinsically into the reasoning by which the discipline reaches its universal conclusions.[149]

To recapitulate, concludes Guevara, mechanics is a practical science that uses geometrical demonstrations to study the quantity of weight of heavy and light bodies and of the forces by which they are moved and sustained, in remarkable and artful ways, to attain goals intended by the engineer. Such a characterization differentiates mechanics from speculative sciences and from other practical sciences: the specification of demonstrations as geometrical separates it from other mathematical disciplines that are not subalternated to geometry; and the remaining specification contracts its difference yet further, showing how it is not the same as perspective, geodesy, or astronomy, since it studies continuous quantity as this is found in the weights and forces that account for the dynamic and static properties of physical bodies.[150] This definition also confers a unity on mechanics as a total science—and here Guevara rejoins the terminological usage of the Collegio professors[151]—and provides the basis for specializations within the discipline that can be viewed as partial sciences, either actual or habitual. Such partial sciences are integral parts of the discipline differentiated on the basis of subject matter; as Guevara sees it mechanics comprises only two principal partial sciences—*Centrobarica*, concerned with studies of the centers of gravity of bodies, and *Machinaria*, concerned with the instruments and machines by which bodies can be moved and supported. Under *Machinaria* he then proceeds to list the various types of machines and devices studied by the ancients, and subdivides these yet further into partial sciences.[152] And in his final animadversions Guevara attributes a special degree of perfection and dignity to mechanics, listing it below pure mathematics but above most practical sciences; granted that some of the latter, such as medicine, he says, study a subject that in itself has greater importance, for example, the human body, their reasoning is so uncertain and inconclusive that they hardly deserve to be called sciences.[153]

Following this introduction, Guevara first discourses briefly on the basic properties of the circle and shows how these can be used in the analysis of simple machines, as set out by Hero and Pappus, and in most recent times by Guidobaldo del Monte.[154] He then turns to a consideration of centers of gravity, likewise pioneered by Hero and Pappus, but also

[149] *Mechanica*, pp. 21-24.

[150] *Mechanica*, pp. 25-26.

[151] Sec. 3.4.a.

[152] *Mechanica*, pp. 26-29.

[153] *Mechanica*, pp. 30-32.

[154] *Mechanica*, pp. 33-54.

perfected by Commandino and Guidobaldo among recent writers. The center of gravity is particularly important, in his view, for analyzing the natural mobility of heavy and light objects—and here he allows that "light" is now being taken in the sense of "less heavy." Natural bodies have a certain aptitude or propensity for motion downward, and this comes from an intrinsic principle that is both active and passive, inclining them to move along a straight line that connects the body with the center of the world. This is obvious from sense experience, he claims, since when restraints are removed bodies immediately tend downward and seek the center of the universe, and they do so not in an oblique direction, nor along a line drawn from one of their extremities, but rather from the point in which all of their gravity seems to be concentrated, namely, their center of gravity.[155] Apart from these natural motions, however, bodies also have a preternatural or artificial mobility that enables them to be moved from without, by force, through an impressed impulse or impetus. This type of motion is called violent or artificial, he observes, but this does not rule out that it is also produced by natural causes, as is seen in the eruptions of volcanoes and various types of explosions, including those resulting from the use of gunpowder. When one combines these two types of mobility, moreover, and is able to ascertain the causes of various motions and the ratios that obtain between their weights and motive forces, as well as the distances through which such forces operate and objects move, one is equipped with principles whereby the science of mechanics can be expanded far beyond the scope envisaged for it by Aristotle in his *Mechanica*. Archimedes, he goes on, contributed greatly to this development, but it is important to realize that he did so by building on Aristotelian principles. Some question this, and here Guevara mentions Guidobaldo by name, but in his view Archimedes clearly employed the same suppositions as did Aristotle in elaborating the principles of his mechanics.[156] Unfortunately Guevara does not enumerate these suppositions, nor does he make reference to the problem of bodies gravitating to earth from different positions on the earth's surface along parallel lines, and so one can only speculate as to the precise *suppositiones* he may have had in mind.

The remainder of Guevara's commentary on the *Quaestiones mechanicae* takes up Aristotle's questions in detail, each of which he explains and supplements with materials drawn from the study of heavy bodies in terms of *gravitas* and *virtus impressa*. In this way he attempts to work out a complete science of mechanics that is consistent with the discussions explained in the earlier sections of this chapter, only presented now in a

[155] *Mechanica*, pp. 64-68.

[156] *Mechanica*, pp. 68-72.

more quantitative way.[157] Guevara makes no reference to Clavius or any of the Jesuit professors with which the foregoing sections have been concerned, but the terminology he employs is so similar, and the conclusions to which he comes so congruent, with those of the Collegio Romano that it is difficult to imagine that he did not have some association with the Jesuits in his earlier years. Be that as it may, his works stand as an exemplar of how the Aristotelian study of local motion, supplemented with materials deriving from Archimedes, Pappus, Jordanus, and others, could be combined with sixteenth-century mechanical investigations to yield a science of mechanics that would have a number of elements in common with that developed by Galileo, as will now be detailed.[158]

[157] *Mechanica*, pp. 73-277.

[158] The importance of Guevara in Galileo's life, and particularly in his relationships with the Roman Inquisition, has generally been overlooked by scholars. A recent study that uncovers and describes Guevara's interventions on Galileo's behalf, which unfortunately did not appear until this volume was in proof, is Pietro Redondi, *Galileo Eretico* Microstorie 7 (Turin: Giulio Einaudi Editore, 1983).

PART THREE

Galileo's Science in Transition

CHAPTER 5

Galileo's Earlier Science (Before 1610)

THE MATERIALS surveyed in the preceding two chapters are very extensive, and it is no simple matter to relate them to Galileo's logical and physical questions and to his other writings in his early and late years. Part of the difficulty stems from the sheer number and extent of Galileo's compositions, which fill the twenty volumes of his collected works in the National Edition. The problem is compounded when one considers the vast literature that has grown up around these works, especially in recent years, when historians and philosophers of every persuasion have set themselves the task of interpreting Galileo and reconstructing him, not infrequently in their own image.

To keep the remainder of this study within manageable proportions, discussion will be restricted to those writings in which Galileo casts light on the concept of science he is employing and the methods he uses to attain it. These writings will be treated in two stages, with the dividing point being the publication of the *Sidereus nuncius* in 1610. The first consideration will therefore focus on Galileo's attempts to develop a science of mechanics or a science of motion prior to his discoveries with the telescope, and will occupy the present chapter. The remaining treatment will continue to focus on the science of local motion, but will also pay attention to Galileo's thought concerning the motion of the earth and the sun as this developed in the course of his controversies over Copernicanism, and will occupy the following chapter.

Perhaps it should be emphasized here that the intent in the remainder of this work is not to provide an intellectual biography of Galileo or to attempt a definitive statement of the philosophy of science that lay behind his work. Its aim is more modest. The first two parts have been devoted to a textual analysis of his early Latin writings and to an elaboration of the context in which these may be properly understood. The remaining part now attempts to trace the influences this conceptual structure could have exerted on Galileo's subsequent work. To this end all of his major writings have been examined to document the use he makes of the logical and physical terminology already sketched. In the area of logic, as detailed in chapters 1 and 3, the main emphasis has been on demonstration and the principles on which it must be based, viz., definitions and suppositions, subjects being examined, ways of demonstrating from cause to effect and vice versa, the kinds of science that are generated by various demonstrations, the methods characteristic of the different sciences and the impediments they must circumvent, and the relationship between sciences—particularly those between mathematics and physics. In the area of natural philosophy, as focused in chapters 2 and 4, the corresponding emphasis has been on its subjects and the possibility of a natural science concerning them, the ways in which causes function in nature, and problems respecting local motion and related topics—such as the continuum, motive powers (*gravitas, levitas,* and *virtus impressa*), and mechanical effects that can be studied using mathematical techniques. All of this material continues to recur in Galileo's writings, both in Latin and in Italian, in surprisingly consistent ways that have hitherto been overlooked, especially by scholars intent on characterizing him as a positivist or a Platonist. Actually, his philosophical stance turns out to be more scholastic, and much more nuanced, than has been suspected. The documentation of the remainder of this study, understandably selective, is designed precisely to make that point.

In the period prior to 1610 Galileo's most important writings, for purposes here, are the various treatises on motion contained in MS 71 and related documents. These were preceded by a number of smaller compositions concerned essentially with problems of statics, and they are followed by other writings on mechanics and cosmography and by manuscripts containing records of his experiments and calculations relating to problems of motion. In what follows these works will be considered in the approximate order of their composition.

1. Archimedean Beginnings

At the time Galileo gave up studies at the University of Pisa in 1585, according to the information analyzed thus far, he would not have written anything that would give indication of his latent abilities as a mathematician. Following the usual course of studies, he by then would have studied Euclid, the *Sphaera,* and the *Theorica Planetarum,* all under the tutelage of Father Fantoni, whose surviving manuscripts show him to be somewhat undistinguished in these subject matters.[1] Thus the commonly accepted story of Galileo's interest in mathematics being kindled

[1] Again see Schmitt, *Studies in Renaissance Philosophy and Science,* especially his Essays 9 and 10, and Essay 5, p. 505.

through contact with Ostilio Ricci, a friend of the family and a competent mathematician—possibly a student of Tartaglia—offers a plausible explanation for the new direction his work would take.[2] In the four years that intervened before Galileo returned to Pisa as a lecturer in mathematics, surviving materials show that he composed a treatise in Italian on a small balance useful for determining specific gravities, *La bilancetta*, written in 1586; a more technical work in Latin entitled *Theoremata circa centrum gravitatis solidorum*, completed in 1587 but probably begun earlier; some tables listing specific gravities of metals and jewels measured in air and water; and a few notations on a work of Archimedes entitled *De sphaera et cylindro*. Collectively these compositions show that he had progressed beyond Euclid and had a good knowledge of Archimedes, so much so that he was capable of doing independent work in statics. His treatise on the center of gravity of solids, in fact, was circulated to a number of distinguished mathematicians, and it was their favorable reaction to the work that gained for him the teaching post vacated at Pisa by Fantoni in 1589.

a. *LA BILANCETTA*

The first of these works, *La bilancetta*, holds little theoretical interest, being of importance mainly for showing Galileo's skill as an instrument maker, which he was to develop extensively in later life. A few points, however, are worthy of comment. The occasion for the work, by Galileo's account, was to investigate the method used by Archimedes to determine that the king's crown was not of pure gold but rather a mixture of gold and silver. The usual understanding of Archimedes' procedure would not yield results of great accuracy, and so he applied himself with diligence to reviewing the demonstration given by Archimedes (*quello che Archimede dimostra*) in his books *De his quae vehuntur in aqua* and *De aequiponderantibus*. Such a study enlightened him (*mi è venuto in mente*) with a way that yielded the desired result with almost unbelievable precision (*esquisitissimamente*), so much so that he became convinced that it was the method Archimedes himself actually employed. To understand it, one must construct a balance such as he is about to describe, and then make use of a principle already demonstrated by Archimedes (*il che da Archimede e stato dimostrato*), namely, "that solid bodies that go to the bottom in water weigh less (*pesano meno*) in water than they do in air by as much as the *gravità* in air of a body the same size is more than that in

[2] See Drake, *Galileo at Work*, pp. 2-5; material on Ricci is sparse, but some details are provided in T. B. Settle, "Ostilio Ricci, a Bridge between Alberti and Galileo," in *XIIe Congrès international d'histoire des sciences* (Paris, 1971), tome IIIB, pp. 121-126.

water" (1:216).[3] He goes on to consider the case of a ball of gold placed in water, and notes that if the ball were made of water it would weight nothing (*non peserebbe nulla*), because water neither moves up nor down in water (1:216-217). Then, using the principle of the balance, he develops a method of calculation that will yield the percentage of gold and silver in a sample given for test, using the instrument he has devised.

From the terminology he here employs, one can see that Galileo had no scruples at this stage of his life about the demonstrative character of Archimedes' reasoning. The concepts he employs are mathematical, but they are not purely such; he speaks also of *gravità* and its effects, namely, weight and motion, and he is convinced that an element has no weight, and thus no motion, in its proper place. The case that interests him is one of static equilibrium, but this should not disguise his knowledge of the physical causes that produce motion and rest. Thus he is pursuing the model of an applied science, in the traditional understanding of *mechanica*, to justify his solution of the famous problem.[4]

The tables in which Galileo lists the specific gravities of various metals and jewels (1:225-228) were obviously made with the aid of the *bilancetta*, and give early proof of his skill in experimentation and measurement. Similarly, his annotations of Archimedes' *De sphaera et cylindro* (1:223-242) were probably done during this period. These were made marginally alongside the Latin of a text that appeared both in Latin and in Greek, printed at Basel in 1544. Galileo's remarks, written in Latin, indicate his concern with the cogency of the argument being developed; he provides numerous inserts and emendations designed to clarify the thought or to supply steps missing in the original. Occasionally his observations have methodological import, as when he notes *suppositiones* or *positiones* employed by Archimedes (e.g., *his supponitur circulus* at 1:234.7, and *positum est enim* at 1:241.13), or various matters demonstrated or shown (e.g., *quae demonstrata est* at 1:234.24, *demonstrabitur* at 1:241.18, and *ostendendum ergo est* at 1:241.25).[5] Here he is obviously working as a pure mathematician and is using the straightforward terminology of classical Euclidean geometry, merely testing his ability, as it were, against the "divine man" (1:215.11) whose thought he so much admired.

[3] *Opere* 1:216 (see note 4 to chap. 1 *supra*). Here and henceforth throughout this and the following chapter the word *Opere* will be omitted and references to volume and page number will be inserted directly into the text. Where a decimal point and a numeral follow the page number in such citations, this indicates the line number on the page cited; thus 1:234.7 refers to vol. 1, p. 234, line 7.

[4] Cf. sec. 4.4.

[5] Cf. sec. 1.3.a-b.

b. *THEOREMATA* ON CENTERS OF GRAVITY

The *Theoremata circa centrum gravitatis solidorum* (1:187-208) is a more systematic treatise than *La bilancetta* and gives better indication of Galileo's considerable ingenuity in devising proofs not found in Archimedes. Apparently having worked through the *De aequiponderantibus* to its last theorem, which showed how to calculate the center of gravity of a plane parabolic section, Galileo conceived the idea of extending the treatise to include a triangular section whose center of gravity would correspond to that of a truncated one. He worked out his solution through a lemma he devised employing mean proportionals and submitted it in 1587 to the University of Bologna in conjunction with his application for a position there as a mathematician. It appears that he also, at about the same time, decided to expand this treatment into a series of propositions and proofs that would serve as more general principles from which the solution would follow, thus supplementing Archimedes' exposition of the center of gravity of planes with another on the center of gravity of solids. Galileo left some propositions of the resulting *Theoremata* with Christopher Clavius when he visited him in Rome in 1587, and also showed them to Guidobaldo del Monte. The favorable judgment of Gioseppe Moleto, professor of mathematics at Padua, and of other mathematicians on the solution of the problem of the truncated cone, together with the correspondence elicited by the other propositions from Clavius, Guidobaldo, and a Belgian mathematician, Michael Coignet, established Galileo's reputation very quickly. During 1588 he continued to work on the *Theoremata* in response to criticisms he received from Clavius and Guidobaldo. It is not known when the work was completed in definitive form; Galileo planned to publish it in 1613 but withheld it at the request of Luca Valerio, who was then revising his own book on centers of gravity. He finally included it as an appendix to his *Two New Sciences* of 1638, some fifty years after its original composition.

Written in Latin, as its title indicates, the *Theoremata* is Archimedean in style as well as in inspiration. It begins with a *postulatum* wherein Galileo petitions (*petimus*) a principle of symmetry, namely, if equal weights are similarly arranged on different balances and the center of gravity of one group of weights divides its balance in a certain ratio, then the center of gravity of the other group of weights will divide its balance in the same ratio. He then proceeds to approximate a triangular distribution of weight along an axis with five groups of small rectangular weights suspended at equal intervals and increasing in number from one to five from one end of the axis to the other. With the aid of a lemma he is then able to demonstrate (*his positis demonstratur*, 1:187.15) that the center of gravity of the weights, and also of the figure they approximate, divides the axis so that the part on the side of the smaller groups of weights will be double the part on the side of the larger groups, or, in modern terminology, that the center of gravity will be one third the distance from the base to the point of the triangle. The reasoning is geometrical but the considerations are also physical, as testified by his use of such expressions as *aeque ponderant* and *suspensae magnitudines*, and thus the demonstrations are those of the classical *scientia media*.[6] Following his initial proposition, Galileo investigates the center of gravity of a parabolic conoid, first approximating it as the mean between the centers of gravity of a series of circumscribed cylinders and of a similar series of inscribed cylinders, and then determining the actual center of gravity by a double *reductio ad impossibile*.[7] He then extends the method to include the frustrum or truncated section of a parabolic conoid. Additional propositions are concerned with determining the center of gravity of a cone, which he likewise attacks through the use of circumscribed and inscribed cylindrical sections, and finally the center of gravity of the truncated cone.

In form, apart from the initial *postulatum*, the treatise is organized into a series of propositions, lemmas, and corollaries, although these are not explicitly identified as such. The problem or proposition to be proved is set out by expressions such as *ostendendum est* and the conclusion is usually indicated as *quod est propositum*, or *quod demonstrandum fuit*, or *quod ostendendum erat*. Propositions proved earlier are used in subsequent proofs, as indicated by such expressions as *his autem praedemonstratis, demonstratur* . . . (1:192.20), thus conveying the impression of an organized system of deductions from simple principles. Occasionally Galileo uses *ratio* or *pactum* to designate previous arguments, as in *eadem ratione demonstrabitur* (1:194.8) and *eodem pacto demonstrabitur* (1:195.3), but he is not adverse to using *eamdem ob causam* in the same way (188.23, 191.17, 198.35), thus showing that he did not exclude causes from mathematical reasoning.[8] Overall the treatise is most impressive, particularly for the mastery Galileo shows in handling limiting cases and methods of approximation and for the simplicity and elegance (1:198.6) of his proofs.

As already noted, Galileo left a draft of the *Theoremata* with Clavius during a visit to Rome, at which time he probably discussed the manuscript with him.[9] On January 8, 1588, he wrote to Clavius that he had been unable to find a better proof for a proposition questioned by the Jesuit mathematician, other than one *per induzione* (10:23.3), and otherwise took the occasion to correct an error in his analysis of the center of

[6] Cf. secs. 3.4.a and 4.4.b.

[7] Cf. sec. 3.2.b.

[8] Cf. sec. 3.4.c.

[9] Sec. 2.4.

gravity of the frustrum of a parabolic cone. The questionable proposition was undoubtedly Galileo's first, which was basic to his entire treatise, for on January 16 Clavius replied that he had no trouble with the postulate (*il supposto*), and that Galileo's method of reasoning had a precedent in Archimedes' *De aequiponderantibus*, but that he thought his first demonstration had the defect *quod petitur principium* (10:24.14). Apparently Clavius wrote to Guidobaldo del Monte that he also experienced difficulty with some of the proofs contained in Commandino's *Liber de centro gravitatis solidorum*, and Guidobaldo later informed Galileo that he had a similar difficulty with the first demonstration. Galileo attempted to answer Clavius's objection in a letter to Rome dated February 25, 1588, wherein he elaborated his proof with the aid of a new diagram; to this Clavius replied on March 5 that he did not have the time to give the matter his full attention but that it still seemed to invoke an arbitrary supposition, *quod est petere principium* (10:29.14). Guidobaldo reopened the matter in a letter to Galileo dated June 17, to which Galileo responded on July 16, admitting that had twice failed to convince illustrious mathematicians on this point, but that he still had not found a better proof. His friend was satisfied with this, as recorded in his letter of July 22, and at that point let the matter rest.[10]

c. GALILEO AND MAZZONI

There are no evidences of other technical activity on Galileo's part in 1588, but it is known that later in the year he was invited to give two lectures at the Florentine Academy on the dimensions of hell as described in Dante's *Inferno*. These literary exercises probably kept him occupied during the fall and winter. He may also have assisted his father Vincenzio with musical experiments, for the latter's adversary, Gioseffo Zarlino, published a work in 1588 attacking Vincenzio's musical theory, and Galileo's skill in measurement would have been a considerable asset in justifying his father's views against those of Zarlino.

Galileo's application for the post in mathematics at the University of Bologna was seconded by a Signor Artani, who certified that Galileo, then "about 26 years of age," had studied under Ricci and had also given public lectures on mathematics at Siena, plus some private tutoring in Florence and Siena (19:36). Another letter from Enrico Cardinal Caetano, written from Rome on February 10, 1588, similarly attested to his suitability for the post (10:26-27)—possibly on the basis of Clavius's rec-

[10] For a full treatment of the matter of the *petitio* and subsequent efforts to tighten up Galileo's proof, see Raffaello Caverni, *Storia nel metodo sperimentale in Italia*, 6 vols. (Florence: Stabilimento G. Civelli, 1891-1900; reprinted New York: Johnson Reprint Corporation, 1972), vol. 4, pp. 101-112.

ommendation. Despite these testimonials, however, the position at Bologna did not go to Galileo but went to a Paduan astronomer named Magini, who took over the post on August 4, 1588. Thus Galileo had to seek another position, which fortunately opened up when Fantoni retired from his chair at Pisa in July of 1589. Galileo was appointed to succeed him, undoubtedly through the influence of Guidobaldo del Monte, and began public lectures on the first book of Euclid's *Elements* on November 14, 1589 (19:43). Joining the faculty slightly before Galileo, but as a professor of philosophy, was Jacopo Mazzoni, a distinguished humanist who in 1587 had published a scholarly work on Dante and who was soon to become Galileo's close friend.[11] Mazzoni entered the philosophy cycle at the point where he would be teaching the *De caelo* in 1589, along with Verinus and Bonamicus, and the *De anima* in 1590, but at a salary higher than those of his colleagues in the cycle. Galileo, on the other hand, received less than a tenth of Mazzoni's salary and only half of what had been given his predecessor in the chair of mathematics.

The question naturally arises about what Galileo was doing in the early part of 1589, preparatory to taking over Fantoni's teaching duties. He could have continued with private teaching, and he may even have taught mathematics at the Monastery of Vallombrosa, since he was engaged there with a course on *Perspectiva* in the summer of 1588.[12] An additional possibility suggests itself, however, and this is in connection with a statement made by Galileo some twenty years later, when seeking a post as mathematician to the Grand Duke of Tuscany, Cosimo II. On May 7, 1610, Galileo wrote to Belisario Vinta asking that, should he get this position, he would wish that "in addition to the name of Mathematician his Highness adjoin that of Philosopher, for I may claim to have studied more years in philosophy than months in pure mathematics" (10:353). Since Galileo was apparently dissatisfied with his instruction in philosophy at Pisa, as about to be noted, one may wonder when he had time to pursue these additional "years" of philosophical study. It is at this point that the materials surveyed in the first two chapters of the present work become immediately relevant. At that time Galileo was showing great promise as a mathematician, but he had already run into a methodological problem in explaining one of his demonstrations to the great Clavius. What better expedient, under the circumstances, than to return

[11] For information on Mazzoni, see Frederick Purnell, "Jacopo Mazzoni and Galileo," *Physis* 3 (1972): 273-294; also N. Badaloni, "Il periodo pisano nella formazione del pensiero di Galileo," preprint of an article to appear in *Saggi* on Galileo and furnished the author by Dr. Carlo Maccagni of the Domus Gailaeana, Pisa, in October of 1972. To the author's knowledge Badaloni's article has not yet been published.

[12] The author is indebted to Thomas B. Settle for knowledge of the course on *Perspectiva*, furnished him in a private communication.

to the study of philosophy, and to go precisely to the point of the difficulty he had encountered with Clavius, namely, to logic, and review all the matter relating to the suppositions required for demonstration and to the tract on demonstration itself. Galileo's relations with Clavius, moreover, were sufficiently cordial for him to have no qualms about seeking the Jesuit's advice about the materials he should study. In fact, in light of the evidence presented in part 1, it is reasonable to believe that this is precisely what Galileo did, and that, in reply to his request, Clavius sent him the set of notes for the logic course just completed by Valla at the Collegio Romano. This would explain how Galileo came to acquire such a recent set of lectures, why he should have been interested in the treatises on *De praecognitionibus* and *De demonstratione*, and why he should have gone to the trouble of writing out such a complete set of notes on these difficult topics at this stage of his academic career.

There is ample reason to believe, moreover, that Galileo was not on the best of terms with the philosophers who had taught him when he was a student at the University of Pisa, namely, Borrius, Verinus, and Bonamicus, so that apart from the newly arrived Mazzoni he had no one to whom he could turn for help in the study of philosophy. The method of teaching at Pisa, where the books of Aristotle were covered in a cycle rather than in systematic order, apparently did not appeal to a mind such as his, with its mathematical cast. In one of the drafts of the *De motu antiquiora*, in fact, Galileo inveighs against philosophers who teach the physics and even the logic in reverse order, beginning with the last books on these subjects rather than the first, asking students to believe that their statements are true merely on the strength of Aristotle's authority, with the result that the students never come to know anything by its causes (1:285). This was a criticism that surely could be directed against his Pisan instructors, whereas it could not be levied against the courses offered at the Collegio Romano, where each professor started his class in philosophy with the first books on logic and then treated all of the philosophical disciplines systematically from beginning to end. Having Valla's notes on logic before him must have been a revelation to the young Galileo, and one can easily believe that, having reviewed the matters on demonstration and the foreknowledge requisite to it, he would again have appealed to Clavius for a more complete set of notes that would also embrace the whole of natural philosophy. Again, it is reasonable to believe that Clavius would have secured for him the requested materials, which then formed the basis for his composing the physical questions along the lines suggested in chapter 2.[13]

With Galileo thus engaged in the intensive study of philosophy, re-

[13] Secs. 1.1 and 2.4.

viewing the whole of the logic and then the eight books of the *Physics*—and probably composing notes on all of this subject matter, now lost but yet referred to in those that still survive—there is no problem accounting for his activity in the years from 1588 to 1590 or for his later claim about the years spent studying philosophy. An additional piece of information strengthens this interpretation, namely, the letter written by Galileo to his father on November 15, 1590, wherein he requests a copy of his *Sphaera* and some volumes of Galen, and then assures his father that he is applying himself to study and is learning from Signor Mazzoni, who sends his greeting (10:44-45). The references to the *Sphaera* and to the Galen, who is cited repeatedly in the portions of the physical questions dealing with the elements, suggests that Galileo had by this date progressed to a detailed study of the *De caelo* and the *De generatione* and was actually composing MS 46 late in 1590. Such an activity coheres well with the courses Galileo would himself be teaching at the university at that time, and the reference to the assistance given by Mazzoni makes further sense, since the latter would have covered the *De caelo* and the *De generatione* in the portion of the philosophy cycle he had just taught in the academic year 1589-1590.[14]

This reference to Mazzoni is of considerable importance for indicating an influence on Galileo during his formative years that was different from those of Ricci and the Jesuits of the Collegio Romano and yet turned out to be compatible with both. Unlike his philosophical colleagues at Pisa, Mazzoni was not a monolithic Aristotelian; although he was undoubtedly competent in expounding the text of Aristotle, he also had Platonic sympathies, and in fact introduced a course in Plato's thought at the university during the summer of 1589.[15] One of his predilections was to compare Plato with Aristotle. He had attempted a concordance of their views in an early work, *De triplici hominum vita*, printed at Cesena in 1576, and then, after mature consideration, published at Venice in 1597 his *In universam Platonis et Aristotelis philosophiam praeludia, sive de comparatione Platonis et Aristotelis*, usually referred to as the *Praeludia*. After the publication of the latter work, Galileo wrote to Mazzoni, remarking how their discussions at the beginning of their friendship had influenced its composition (2:197.14-17). The work is therefore valuable for the light it sheds on Galileo's studies in 1590, especially since a number of the topics taken up in the *Praeludia* are closely related to matters discussed in the logic and physics notes deriving from the Collegio Romano. Despite his erudition, Mazzoni appears to have oversimplified the positions

[14] See sec. 2.4.

[15] On Mazzoni's Platonism, see Galluzzi, "Il 'Platonismo' nel tardo Cinquecento e la filosofia di Galileo," pp. 65-79.

of Plato and Aristotle, giving all credit to the former for using mathematics in natural philosophy and depicting the latter as adamantly opposed to such use. This view, which was advanced by Pererius at the Collegio Romano and had to be actively combatted by Clavius (both Pererius and Clavius are cited by Mazzoni), would help explain the first anti-Aristotelian statements of Galileo in compositions dating from this period.

Noteworthy in Mazzoni's treatment is the favorable consideration he gives to scholastic thinkers, in contrast to his Aristotelian colleagues who dwelt almost exclusively on Greek and Arab commentators. Mazzoni is especially partial to St. Thomas Aquinas and to Cajetan, although he mentions other Thomists such as Capreolus, Herveus Natalis, and Domingo de Soto; he also references Scotus, Henry of Ghent, and nominalists such as Gregory of Rimini. When discussing theories of the division of the sciences he is aware of the extreme positions of Antonius Mirandulanus and of Scotus, but adopts nonetheless that of St. Thomas.[16] He argues, against Copernicus, that the earth is at the center of the universe, apparently relying on Clavius, much along the line pursued in Galileo's physical questions.[17] Another topic having affinity with Galileo's composition is Mazzoni's raising the question as to whether primary qualities are the forms of the elements, which occurs in a context where he is discussing how accidents can be superior to substance, a discussion not without interest for those promoting the acceptance of mathematics as a true science.

The most important thesis advanced by Mazzoni, however, is developed in the context of his inquiring about the relative merits of Plato and Aristotle for removing impediments (*impedimenta*) to learning. Such impediments interested Galileo throughout his life, and it is likely that his studies with Mazzoni were seminal in their regard.[18] Generally Mazzoni is much more partial to Plato than he is to Aristotle in working out this thesis, but there is one important exception. On the matter of physics or natural philosophy being either a science (*scientia*) or mere opinion, the former being Aristotle's position and the latter Plato's, Mazzoni sides categorically with Aristotle, and cites Aquinas's explanation that there can be a science of nature as persuasive in this matter.[19] While seeing this defect in the Platonic position, however, Mazzoni favors Plato for his

[16] *Praeludia*, pp. 156-157.

[17] For a sketch of Galileo's arguments and their dependence on Clavius, see the author's "Galileo's Early Arguments for Geocentrism and His Later Rejection of Them," in *Novità Celesti e Crisi del Sapere*, Paolo Galluzzi, ed. (forthcoming).

[18] Secs. 1.3.a, 5.2, 5.3, 5.4.a, 6.3.b, and 6.4.a.

[19] *Praeludia*, pp. 183-185.

view that mathematics is superior to physics and shares more in the prerogatives of metaphysics. To make his point he relies heavily on the anti-Aristotelain polemics contained in Benedetti's *Diversarum speculationum mathematicarum et physicarum liber*, which had been printed at Turin in 1585. In this work, whose tone is so similar to that of drafts of the *De motu antiquiora* that it is difficult to believe Galileo was not influenced by it, Benedetti uses Archimedean principles to refute a number of dynamical statements found in the text of Aristotle, and blames the latter's errors on his failure to make proper use of mathematics in elaborating his physics.[20] The obvious import of Mazzoni's discussion, wherein he exposes most of Benedetti's dynamical teachings, is that the light furnished by mathematical reasoning can illuminate, and thus remove, many of the obstacles and impediments the human mind encounters in its attempts to understand the physical universe. The appeal of such a viewpoint to Galileo, who by then was already convinced of the value of Archimedean methods and who, like Benedetti, was beginning to approach physical problems using the techniques of the mathematician, can hardly be overestimated.

2. The *De motu antiquiora* (MS 71)

If the thesis being investigated in this study is correct, namely, that Galileo's early treatises on motion were written in continuity with his logical and physical questions,[21] one would expect some evidence at this point linking MS 46, containing the physical questions, with another codex containing drafts of his early works on motion, namely, MS 71. Such evidence is clearly present on the last eight folios of MS 46, for these contain an extensive series of memoranda, all pertaining to the general subject of motion, and most of which can be tied to specific passages in MS 71.[22] Because of the place at which these notations appear in MS 46, viz., that at which Galileo had just finished the alterative qualities of the elements and would next treat of their motive qualities, one can see here the program he followed. Rather than continue simply abbreviating the systematic treatises of the Jesuits, since he had reached

[20] Selections from Benedetti that make this point have been translated by Drabkin in Drake and Drabkin, *Mechanics in Sixteenth-Century Italy*, pp. 166-223.

[21] Sec. 2.4.

[22] The memoranda are translated into English, by I. E. Drabkin, in Drake and Drabkin, *Mechanics in Sixteenth-Century Italy*, pp. 378-387; an exhaustive study of their correspondences, with the materials in MS 71, is given in Fredette, "Bringing to Light the Order of Composition of Galileo Galilei's *De motu antiquiora*."

the subject that had interested him at the outset, he could now compose his own tract on the motion of heavy and light objects, which would continue on with the materials covered in *La bilancetta* and the *Theoremata*. His predecessor at Pisa, Filippo Fantoni, had composed a *Quaestio de motu gravium et levium* while teaching mathematics there, so apparently there would be no objection from the philosophers on the faculty against his doing the same.[23] To do so, however, he would make jottings, or take notes on the positions adopted in the *reportationes*, supplement them with reflections of his own and with materials from other sources, including those referenced by the Jesuits, and then use these memoranda as the basis for his own composition.

The fruits of Galileo's labors in constructing his *De motu*, filling out the outline thus provided by the memoranda, are preserved in MS 71. Unfortunately, the curator who bound together the folios making up this codex got them slightly out of order, and his error was compounded by Favaro's rearranging its components in a different order still in the National Edition. If one follows the order in which the memoranda appear in MS 46, however, one can discern the way in which Galileo wrote out the successive drafts and thus can gain a good idea about the development of his thought.[24] His first attempt was to lay out a plan for the work, preserved in folio 3v, which mentions most of the topics he wanted to treat, although not in the order he finally adopted (1:418-419). Then came a fairly well-developed treatment of these problems cast in the form of a dialogue, on folios 4r through 35v, which was never completed (1:367-408). Following this, Galileo apparently decided to adopt the more formal style of a *tractatio* and wrote out a first version of this, in twenty-three unnumbered chapters, now found on folios 61r through 124v (1:251-340). Dissatisfied with the first two chapters of this, he then attempted a revision of these, only to discard them also; they constitute a brief second version and are preserved on folios 133r-134v (1:341-343). Finally he undertook a more thorough overhauling of the first six chapters, which he replaced by ten alternate chapters, found on

[23] See Schmitt, *Studies in Renaissance Philosophy and Science*, Essay 10, p. 61.

[24] This mode of reasoning has been pursued by Fredette in his various studies of MS 71; see his "Galileo's *De motu antiquiora.*" *Physis* 14 (1972): 321-348, which is a summary of his Ph.D. dissertation, "Les *De motu* 'plus anciens' de Galileo Galilei: Prologomènes" (Montreal, 1969). Stillman Drake has proposed a different ordering of the contents of that manuscript in his "The Evolution of *De motu* (Galileo Gleanings XXIV)," *Isis* 67 (1976): 239-250, which unfortunately does not take account of Fredette's findings as reported in his Blacksburg paper of 1975. Although some elements of Drake's ordering can be used to make a stronger case for Jesuit influences on Galileo, in what follows preference will be given to the ordering worked out by Fredette.

folios 43r-60v (1:344-366); these, taken with the remaining seventeen chapters of the original version, go to make up the third and definitive version. In the end this also proved unsatisfactory and was never published by Galileo. He preserved all of these versions, however, and at some time later added to them a brief composition, *De motu accelerato*, which formed the basis for his treatment of falling motion in the *Two New Sciences*; this is contained along with the other drafts in MS 71 on folios 39r-42v (2:261-266).[25] Favaro transcribed the memoranda of MS 46 and put them not at the end of the physical questions (1:15-177) where they occur in the manuscript, but rather at the end of the materials transcribed from MS 71 (1:409-417). Perhaps it needs be mentioned that these memoranda were probably not composed all at one time; they seem to have been an ongoing exercise to which Galileo kept adding materials and ideas as these occurred to him, and which thus formed the framework around which he constructed the dialogue and the successive versions of the treatise.

a. THE DIALOGUE ON MOTION

The dialogue on motion, written in Latin as is the remainder of the manuscript, in untitled and is far from being a finished piece of work. Set at Pisa toward the end of 1590, it records discussions on the motion of heavy and light bodies between Alexander, obviously Galileo, and Dominicus, unidentified but possibly Dominicus Pisanus, to whom Benedetti had addressed a letter on the importance of mathematics as a part of philosophy.[26] In tone it is critical of Aristotle, but not stridently so. Dominicus sets the stage by saying that he would like to discuss motion, but that to do so in general terms, to determine its essence and properties, would require working one's way through all of Aristotle's *Physics* and that he would prefer now to confine himself *ad motum gravium et levium* (1:367.25)—a possible hint at the enterprise in which Galileo himself had recently been engaged.[27] He then outlines six problems he would like to have solved, all of which require mathematics for their solution and on which his interlocutor can help him because of his familiarity with "the divine Ptolemy and the most divine Archimedes" (1:368.32-33). The problems are: (1) whether there must be rest at the turning point of motion; (2) why wood falls faster than iron when dropped from a height, if it truly does so; (3) why falling motion is swifter at the end than at

[25] Sec. 5.4.d.

[26] This is printed in Benedetti's *Diversarum speculationum mathematicarum et physicarum liber* (Turin, 1585), p. 298.

[27] Sec. 5.1.c.

the beginning, and forced motion the reverse; (4) why a given body falls more swiftly in air than in water; (5) why cannons shoot farther when inclined to the vertical than when pointed horizontally; and (6) why they shoot heavier balls more swiftly and farther than light ones, though the latter are easier to move. The replies to these are poorly organized, and by the time the dialogue breaks off only three have been answered, namely, the fourth, the first, and the third, in that order, and these in rambling fashion.

Alexander (i.e., Galileo) begins by introducing the topic of heaviness and lightness, with which much of the dialogue is concerned, but the discussion turns immediately to the projectile problem, and here the main classical and medieval solutions are rehearsed along the lines of Menu's and Valla's lecture notes. The notion is even introduced that a nonviolent motion, in the absence of resistance and contrary forces, would endure forever (*perpetuo duraret*, 1:373.30–31), suggesting a knowledge of Vitelleschi's position on this point.[28] The question is asked why heavy and light bodies are arranged as they are, and to this only general answers are given similar to those in the Collegio notes.[29] The idea is suggested, however, that heavy and light may be said only comparatively and not absolutely,[30] and here a revision of the original version of the dialogue is attempted; the revision is contained in the memoranda and is marked for insertion at this point in the original. The point it makes, namely, that Aristotle was wrong in holding for lightness and heaviness as absolute qualities, since they merely effect motions of one element relative to another, permits an easy transition to the use of Archimedean principles in discussing upward and downward motions. Return is then made to the original manuscript, where Galileo calls attention to his work on the *bilancetta* (1:379.12–23) and notes that he has already proved all the theorems necessary for its understanding. These theorems, he says, though not different from those demonstrated by Archimedes, will be supported by proofs that are less mathematical and more physical (*demonstrationes minus mathematicas et magis physicae . . . afferam*, 1:379.21), and they will be based on *positiones* that are clearer and more manifest to the senses (*positionibus utar clarioribus et sensui manifestioribus*, 1:379.22) than those employed by Archimedes. There follows a series of demonstrations on the flotation and submergence of bodies in water, accompanied by diagrams that picture bodies immersed in pie-shaped segments of the me-

[28] Secs. 4.1.d and 4.3.

[29] Sec. 4.2.a.

[30] The same idea resurfaces in Guevara's *Mechanica* of 1627, pp. 10 and 66, as noted in chap. 4 *supra*, sec. 4.4.b.

[31] Secs. 3.2.b and 5.1.b.

dium, thus seemingly taking into account the curvature of the earth's surface. Galileo's method of proof frequently employs a double *reductio ad impossibile* such as is found in the *Theoremata*;[31] for example, he shows that a volume cannot be greater than a given amount and that it cannot be less than the same, and therefore must have the value he claims for it. The discussion concludes with an airing of the problem whether elements have weight in their proper place, and why men do not sense the weight of air or water in which they may be immersed, for which answers similar to those in the Collegio lectures are given.[32]

The dialogue then moves to a consideration of the turning point of motion, and here Galileo's methodological statements are again noteworthy.[33] Starting from four *suppositiones* expressed in quantitative form relating to the motive forces and resistances encountered by an object thrown upward, he argues that the object's motion will be continuous and that no rest will intervene when it turns to begin its downward course. Dominicus is most impressed by this procedure, and says that he cannot but agree that these demonstrations conclude necessarily, since they depend on the most manifest and most certain principles, which cannot possibly be denied (*demonstrationes istas necessario concludere, cum ex principiis manifestissimis et certissimis, quae nullo modo negari possint*, 1:391.21-23). Galileo then proceeds to contrast his method with that of Aristotle in his treatment of the void, where the latter employed a kind of geometric demonstration (*demonstratione quasi geometrica est usus*, 1:396.6) and yet come to an erroneous result. The form of argument Aristotle used was correct, says Alexander, speaking for Galileo, and it would have led to a necessary conclusion, if Aristotle had demonstrated what he had assumed or if, at least, his suppositions were true even though not demonstrated (*si ea quae assumpsit demonstrasset Aristoteles, aut, si non demonstrata, fuissent saltem vera*, 1:397.11-13). He then argues that Aristotle regarded his assumptions as axioms, whereas in fact they are not obvious to the senses, nor have they ever been demonstrated as true, nor are they even demonstrable (*non sunt sensui nota, verum nec unquam demonstrata, nec etiam demonstrabilia*, 1:397.14-16), and proceeds to show their falsity. These remarks show a sophisticated awareness of the matters contained in the logical questions, and especially of Galileo's concern over the truth of premises on which arguments are based, possibly triggered by his first encounter with Clavius.[34]

After thus exposing the defects in Aristotle's argument, Alexander (i.e., Galileo) voices an aside (1:398.4-19) in which he complains that contemporary philosophers will not listen to his reasoning; this aside has

[32] Sec. 4.2.c.

[33] Sec. 4 1.d.

[34] Secs. 1.3 a-b and 3.1-3.

a counterpart among the memoranda (1:412.19–22), possibly written later, wherein Galileo complains that those who read his writings will not even attempt to understand them but will reject them as false out of hand. Such expressions suggest that Galileo had already had unpleasant experiences with the philosophy professors at Pisa, and thus confirm that the time of their writing should be dated late in 1590 or in 1591. The context of these remarks is a discussion about what determines the speed of a natural motion, wherein Galileo's claim is that the true cause (*vera causa*, 1:400.9) is the lightness or heaviness of the medium with respect to that of the moving body. From this he deduces that the arguments of Aristotle based on geometrical ratios are erroneous, and that arithmetic differences should be used in the calculations. This result is applauded by Dominicus, who sees in it confirmation that one cannot "attain philosophy without a knowledge of divine mathematics" (1:401.23)—a conclusion to which Benedetti undoubtedly had wanted Dominicus Pisanus to come.

The concluding portions of the dialogue address the problem of the acceleration or of the *causa intensionis* (1:405.31) of falling motion, and make use of the preceding discussion to show that the increase of speed results only from the continual decrease of a force previously impressed on the falling body. Archimedean notions of specific gravity are introduced into this exposition to account for the effect of the medium through which the body falls. It is noteworthy that Galileo here refers to specific gravity as *propria gravitas* (1:402.5, 406.25, 408.20 and 30), a characterization possibly deriving from the Jesuit *reportationes*, wherein the expression *propria forma* is frequently used to describe the specific or substantial form of a physical body.[35] Another point of similarity is Galileo's pointing out the difficulties of verifying how fast bodies fall when they are observed from below, which resonates with Vitelleschi's observations on the problems of obtaining precise data in experiments with falling bodies (1:407.8–31).[36]

b. THE FIRST VERSION OF *DE MOTU*

The reformulation of the materials contained in the dialogue goes to make up the major portion of the *De motu antiquiora*, of which three versions exist. The first of these is the most extensive, and a goodly part of it remains substantially unchanged even after the revisions proposed

[35] The expression *propria anima* was also used by the Jesuits to designate a soul that served as the substantial or specific form of an animated body; it is noteworthy that Galileo used this terminology in PQ3.6.11–12 when discussing whether the heavens were animated by a "proper" intellective soul; for the Latin, see *Opere* 1.105.17, 19, and 24.

[36] Sec. 4.2.d.

in the second and third versions are taken into account. This being so, major attention will be given in the following to the twenty-three chapters constituting the preliminary version, after which some brief comments will be made about the other two.

The key to the organization of the first treatise or essay *De motu* is contained in an item indicated on the plan with which MS 71 begins, viz., one stating that in motion three things are to be considered—the movable object, the medium, and the mover or the motive force (1:418.15). Internal evidence indicates that the essay is divided into two books, the first made up of thirteen chapters and the second of the remaining ten, and a further perusal of the contents shows that the first book treats mainly the movable object and the medium through which it moves, whereas the second book concentrates on the movers or motive powers that effect the motions. The work thus shows an orderly progression through complex subject matter, quite different from the rambling structure of the dialogue, but with a resumption and completion of all the topics previously discussed. More specifically, Book One gives evidence of being divided into three sections, the first, comprising chapters 1 through 6, is concerned with *gravitas* and *levitas* in general; the second, made up of chapters 7 through 9, treats of the motion of heavy objects considered from the viewpoint of the movable body itself; and the third, chapters 10 through 13, considers the effect of the medium on the body's motion. Book Two then takes up problems arising from the type of force that initiates the motion, viz., whether internal and thus giving rise to natural motion, or external and so producing forced or violent motion. Its first section is concerned with delineating difficulties associated with the two types, chapters 14 through 16 treating of the movers involved in natural motion and chapters 17 and 18 concerned with those in forced motion. Its second and concluding section then examines quantitative characteristics or mathematical properties of the two kinds of motion, chapters 19 through 22 being concerned with falling motion and chapter 23 with projectile motion.

The first six chapters essentially duplicate the treatment of heaviness and lightness in the dialogue, admitting the existence of both types of motive quality, *gravitas* and *levitas*, but now introducing the concept of specific gravity (*propria gravitas*, 1:251.2) at the outset. A more extensive examination of the reasons for the arrangement of the elements in the universe and their quantitative distribution is given, along lines similar to those found in the Jesuit *reportationes*.[37] Such general considerations aside, the first demonstration proves a proposition basic to the treatise

[37] Sec. 4.2.a.

as a whole, namely, that bodies of the same heaviness as the medium in which they are situated move neither upward nor downward. The proof takes the form of a *reductio ad inconveniens* (1:256.13), and the diagram used to illustrate it and subsequent demonstrations employs rectangular segments (rather than the pie-shaped segments of the dialogue) to represent bodies and the media in which they are immersed; by this time, therefore, Galileo had apparently ceased to be concerned about the curvature of the earth's surface, for reasons to be discussed shortly. A second demonstration is concerned with showing how and why motion upward results from *levitas* (1:257.1), a conclusion that will be revised in subsequent versions of the treatise. Toward the end of the last chapter of this section there is a copying break in the manuscript (at 1:259.27), which is a fairly good sign that, although written in Galileo's hand, MS 71 contains mainly clean copies of drafts he had prepared earlier—a circumstance that can serve to explain why the hand of MS 71 sometimes appears different from that of MSS 27 and 46, although the contents of all three manuscripts date from the same period.

The second section of the first book, which focuses on the movable body in its natural motions, raises more interesting questions and goes beyond the quantitative methods of the dialogue to parallel those used by Benedetti in his *Diversarum speculationum . . . liber* of 1585. The similarity is so striking, in fact, and the anti-Aristotelian tone of Benedetti's work becomes so much more explicit in Galileo's composition, that it seems likely that at the time of its writing he and Mazzoni were reading extensively in Benedetti and absorbing the latter's mentality.[38] It is here also that problems begin to surface relating to the closeness of fit between mathematics and physics that were to occupy Galileo until the end of his life.

There can be no doubt that Galileo is very much concerned with a casual analysis of natural motion, for the section starts out by inquiring *unde causetur celeritas et tarditas motus naturalis* (1:260.5); yet Galileo soon admits that though he is seeking the causes of effects, such causes are not really found in experience (*quaerimus enim effectuum causas, quae ab experientia non traduntur*, 1:263.26). In other words, natural causes are in large part hidden causes, and they can only be discerned from a careful study of the effects they produce—a standard Aristotelian and scholastic doctrine.[39] When inquiring about the speed of fall of bodies in relation to their *gravitas*, however, Galileo shows his awareness of the difference

[38] Sec. 5.1.c. On Benedetti's anti-Aristotelianism, see Carlo Maccagni, " 'Contra Aristotelem et omnes philosophos,' " in *Aristotelismo Veneto e Scienza Moderna*, Luigi Olivieri, ed., vol. 2, pp. 717-727.

[39] Secs. 3.2.a., 3.3.b, and 3.4.c

between a *causa per accidens* and a *causa per se* (1:266.8-9) and the way in which this complicates the search for causes. Earlier, in the dialogue, he admitted the possibility that a *causa per accidens* might introduce an interval of rest at the turning point of motion (1:392.32), but had brushed this aside on the basis that such rest would then not be necessary. Now, too, he seems confident that the essential factors that affect the ratios of speeds of fall of different bodies in the same medium can be isolated, mainly through the use of specific or *propria gravitas* and the Archimedean principles it allows one to import into the investigation. Yet accidental factors are not so easily disposed of, for Galileo returns at this juncture to the second problem with which he began the dialogue, namely, why is it that a lighter body falls more swiftly than a heavier one at the beginning of its motion (1:273.27; see also 1:267 n. 1). He accepts this as a fact that contravenes the general rules (*universales regulae*) governing the ratios of fall he has calculated, and frankly admits "that a great difficulty arises at this point, because these ratios will not be observable by one who makes the experiment" (1:273.22-23). The remark is very important, for it shows that the experimental cast of Galileo's thinking, already evident in *La bilancetta*, had not been obliterated by his work in mathematics. Apparently he thinks that the departures from expected results will be explicable, for he returns to them later, but for the moment he ascribes them to accidental causes; they are to be treated as quasi monsters (*quodammodo prodigia*) that arise in nature but from unnatural causes, a remark that reflects a sound knowledge of the second book of Aristotle's *Physics*.[40]

The third and concluding section of this first book of the *De motu* contains four chapters, all of which indicate in their titles that they are written "in opposition to Aristotle" for his teachings on the void and on the medium through which motion occurs. The tone is more strident than that in the dialogue, but it is noteworthy that Galileo criticizes the methodology Aristotle employs to dispose of the void in exactly the same terms he uses in the dialogue. Aristotle's conclusions would have been necessary, he writes, if his suppositions were either demonstrated or, if not demonstrated, at least true (*si ea quae assumpsit demonstrasset Aristoteles, aut, si non demonstrata, fuissent saltem vera*, 1:277.33-34), but he erred in taking as axioms assumptions that are not manifest to the senses, nor have they ever been demonstrated to be true, nor are they even demonstrable (*tanquam nota axiomata assumpsit, quae non solum non sunt sensui manifesta, verum nec unquam demonstrata, nec etiam demonstrabilia* (1:277.35-278.2). The reassertion is of significance methodologically, for it shows unambiguously that at this stage of his life Galileo was aware

[40] Aristotle treats of accidental causes in nature and the ways in which they produce monstrosities in chap. 8 of that book; see especially 199b5.

that the principles on which sound demonstrations are based must either be manifest to the senses or be true, whether they have been already demonstrated as true or will be so demonstrated in the future.[41] On a separate folio prepared for insertion into the chapter where he is discussing these matters (c. 10), Galileo makes the statement that no one up to now has dared to question the rule Aristotle used for calculating speeds of fall in various media (*hanc nullus hucusque negare ausus est*, 1:284.15-16). Apparently by this he wished to assert the independence of his own discovery, for many authors (including those surely known to Galileo) had previously questioned and rejected precisely this rule.[42] In another chapter of the section (c. 11), which he begins by criticizing the method of instruction in philosophy at Pisa, already noted, he reviews all of the arguments proving that air has no weight in air, including many touched on in the lectures given at the Collegio Romano.[43]

Book Two of the *De motu* breaks new ground, for it begins by taking up a topic mentioned in the plan for the treatise, namely, the ratios of motions along inclined planes (*considerandum est de proportione motuum super planos inclinatos*, 1:418.25), which had not been discussed in the dialogue. Galileo prefaces his treatment with a claim similar to that just seen, namely, that this topic has not been taken up by any philosophers so far as he knows. What he intends to safeguard by this statement is difficult to ascertain, for the treatment of weights on inclined planes was quite common in the mechanical treatises of the time, including those with which Galileo must have had an acquaintance, viz., the works of Guidobaldo del Monte and Tartaglia, the latter having made available the *De ponderibus* of Jordanus Nemorarius in a sixteenth-century edition, wherein problems relating to positional weights are thoroughly investigated.[44] The original aspect of Galileo's analysis would seem to lie in his claimed success in having derived the ratios of motions along such planes from principles of nature that are known or manifest (*ex notis et manifestis naturae principiis*, 1:296.20-21). He puzzled over the question, he says, why a heavy body descends faster along a plane that is more inclined to the horizontal, seeking to resolve the demonstration of this fact to its proper principles (*eiusque demonstrationem in sua principia resolvere*, 1:296.18-19), namely, those that initiate a natural motion. Galileo's

[41] Secs. 3.1.a and 3.2.a.

[42] Such rules, in fact, were much discussed by medieval and Renaissance commentators from the fourteenth through the sixteenth centuries, especially those in the "calculatory" tradition; see part 2 of Wallace, *Prelude to Galileo*, particularly pp. 78-90. Soto, of course, had rejected the rule referred to by Galileo, and his alternative was endorsed by Jesuit professors such as Rugerius; see ibid., p. 312.

[43] Sec. 4.2.c.

[44] Sec. 4.4.a.

use of *resolvere* here signals his use of a resolutive method, common practice for the Aristotelians of his day, to reduce effects to their essential causes.[45] The cause that accounts for the swiftness of descent he then identifies as the *gravitas* the body has by reason of the incline on which it is situated, which leads him to investigate what such weight (*talis gravitas*, 1:297.16) will be. His calculations then yield the famous *De motu* theorem, one of the cornerstones on which the *Two New Sciences* would be built, namely, "the same heavy body will descend vertically with greater force (*maiori vi*) than on an inclined plane in proportion as the length of the descent on the incline is greater than the vertical fall" (1:298.28-30).[46]

Immediately after asserting this remarkable proposition, Galileo qualifies it by pointing out that his demonstration is valid only on the supposition that there are no accidental factors present to perturb the result. One must suppose (*supponendum est*, 1:298.32) that the plane is in some way incorporeal or at least exactly level (*quodammodo incorporeum, vel saltem exactissime expolitum*, 1:298.33) and that the ball is perfectly spherical (*perfecta sphaerica*, 1:299.1). Under these suppositions, he goes on, one can even show that any body on a plane parallel to the horizon will be moved by a minimal force, and indeed, "by a force less than any given force" (1:299.3-4). He proceeds to prove this by mathematical arguments made with the express assumption that no accidental resistances are present (1:299.25), and concludes that the motion of such a body would be neither natural nor forced (*nec naturaliter nec violenter moveatur*, 1:299.29). In a marginal addition he speculates how such motion should properly be described, and prefers to call it neutral motion rather than mixed motion (*iste melius dicetur neuter quam mixtus*, 1:300 n. 1)—a characterization similar to Vitelleschi's *motus medius* or intermediate motion for an analogous case.[47]

Here Galileo apparently senses a need to defend his *suppositiones*, for it may appear, he writes, that he has used false propositions to demonstrate a true result (1:300.15)—precisely the charge he had directed against Aristotle. It is in this context that Galileo takes his oft-cited refuge under "the protecting wings of the superhuman Archimedes," whose name he never mentions "without a feeling of awe" (1:300.18-19). Archimedes, he recalls, made precisely the same type of supposition in his *Parabolae quadratura*, for there he treated weights suspended from a balance as making right angles with the balance even though they do not exert

[45] Sec. 3.3.a-b.

[46] The genesis and use of this theorem is discussed in detail by Wisan, "The New Science of Motion," pp. 152 ff.

[47] Sec. 4.1.d.

their force in parallel lines but actually converge toward the center of the earth. One could maintain, he now says, that the angles are right angles, or that this is an immaterial consideration since all that is necessary for the proof of the balance theorem is that the angles be equal; an alternative defense of Archimedes, he goes on, would be that he had simply employed geometric license (*geometrica licentia*, 1:300.27), as he had quite clearly done in other situations, such as when supposing that surfaces have weight, or that one surface is heavier than another, although in point of fact a surface can have no weight. Galileo is back on familiar ground when making such statements, for here he has rejoined his earlier project in the *Theoremata* of extending Archimedes' analysis of the centers of gravity of surfaces to include the centers of gravity of solids. What is remarkable about his defense of Archimedes, however, is that in it Galileo has signalled an important break with Benedetti and Guidobaldo del Monte. Both of the latter had examined Archimedes' proof of the balance theorem and had rejected it for its lack of rigor; in their view the suppositions employed exclude statics from the realm of strict science. Tartaglia, however, and Jordanus had both contenanced the use of this type of supposition to generate a mixed science, and Galileo now sides with them in his justification of Archimedes.[48] This may serve to explain why he changed from pie-shaped to rectangular figures in his drawings; for the former would indicate the necessity of taking account of the earth's spherical shape whereas the latter would indicate that this shape is immaterial, since it can be neglected on the basis of appropriate suppositions.

The inspiration behind Galileo's attitude could well have been Mazzoni, who was not disposed to follow Benedetti in all aspects of his thought.[49] A clue to this comes from Galileo's further remarks on external resistance and on the fact that no plane on the earth's surface would be horizontal in the strict sense, two factors that would prevent anyone from experimentally verifying that a sphere could be moved horizontally by a minimal force (1:301.1-10). Passing from the case of the horizontal surface to that of the inclined plane, he further concedes that, just as in the cases of vertical and horizontal motions, so too the ratios he has calculated for motion down an incline are not observable (*etiam in his motibus super planis accidit non servari has proportiones quas posuimus*, 1:302.3-4). The reason for all these deviations, he now sums up, is that his demonstrations generally are based on the supposition that there are no extrinsic impediments (*hae demonstrationes supponunt nulla esse extrinseca impedimenta*, 1:302.8-9). Such *impedimenta* are all so many accidental causes

[48] Sec. 4.4.a.

[49] Some instances of disagreement between Mazzoni and Benedetti are sketched by Purnell, "Jacopo Mazzoni and Galileo," pp. 290-293.

for which rules cannot be expected to account, since they can occur in countless ways, and so they invariably affect experimental accuracy (1:302.12-13). Vitelleschi's remarks about the extreme difficulty of deciding which ratios are correct by experiment could well be in the background here.[50] More significant, it would appear, are the lengthy discussions of *impedimenta* to knowledge that are found in Mazzoni's *Praeludia*, for it is in their context that Mazzoni sees Plato's esteem for mathematics as making his philosophy superior to Aristotle's.[51] Galileo was now clearly convinced that the *impedimenta* and the *accidentia* found in the universe of sensible matter would have to be transcended if one was to arrive at a science of mechanics, and the way was clearly marked out. Mathematical demonstrations would supply the answer, but they would have to be made *ex suppositione*, with a clear knowledge of the many impediments their suppositions would be designed to eliminate.[52]

Further discussion of circular motion in succeeding chapters amplify this conclusion. Aristotle was wrong in saying that rectilinear and circular motions cannot have a ratio to each other, for Archimedes shows in his work on spiral lines how such a ratio can be calculated. More interesting is the physical question whether circular motions are natural or forced, for here Galileo suggests the possibility that they could be neither, citing as an example a sphere rotating with its center at the center of the universe (1:304.19-21). It is in this setting that he distinguished between motions that are *praeter naturam* and those that are *secundum naturam* (1:305.11-12), employing the terminology already found in the Collegio notes.[53] This leads to the question whether the rotation of such a sphere, once started by an external mover, would go on forever (*perpetuo moveretur nec ne*, 1:305.10); Galileo postpones his answer, then never returns to it, although, as has been seen, Vitelleschi replies to a similar question in the affirmative.[54] As part of his discussion of circular motion, however, he again introduces the distinction between *causae per se* and *causae per accidens*, and points out that the accidental resistance impeding the rotation of a sphere can be reduced as much as one likes by making the ends of its axis finer and finer. If one imagines them as points, he says, implicitly letting their fineness go to the limit, they will not be the source of any resistance at all (1:307.4-15).

The foregoing problems have been concerned mainly with the moving forces involved in natural motions; in the remaining chapters of this section Galileo turns his attention to the agent causes involved in projectile motion. Here he expounds at greater length the teaching on *virtus*

[50] Sec. 4.2.d.

[51] Sec. 5.1.c.

[52] Secs. 3.1.a, 3.2.a, and 3.4.c.

[53] Secs. 4.1.b and 4.1.d.

[54] Secs. 4.1.d and 4.3.c.

impressa already touched on in the dialogue.[55] His arguments in support of impetus are those of the late medievals and the scholastics, and Galileo presents them as anti-Aristotelian arguments, even though they were all developed in Aristotelian commentaries and based on Aristotelian principles. He employs analogies between local motion and alterative motion (*motus alterativus*, 1:307.7) to strengthen his case. The most interesting development is Galileo's attack on the problem of what happens to the natural weight of a body that is thrown upward, since the force impressed on it seems to endow it with lightness rather than heaviness. His solution is not as detailed as those found in the lectures given at the Collegio Romano, though it is consonant with them.[56] The lightness impressed is really extraneous to the body (*levitas extranea*, 1:312.20), and since it exists there *praeternaturaliter* (1:311.34), the body still retains its natural gravity while being impelled upward. With regard to the permanent or transient character of the *virtus impressa*, Galileo clearly opts for the position that the *virtus* is self-expending, diminishing continuously from the moment the projectile loses contact with the impelling agent.

The concluding section of Book Two is concerned with the mathematical properties of falling and projectile motion, and in it Galileo applies the principles he has been developing to refute some of the commonly held opinions of the Aristotelians of his day. The major problem with falling motion is explaining why it accelerates toward the end of the fall; his adversaries, he says, err for a variety of reasons, among which is their confusing *causae per accidens* with *causae per se* (1:317.13-14). His preoccupation will be uncovering the true cause (*vera causa*, 1:318.3-4) of the acceleration, and to do this he will employ a resolutive method (*resolutiva methodo*, 1:318.4). Since he has already shown that the velocity is a function of the *gravitas* and *levitas* of the moving object, he will consider the problem solved if he can show how and why the falling body is less heavy (*minus grave*) at the beginning of its fall. And since its natural gravity must remain unchanged, the only reason why it could be *minus grave* would be because the diminution was *praeternaturalis* (1:318.15) and introduced from without. The explanation is easily seen in the case of a heavy object thrown upward, for in this case the *virtus impressa* overcomes the body's weight all during its upward course, though it diminishes gradually until it equals the downward force at the top of the body's trajectory; after the turning point, however, the externally induced force diminishes yet further and the natural weight becomes more and more felt. The gradual predominance of the natural weight over the extraneous lightness thus explains the body's downward acceleration. Two months

[55] Sec. 5.2.a.

[56] Secs. 4.2.a and 4.2.c-d.

after working this out for himself, Galileo writes, he discovered that the Greek philosopher Hipparchus had proposed essentially the same solution, but he thinks he has improved on Hipparchus's explanation by applying it not only to the case of bodies that fall after having been thrown upward but also to the case of bodies that fall from rest (1:319.23-320.7). The fact that Hipparchus had actually discussed both cases is an indication that Galileo got his information from a secondary source, such as one of the Collegio *reportationes*; Rugerius, as already observed, describes Hipparchus's theory and misrepresents it in the precise way it was understood by Galileo.[57]

Following some mathematical calculations illustrating how his results should be applied, Galileo summarizes his position as one invoking a combination of local and alterative motions wherein *propria gravitas* and *extranea levitas* become the principal explanatory forces, and wherein he will henceforth refer to the *virtus impressa* as *levitas* (1:322.23-26). The principle on which his solution is based is essentially that subscribed to by the Jesuits, who generally taught that the speed of the body's fall was regulated by the difference between the downward force of the body's natural *gravitas* and the resistance it would have to overcome.[58] Some particulars wherein the mechanism employed by Galileo differs from that proposed in the Collegio lectures will be discussed below.[59]

The next topic Galileo addresses, which fits naturally in this context, is the problem whether a period of rest intervenes at the turning point of motion, and here his solution parallels that already given in the dialogue, with approximately the same *suppositiones* being stressed. Following this Galileo extends his previous explanation of falling motion to show that the acceleration is only temporary, since when the extrinsic lightness is overcome the body's fall is regulated by its *gravitas*, which remains constant thereafter—a position he would later revise as he prepared the groundwork for the *Two New Sciences*.[60] The final problem relating to falling motion is the one to which he has already referred several times, namely, why lighter bodies fall more swiftly than heavier ones at the beginning of their motion. Here Galileo attempts an explanation that is somewhat implausible, but wherein he hints at experiments

[57] Sec. 4.2.d and note 87. For the misrepresentation in both Galileo's and Rugerius's explanations, see Drabkin's notes to the English translation *On Motion*, p. 90, notes 6-7, where he outlines Simplicius's and Alexander's understanding of Hipparchus's theory. From his examination of these texts Drabkin concludes that "it is questionable that Galileo knew the material at first hand. . . ." The presence of a similar error in the Jesuit lecture notes suggests these as the likely source of Galileo's information here and in the memorandum in which Hipparchus is mentioned (*Opere* 1:411.1-5).

[58] Sec. 4.2.d.

[59] Sec. 5.2.c.

[60] Sec. 5.4.d.

he has already performed with pendulums of equal length but whose bobs were made of wood and lead respectively (1:335.22-26)—another indication of the experimental bent of his thought at this period. More striking is his statement that he has repeatedly let wooden and lead objects fall from a high tower, and has found that the wood moves more rapidly than the lead at the beginning of the motion, but that later the lead accelerates and leaves the wood far behind (1:334.12-18). There is no reason to suspect that Galileo did not perform such experiments while at Pisa, and they could well be the basis for the famous legend of the Leaning Tower, originated by Viviani over sixty years later (19:606.210-218).

The final chapter of the treatise is concerned with the paths followed by projectiles that are propelled at various angles to the horizon, and attempts to account for phenomena associated with cannons and cannon balls in terms of principles previously developed, especially those relating to inclined planes. The treatment is primitive, not at all like that later taken up in the *Two New Sciences*, although it is noteworthy that Galileo had already begun to consider the mechanics of projectiles in this early essay.

c. REVISIONS OF *DE MOTU*

A clue to the purposes of the two revisions of the *De motu* treatise preserved as the second and third versions, respectively, is given in the first chapter of the second version, which is quite obviously a reworking of the same chapter in the first. In the reformulation Galileo is explaining how greater and less *gravitas* can serve to explain the natural motions of bodies; having stipulated what it means for one body to be more heavy (*gravior*) than another, he proceeds to treat of the less heavy in the following words: *Converso demum modo de levibus minus gravibus est statuendum* (1:341.23-24). Here it is noteworthy that the term *levibus* is not crossed out in the manuscript, whereas it is clearly superfluous and should have been. What Galileo apparently had in mind was a reworking of the material he had already written wherein he would replace every occurrence of *leve* with *minus grave*, and correspondingly every occurrence of *levitas* with *minor gravitas*. In the passage cited, however, he wrote *de levibus* from force of habit, then wrote *minus gravibus* after *levibus* with the intention of crossing out the latter, but forgot to do so and continued on, unaware of the redundancy.[61] Such a *lapsus calami* was not unusual for Galileo—his manuscript containing the physical questions is replete

[61] The author is indebted to Raymond Fredette for pointing out this passage in the revision of the *De motu* treatise.

with them—but this one has special interest because of the situation in which it occurs.

The project of revising the *Tractatus de motu* in order to replace the two natural motive powers of Aristotle with only one, *gravitas*, would not be unusual at this time, considering that the possibility is discussed in the notes from the Collegio Romano.[62] Those using Archimedean principles in calculating ratios of motions, as were Galileo and Benedetti before him, might see some economy in having to employ only one motive quality, although, as Guevara was to indicate years later, from a mathematical point of view the choice had little significance.[63] What might have interested Galileo in developing the alternate versions is the opportunity it would afford him of distancing himself yet further from Aristotle, and this would be very much in the spirit of the *De motu*. Thus it appears likely that Galileo attempted such a version, and did so in two stages. The first, thus the second version, consists of only two brief chapters similar to the opening ones of the original version, except that explicit references to *levitas* are replaced by references to less intense degrees of *gravitas*. Only one of the memoranda contained at the end of MS 46 relates to this version; it is similar to a portion of the second chapter, and the fact that it is late in the sequence of memoranda (twenty-second in a total of forty-two), whereas most of the memoranda that precede it can be connected to the version already discussed, argues for its later composition. The third version, on the other hand, represents a more substantial reworking of the first part of Book One in ten alternate sections or chapters, wherein a more thorough attempt is made to be rid of *levitas* as a motive principle. Most of the memoranda after the twenty-second relate to materials in its chapters, thereby providing evidence of the lateness of its composition.

Interestingly enough, two passages in the third version can be tied fairly directly to the latter part of the physical questions contained in MS 46, namely, that dealing with the elements and their qualities. In paragraph PQ5.6.7, contained in a question on the number and quantity of the elements, Galileo mentions the figures given by Aristotle for the volumes of space occupied by the different elements and says, with respect to this calculation, that he "will show elsewhere by mathematical demonstrations that it is false" (1:138.8-9). The locus he cites for Aristotle's teaching is the first book of the *Meteors*, chapter 3. The first chapter of the third version contains such calculations and also gives as a marginal reference *primo Meteororum, capite tertio* (1:346 n. 1). In a later question dealing with the alterative qualities of the elements and inquiring whether

[62] Sec. 4.2.a.

[63] Sec. 4.4.b.

they are all positive or some only privative, Galileo invokes the principle *effectus positivi debet esse causa positiva* (PQ6.2.9, 1:161.22). The same principle is recorded in the thirtieth memorandum, which begins *Effectus positivi causa debet esse positiva* (1:416.4), and it is noteworthy that exactly the same words are recorded in a passage, later cancelled, within the ninth chapter of the third version (1:362 n. 1). Such associations would suggest that all of the materials contained within MS 71, regardless of their internal ordering, stand in reasonably close temporal proximity to those contained in the latter folios of MS 46.

With regard to detailed correspondences between Galileo's teachings on falling motion and its acceleration and those contained in the Collegio notes,[64] any definitive judgment is impossible because of the late date of the available Jesuit manuscripts in which speed of fall in analyzed. Of the four sets of lectures that have been discussed, *viz.*, those of Menu, Valla, Vitelleschi, and Rugerius, only the last two treat this topic. Vitelleschi, who has the most complete analysis, provides this in lectures that were not delivered before August of 1590, whereas Rugerius explains his views in a set of lectures completed on August 7, 1591. Allowing time for the copying and distribution of the notes, it seems unlikely that Galileo would have had Vitelleschi's *reportationes* in hand by the end of 1590, and it is impossible that he would have had Rugerius's. Menu does not take up this topic in his lectures, which were not recent anyway, dating from 1578, nor does Valla in the portions of his lectures that have been preserved, though he might have done so in his course on the *De caelo*, probably completed in August of 1589. Since Valla is the most likely candidate for being the source on which Galileo's physical questions, in the latter portions, are based, the ground for arguing similarities between Galileo's treatment and those of the Jesuits would have to be that elements of Valla's exposition were preserved in the subsequent lectures of Vitelleschi and Rugerius. Such a ground is certainly adequate for showing that the conceptual framework that underlies all the analyses is the same, but it is less safe for discussing the details of particular mechanisms.

As already noted, Galileo never published any of the *De motu antiquiora*, and this is somewhat remarkable considering the confidence with which he develops the essay versions and the strength of his claims against his adversaries. Collectively taken the works contained in MS 71 are truly remarkable. They show Galileo already at the peak of his powers, having mastered the mathematical techniques of Euclid and Archimedes and showing a sophisticated awareness of the logical and methodological

[64] Sec. 4.2.d.

problems that would have to be surmounted if one were to develop a science of motion. Indeed, one could say that all of the difficulties that would have to be overcome to write the *Two New Sciences* had already been sensed or identified by him by the time he had finished their writing. His unwillingness to publish them might be traceable, as has been proposed, to the difficulties he experienced with the concept of *levitas*, especially considering that when he thought himself rid of it, it still showed up in the form of the *virtus impressa* (which he himself admitted is equivalent to *levitas*, 1:322.25-26).[65] More likely, however, it was connected with a more serious difficulty he encountered in his attempts to explain the natural acceleration of bodies, and this was his failure to verify in experience or experimentally the ratios he had calculated.[66] The program on which Mazzoni had embarked him was satisfactory from the mathematical point of view; what required more work was the identification of the physical *accidentia* and *impedimenta* that seemed to invalidate his claims not only for vertical fall and descent along inclined planes but for horizontal motion as well. Thus one may suspect that it was Galileo the experimentalist who held back the manuscript, not Galileo the mathematician. But he carefully conserved all its folios, and years later, when he had obtained the experimental data that enabled him to alter his explanation of acceleration and show its agreement with experience, he inserted the new explanation, labeled *De motu accelerato*, among the original folios, where it is still to be found. By the end of 1591, however, he was still a long way from the *scientia* and the *demonstrationes* concerning local motion that he had hoped to provide in his first major composition.

3. Other Scientific Treatises

The death of Galileo's father in 1591 placed heavy obligations upon him, and this circumstance, plus growing tensions at the University of Pisa, led him to seek the post of mathematician vacated by Moleti at Padua. By that time his reputation had grown and in 1592 he was successful in obtaining this prestigious position at a salary three times what he was receiving at Pisa. His inaugural lecture at Padua was delivered on December 7, 1592, and with it he began one of the most productive periods of his life. Paradoxically, however, Galileo wrote few systematic treatises during his years at Padua: his time was undoubtedly spent in

[65] This is basically Fredette's argument in his dissertation and in the summary given in *Physis*, note 24 *supra*.

[66] Drake makes this point in his article, "The Evolution of *De motu*," p. 249, cited in note 24 *supra*.

lecturing and tutoring, in building instruments, and, as will be seen, in extensive experimentation relating to the problem of motion. The essays that do survive seem related to his teaching activities in applied mathematics, specifically in the areas of mechanics and astronomy, and these show little change in the ideals of science he had pursued while at the University of Pisa.

a. THE EARLY TREATISE ON MECHANICS

Galileo's course in mechanics, based on the tradition of the Aristotelian *Quaestiones mechanicae*, survives in two early versions, one probably dating from 1593 and the other certainly from 1594, and in a finished version that is more difficult to date but was probably written before 1602.[67] The versions of 1593-1594 register few points of disagreement, and thus may be conveniently treated as one. The work's title indicates that it is a mechanics of machines or instruments (*meccaniche dell'istrumento*), thus practical mechanics, the second part of the discipline that, as Blancanus and Guevara attest, would usually follow the more theoretical treatment of centers of gravity.[68] Its opening chapter paraphrases the introduction to the *Quaestiones mechanicae*, and in this is not different from similar treatises on the subject dating from the same period. The science of mechanics (*la scienza delle meccaniche*), writes Galileo, teaches the reasons and furnishes the causes of the marvelous effects (*le cause degli effetti miracolosi*) we see coming from various instruments, moving and raising great weights (*pesi*) with the slightest force (*forza*, M270.7-10). Wishing to present an orderly treatise on this subject, he will first examine the nature of the primary and simpler instruments and then show how compound machines may be reduced (*si riducano*) to them. All of them, he assures the reader, can be reduced to the balance, and thus its understanding is basic to all the rest (M270.14-17).

The explanation of the balance that follows is similar to that in *La bilancetta* of 1586.[69] Galileo treats first in chapter 2 of the balance with equal arms, then extends the explanation in chapter 3 to that with unequal arms, saying that his results are demonstrated by experience (*l'esperienza ci dimostra*) and by reason also (*con la ragione ancora*), especially as Ar-

[67] The finished version is published in the *Opere*, vol. 2. For the earlier versions, see Stillman Drake, "Galileo Gleanings V: The Earliest Version of Galileo's *Mechanics*," *Osiris* 13 (1958): 262-290. From these Drake has constructed a composite text, published as an appendix to that article, which will be cited in what follows by its page number, prefixed by the letter M, and followed by the line number after a decimal, according to the convention being used for citations from the National Edition.

[68] Sec. 4.4.b.

[69] Sec. 5.1.a.

chimedes has demonstrated them in his book *Delle cose che pesano ugualmente*—the Italian title for *De aequiponderantibus* (M271.39–45). The propositions established here are applied to the lever in the next four chapters, and it is in this application that Galileo returns to an idea he had already mentioned in the *De motu antiquiora*, namely, that of a minimum force or force smaller than any assigned force (*vi omnium minima . . . vi quae minor sit quacunque vi proposita*, 1:299.30–33).[70] The context is a clarification of the statement that a force of 200 will move a weight of 2000 if applied with a leverage of ten times the distance of application; Galileo immediately qualifies this to state that such a force will merely sustain the weight and thus it is not absolutely true (*non è assolutamente vero*, M272.78) to say that the force will move it. Yet considering, he goes on, that any minimal moment (*ogni minimo momento*, M272.79) added to the counterbalancing force will produce a displacement; by not taking account of this insensible moment (*non tenendo conto di quel momento insensibile*, M273.1), one can say that motion will be produced by the same force (*forza*) as sustains the weight at rest. This is a very important observation, for it shows that Galileo was continuing to reject the rigorist mathematical position of Guidobaldo del Monte, which would not allow this minimal force to be neglected, and thus was permitting dynamic and static cases to be treated by the same mathematical principles.[71] Effectively Galileo had here bridged the gap between Archimedean statics and the Aristotelian tradition of the *De ponderibus*, and was moving in the direction of a unified science of mechanics.

The same topic is rejoined in chapter 12, where Galileo reduces the operation of the screw (*la vite*) to the principles that govern the inclined plane, apparently on the basis that a screw is nothing more than an inclined plane wrapped around a column or a cylinder (cf. M277.214–215). He observes that a heavy body descends by its natural tendency, whether it falls directly downward or works its way along any surface inclined even slightly to the horizon (M276.170–175). On the other hand, given a plane without any inclination at all, heavy objects placed on it will remain at rest; it is also true, he goes on, that a minimal force (*minima forza*) in such circumstances will suffice to move them from their place (M276.176–178). This is a repetition of the teaching contained in the *De motu antiquiora*, just referred to. Galileo again states it explicitly: a body on a level surface can be moved, not by itself, but by a minimal force (*minima forza*) applied to it from without (M277.190–191). With regard to the calculation as to the amount of force necessary to move it along the incline at various elevations, Galileo observes that to determine this

[70] Sec. 5.2.b.

[71] Sec. 4.4.a.

demonstratively would be somewhat more difficult. Thus he will pass over the matter at this point, merely noting the conclusion, namely, that the weight to be moved has the same ratio to the force moving it as the length of the inclined plane to the perpendicular height to which the weight will be raised (M277.198-204).

The remainder of the treatise is concerned with the capstan, the pulley, the water screw, and the wedge, which are explained by similar principles. To these Galileo adds an account of a few compound machines, including one made of a combination of gears and another, called the perpetual screw (*la vite perpetua*, M285.445) and resembling the modern rack and pinion, which he analyzes into its principal components, the wheel and axle, the screw, and the capstan. He continues to make reference to his proofs as demonstrations, and in his discussion of the force of percussion in chapter 18, related to the analysis of the wedge, he asserts that he is searching for the true cause (*trovare la vera causa*, M287.485) of percussive phenomena.

b. *LE MECCANICHE*

The more fully developed treatise on mechanics, titled *Le meccaniche* in some manuscripts, dates from around the turn of the century; written in Italian, it lacks the introductory paragraph of the 1594 version—also missing in that of 1593—with its brief definition of mechanics. It preserves the same tone and orientation, however. The reason Galileo adduces for writing it is that he wishes to correct the erroneous impression created by most mechanicians, who think they can employ their machines to cheat nature (*ingannando . . . la natura*, 2:155.15), overthrowing its principle that no resistance may be overcome by a force that is not more powerful than it is. He will make the falsity of their view most evident by true and necessary demonstrations (*con le dimostrazioni vere e necessarie*, 2:155.18-19). To do so he will have to take under consideration four things: the weight (*il peso*) to be transferred from one place to another; the force or power (*la forza o potenza*) required to move it; the distance (*la distanza*) over which the motion occurs; and the time (*il tempo*) the movement takes, which is equivalent to the swiftness and speed (*prestezza e velocità*) of the motion (2:156.8-15). What is gained in one of these parameters, he implies, is lost in another, and it is for this reason that nature is not cheated when a heavy weight is moved by a force many times smaller than itself. Those who maintain the contrary do so because they do not understand the nature of mechanical instruments and the reasons why they produce their effects (*le ragioni delli effeti loro*, 2:157.19).

Following a brief discourse on the advantage and utilities provided by instruments and machines, Galileo then begins his scientific treatment and models it on the standard mechanical treatises of his day, viz., those of Tartaglia, Commandino, and others. As in all the demonstrative sciences (*tutte le scienze demostrative*), he writes, it is necessary to begin with definitions (*diffinizioni*) and with primary suppositions (*prime supposizioni*), from which will spring the causes (*le cause*) and true demonstrations of the properties (*vere demostrazioni delle proprietà*) of all mechanical instruments (2:159.5-10). The definitions he then gives are those of *gravità, momento* (revised from his usage in the earlier versions to now mean a propensity to move downward), and *centro della gravità*—the last of which he takes from Commandino. His suppositions are likewise three in number, and all are concerned with various aspects of the center of gravity: the first is that a heavy body will move downward along a line joining its center of gravity to the universal center of heavy things; the second, that a heavy body not only gravitates on its center of gravity but also receives impressed forces at that center; and the third, that the center of gravity of two equally heavy bodies is in the middle of the straight line joining their respective centers (2:160.6-18). In explaining the third supposition Galileo remarks that, if two equal bodies are suspended at equal distances from a point, they will have their point of equilibrium at this common juncture, provided the equal distances are measured with perpendicular lines drawn from the weights to the common center of heavy things. Here again he has sided with Archimedes and Tartaglia against the more stringent position of Commandino and Guidobaldo del Monte, who would not admit the possibility of perpendiculars drawn to a common center because such lines would obviously not be parallel.[72]

Using his third supposition Galileo proceeds to demonstrate the general principle of the balance, namely, that unequal weights hanging from unequal distances will weigh equally (*peseranno egualmente*) whenever the said distances are inversely proportional to the weights (2:161.5-9). This, he then claims, is not only demonstrated to be true in the way in which we are certain of the truth of the third supposition, but that it actually states the same thing (2:161:12-16). The proof he offers is a generalized one that does not assume a uniform shape for bodies distributed along the beam of a balance—an assumption Galileo justifies with the statement that shape (*figura*) pertains to the accident of quality and is thus powerless to alter weight (*gravezza*), which derives from the accident of quantity. Such a statement obviously implies a sophisticated knowledge of the

[72] Sec. 4.4, with the background provided in sec. 3.4.

Aristotelian categories.[73] He goes on to make the same point he had stressed in the earlier versions, namely, that although weights counterbalance each other in a position of rest, the least moment of gravity (*un minimo momento di gravità*) will suffice to initiate movement, and on this account he will ignore this insensible amount and not differentiate between the power of one weight to sustain another and its power to move it (2:163.30-164.4).

Galileo then applies these principles to the steelyard and the lever, and reiterates that in their operation these tools do not cheat nature, since they accomplish in one motion what would be accomplished in a series of smaller movements by the same force acting over the same time—an implicit statement of the modern principle of conservation of work. In conjunction with his analysis of the windlass he again observes that one weight will exactly sustain another, and, with any little additional moment (*con ogni piccolo momento di più*, 2:168.17), will also be able to move it. When treating of pulleys, he first shows how they can be used to double a force and then proceeds to demonstrate (*dimostriamo*, 2:175.12) the method of multiplying it to any degree, which he does in two steps to illustrate multiplication by even and odd numbers, respectively.

It is in his treatment of the screw, however, that Galileo resumes the themes touched on in the earlier versions and becomes yet more explicit on the way he sees mechanics preserving its scientific character. He begins with an insight (*speculazione*, 2:179.6) that he recognizes as somewhat remote from the study of the screw but which he feels is fundamental to an understanding of the instrument. This is that any body retaining its heaviness has within itself a propensity, when free, to move toward the center, and to do so not only when falling perpendicularly but also, when unable to do otherwise, by any possible motion toward the center. Thus, given a surface polished like a mirror and a perfectly round ball (*una palla perfettamente rotonda*, 2:179.24), the ball will move down the surface if the latter has some tilt, even the slightest; if the surface is exactly level, however, and equidistant from the plane of the horizon, the ball will remain still, though it will retain a disposition to be moved by any force no matter how small (*da ogni picciolissima forza*, 2:179.32). To make this point even more dramatically, Galileo reiterates that if the surface is tilted only by a hair (*un capello*), the ball will spontaneously

[73] The Aristotelian nuances of this passage (2:163.13-14) are unfortunately lost in Drake's English translation (*On Mechanics*, p. 155), where *accidente di qualità* is rendered as "qualitative circumstance." This obscures the fact that quantity and quality are, in the accepted terminology of Galileo's day, two distinct categories of being, and thus one can be varied without necessarily affecting the other.

move down it or, conversely, resist being moved up it. Only on a perfectly flat surface will the ball be indifferent to motion or to rest; yet when there, any slightest force (*ogni minima forza*) will suffice to move it, and conversely, any slightest resistance (*ogni pochissima resistenza*) will be capable of holding it still (2:179.33-180.6). This line of reasoning leads Galileo to what he regards as an indubitable axiom (*assioma indubitato*), namely, that if all extraneous and accidental impediments (*tutti l'impedimenti esterni ed adventizii*) are removed, a heavy body can be moved in the plane of the horizon by any minimal force whatever (*qualunque minima forza*, 2:180.7-10). Apparently he did not wish to set himself in explicit opposition to his patron, Guidobaldo del Monte, on this point, so Galileo directs the reader's attention to Pappus and says that the latter failed in his attempt to solve problems of this kind by his supposition that a precise force would be required to move a heavy body on a horizontal plane. But this supposition is false, Galileo insists, for no sensible force (*forza sensibile*) is so required when all accidental impediments (*impedimenti accidentarii*)—which are not the concern of the theoretician (*theorico*) anyway—have been removed (2:181.6-12).

With this axiom as a basis, Galileo proceeds to attack the problem of the force required to move an object up an inclined plane. In the earlier versions he had passed over this exercise as somewhat too difficult, but here he proceeds to solve it through the use of geometrical methods. The only physical principle he imports into his analysis is the one he has repeatedly stressed in the preceding discussion, namely, that the force required to move a weight need only insensibly (*insensibilmente*) exceed the force required to sustain it. From this, plus his geometrical analysis of various paths of possible descent, he is able to derive his general conclusion: the force required to move a weight along an incline has the same proportion to the weight as the vertical height of the incline has to its total length (2:183.20-24).

The remainder of the treatise is not as detailed as the earlier versions. At one place Galileo makes a statement that implicitly poses a problem for the concept of neutral motions he had introduced in the *De motu antiquiora*, for he remarks that "heavy bodies do not have any resistance to transverse motions except in proportion to their removal from the center of the earth" (2:186.2-4), but he does not explore its implications nor does he relate it to his previous discussion in the *De motu*.[74] The last instrument he analyzes is the water screw, and then he concludes with another causal investigation of the force of percussion; this he recognizes as leaving difficulties and objections unanswered, but he promises to solve them in the *Problemi meccanici* he will append to the treatise.

[74] Sec. 5.2.b.

c. THE *TRATTATO DELLA SFERA*

Much of Galileo's teaching at Pisa and at Padua was concerned with the elementary astronomy contained in the *Sphere* of Sacrobosco, and this being the case it is not surprising that the other surviving treatise from his years at Padua is the *Trattato della Sfera*, alternately titled the *Cosmografia*. No autograph version of this work is extant, and it is difficult to date its composition precisely. One version appears to be that of a student who studied under Galileo at Padua in 1604-1605, and this bears the notation that it was copied from an original in Galileo's own hand, which indicates that the treatise was completed by that time. Three other manuscript versions exist, one dated 1606, and all giving substantially the same reading as the first. A fifth manuscript has also been discovered which varies in some particulars from the other four, and may represent an intermediate stage in the development of the treatise.[75] Some of the material contained in it is already present, in highly abbreviated form, in questions PQ3.1 and 3.2 of MS 46—clearly based on Clavius's commentary on the *Sphaera*. Thus it is quite likely that the *Trattato della Sfera*, like *Le meccaniche*, developed out of Galileo's teaching at Pisa and Padua, and indeed is an even stronger indication of the continuity of his thought as he passed from one university to the other than is the course on mechanics.

Written in Italian, the *Sfera* is a clear and elementary presentation of Ptolemaic astronomy, as is the book of Sacrobosco it purports to summarize, leaving out all of the technical details of planetary theory and its calculation of eccentrics and epicycles.[76] Among the topics discussed are the heavenly sphere and the various circles that serve to divide it, the earth and its location with respect to the heavenly sphere, the various circles on the earth's surface that are useful for purposes of measurement, and a few observations on the sun and the moon (including how they are eclipsed) and on the motion of the eighth sphere. One can gather from the exposition that Galileo must have been an excellent teacher, for he is orderly and precise in his presentation and does not clutter it with detail and digressions. The level of instruction also shows, however, that his students were beginners in this subject matter and probably would not have benefited from an advanced, more mathematical, treatment.

Perhaps the most interesting part of the *Trattato* is its introductory section, wherein Galileo explains the scientific character of the work and the general methodology that underlies it, utilizing the logical termi-

[75] See Stillman Drake, "Galileo Gleanings VII: An Unrecorded Manuscript Copy of Galileo's *Cosmography*," *Physis* 1 (1959): 294-306.

[76] For a fuller description, see the author's "Galileo's Early Arguments for Geocentrism and His Later Rejection of Them."

nology deriving from the Collegio Romano. In this science, as in all other sciences (*si come in tutte l'altre scienze*), he begins, it is first necessary to advert to its subject (*suo suggetto*) and then to touch on some matters pertaining to the order and method (*qualche cosa dell'ordine e metodo*) to be followed in the discipline (2:211.1-4). The subject of cosmography (*cosmografia*) is, as its name indicates, the world or the universe understood in the sense of cosmos, and yet not all of this is the exclusive concern of the cosmographer, for he considers only the number and organization of its parts, including their size, shape, distances from each other, and especially their motions, leaving to the natural philosopher the study of their substance and qualities (2:211.5-14). As to method, this involves four steps: the first is the study of the appearances (*apparenze*) or phenomena (*fenomeni*), which are observations of the senses (*osservazioni sensate*); the second is the formation of hypotheses (*ippotesi*), which are nothing more than particular suppositions (*alcune supposizioni*) relating to the structure of the heavenly spheres that correspond (*rispondino*) to the appearances; third come geometrical demonstrations (*dimostrazioni geometriche*), which make use of properties (*le proprietà*) of circles and straight lines to demonstrate particular accidents (*particolari accidenti*) following from the hypotheses; and finally, these geometrical demonstrations furnish materials for arithmetical calculation (*con operazioni aritmetiche calculando*) whereby tables (*tavole*) can be constructed that are useful for finding the heavenly bodies at any time (2:211.14-212.15). This is quite clearly the description of a *scientia media* in the understanding of the scholastics of the day. Cosmography is a *scientia*, for it demonstrates conclusions of its subject matter, but it does so on the basis of *suppositiones* that serve as its first principles and therefore it generates a *scientia ex suppositione* or *secundum quid.*[77] Galileo himself supports this view, for immediately after sketching the order in which the discipline develops, he states that he will be concerned only with the first principles of this science (*primi principii di questa scienza*), skipping the more difficult calculations and demonstrations, and occupying himself only with its hypotheses, striving to establish and confirm them (*confermarle*) in relation to the appearances (2:212.16-18).

Other clues that substantiate this interpretation and show its coherence with the teachings of the Collegio and Galileo's other writings are not difficult to find.[78] Taking the total subject (*totale soggetto*) of the science, here using a term discussed in the logic course, Galileo distinguishes it into two principal parts, the celestial and the elementary regions, following the line of thought developed in PQ1.1, 1.2, and 3.3 (2:212.19-

[77] Secs. 3.2.a and 3.4.

[78] Sec. 3.4.a.

35). The elementary region he further subdivides on the basis of the four elements, summarizing materials in PQ5.1 and subsequent parts of the tractate on the elements (2:213.1-14); with regard to the celestial region, he is clear that its subdivisions are based on two suppositions, the first being that a simple body can have only one proper motion, and the second, that stars are fixed to their proper spheres (cf. PQ3.1.21) and thus there will be as many spheres as there are proper motions (2:213.23-32). The basic problem then becomes one of discerning the number and the order of the heavenly spheres from the motions observed in the heavens—precisely the matter discussed from a more philosophical point of view in PQ3.1 and 3.2. An interesting comment that shows Galileo's continued interest in the relationships between mathematics and physics occurs when he is introducing his proof that the earth and the water surrounding it go to make up a perfect globe. Note, he says, that when we speak of this as a perfect sphere (*una perfetta sfera*) we do not mean this in the sense of precise mathematical perfection (*esquisita perfezione matematica*), since there are many heights and depressions on the earth's surface, granted that these are practically insensible (*quasi che insensibile*) compared to its size; yet one can only say that the earth is spherical with regard to sense (*quanto al senso*) but not with regard to certain mathematical judgment (*quanto al securo giudizio matematico*, 2:217.2-8).

The source Galileo used in preparing the *Trattato*, as already remarked, seems quite clearly to be Clavius's commentary on the *Sphaera*. This would square with Galileo's request to his father for a copy of the *Sphaera* in the letter of November 15, 1590 (10:44.31), and also with Sagredo's remark in the *Two New Sciences* to the effect that he had been studying the *Sphaera* with the help of a learned commentary (8:101.20-22). A detailed comparison of the contents of the *Trattato* with the materials contained in Clavius's commentary gives support to this likelihood, as can be seen in table 7. The upper section of the table (A) gives on the left the page numbers of the *Trattato* in the National Edition where sections following the introduction begin, and on the right the page numbers of the corresponding treatment in the 1581 edition of Clavius. The asterisked page numbers indicate a Table of Climes, in Latin, found in one of the manuscripts of the *Trattato*, and this seems copied verbatim from Calvius. The remaining exposition of the *Trattato* shows no evidence of direct copying, as do the physical questions. Rather it would seem to be a free Italian composition explaining the basic reasoning that appears to justify a Ptolemaic structure for the universe. That the outlines of such reasoning are already contained in Galileo's *Tractatio de caelo* of about 1590, however, is shown in the lower section of the table (B).

TABLE 7

Textual Correlations for The *Sphaera*: Galileo and Clavius

A. Textual Parallels Between Galileo's *Trattato della Sfera* and Clavius's *Sphaera* (1581)

Trattato della Sfera, *Opere*. vol. 2 (pages)	Clavius *Sphaera*, 1581 (pages)
215	46-55, 75-80, 104-108
217	117-134
220	135-143
221	143-148
223	194-198
224	40, 51
226	13-17, 212
227	23, 295
228	23, 225-232, 274, 276
229	233
232	23, 249, 303, 307-314
233	307-308, 321-322, 337-338
235-236	383
241	284
243	407-408, 429
*244-245	*429-430
246, 250	250, 460-461, 464
251	190-191
253	56

* Table of Climes

B. Textual Parallels Between Galileo's *Tractatio de caelo* and his *Trattato della Sfera*

Tractatio de caelo Physical Questions, MS 46 (pars.)	*Trattato della Sfera* (*Opere*. vol. 2) (pages and lines)
PQ3.1.2	214.1, 225.18-20
PQ3.1.3	214.27
PQ3.1.4-5	214.31-33, 253.16-22, 254.27-30
PQ3.1.7	253.26-30, 255.18-20
PQ3.1.12	224.32-33, 225.11-13
PQ3.1.14	253.30, 255.18-20
PQ3.1.20	213.30-32, 214.34
PQ3.1.23	254.26
PQ3.1.30	225.15-18
PQ3.1.31	253.13-16
PQ3.1.33	212.22-23, 213.30

TABLE 7 (*cont.*)

Tractatio de caelo Physical Questions, MS 46 (pars.)	*Trattato della Sfera* (*Opere*. vol. 2) (pages and lines)
PQ3.2.2	220.3
PQ3.2.3	221.17-18
PQ3.2.4	221.25-27
PQ3.2.8	216.2-6, 222.13-14
PQ3.2.10	220.7-14
PQ3.2.11	220.15-23
PQ3.2.12	220.24-221.4
PQ3.2.13	221.5-6
PQ3.2.14	215.4-6
PQ3.2.16	215.6-9
PQ3.2.21	215.1
PQ3.2.23	247.5-6
PQ3.2.25	247.7-10, 250.9-12
PQ3.2.32	250.5-7

Here textual correlations between a significant number of paragraphs of PQ3.1 and 3.2 and corresponding passages in the *Trattato* are indicated. Although Galileo's expression has changed, his conceptions are still fundamentally those of the physical questions and the source from which they were derived.

The late dating of this treatise on cosmography that locates it within the first decade of the seventeenth century poses a problem because of the express teaching it contains to the effect that the earth is immobile and at the center of the universe—and this in view of Galileo's writing in 1597 to both Kepler and Mazzoni that he favored the Copernican system over the Ptolemaic.[79] The difficulty is not a serious one when consideration is given to the type of instruction the *Sfera* was intended to provide. The terminology used to describe the celestial spheres even in the present day is largely earth-centered, and *a fortiori* the description of the earth's surface is unaffected by which cosmological system one prefers. There is little evidence, moreover, that Galileo was seriously interested in astronomical problems during his years at Pisa and Padua. The flurry of interest caused by the nova of 1604 elicited some response from him, but this was not appreciably different from the reaction of

[79] Again see the author's "Galileo's Early Arguments for Geocentrism and His Later Rejection of Them."

Clavius, Vitelleschi, and Rugerius to the nova of 1572. In fact, Galileo's failure to detect parallax in 1604 may well have caused him to lose interest in Copernicanism at that time.[80] Certainly the way in which he handles *supposizioni* in his cosmography is notably different from the interest he shows in the suppositions that underlie his science of mechanics. He simply indicates that they are the principles from which calculations are deduced, and is intent on clarifying how they are used without entering into any discussion about their justification. Quite different is the close scrutiny given to the assumptions required to justify his analyses of motion and forces in the other writings of this period.

In summary, then, as Galileo approached his fortieth year, there is no evidence that he had abandoned the logical and methodological framework of his earliest writings. There are hints at a development in his thought, however, and especially in the way he would use the doctrine of *suppositiones* in his astronomical and mechanical treatises. In the latter, as has just been seen, he would tend to equate *supposizioni* with *ippotesi* and to treat them as principles that would merely save the appearances, after the fashion of medieval astronomers. Although the earth's departure from perfect sphericity is *insensibile*, this does not permit one to neglect the differences between physical and mathematical reasoning in developing a cosmography. With the study of mechanics in both its static and its dynamic aspects the case is different. *Supposizioni* here have the status of *positiones*; like *definitiones* they must be seen and accepted as true if one is to have *scientia* and certain demonstrations. Both usages have adequate foundation in the logic notes of the Collegio Romano.[81] The novelty lies in the way Galileo would go about justifying principles taken in the second sense in the context of his investigations of motion and rest and the causes that produce them. The applied mathematician in him, symbolized by Archimedes, tells him that differences that are *insensibile* can be neglected and that in this domain it is possible to have a true mathematical physics. The inspiration for this move may have come from Mazzoni, but it could equally have come from Clavius, whose thought—particularly as developed by Blancanus and possibly Guevara—was sympathetic to just this type of development.[82] The problem lay in what precise *suppositiones* to employ in order to eliminate the *accidentia* and *impedimenta* that would make such a *scientia* possible. That this was a

[80] Drake is of this opinion; see his *Galileo at Work*, p. 110, where he states: "It therefore appears likely that Galileo, confronted with failure of the first possible confirmation of Copernicanism by direct observation, lost faith in it from 1605 until 1610." See also his *Galileo Against the Philosophers* (Los Angeles: Zeitlin and Ver Brugge, 1976), for further confirmation of this view.

[81] Secs. 3 1.a and 3.2.a.

[82] Secs 3.4.b–c, 4 4.b, and 5.1.c.

dominant concern of Galileo for well over a decade can hardly be doubted in face of the evidence already presented.

4. Experimentation and Reformulation

The obvious difference between the principles of astronomy and those of mechanics is that the former are very remote from sense observation whereas the latter are close to experience and can be subjected to test. Now a good part of Galileo's time at Padua was spent in instrument making, and his skill in measuring and performing experiments then offered just as much promise as did his abilities as a mathematician. The remarks he makes in the *De motu antiquiora*, and the impasse to which he came in working out its main theses, reveal a continuing concern over the failure of the calculations made therein to agree with tests he had devised.[83] All of this suggests that, as just intimated, the different attitude Galileo entertained toward *suppositiones* in astronomy and those in mechanics had to do with the possibility of establishing, by direct or indirect recourse to experience, the truth of the latter, coupled with a somewhat disinterested resignation over the impossibility of doing the same with the former. What is more, by the early 1600s he was at a place where he could employ all of his inventive skills to circumvent the *accidentia* and the contingencies that clothe nature's operations, and go on to formulate a science of motion as true and rigorous as any *scientia* taught in the schools.

To corroborate such a thesis there is abundant evidence that just about this time Galileo embarked on an extensive program of experimentation that really had no precedent among the mechanicians of his day, and yet, within less than a decade, was destined to achieve precisely this goal. The uncovering of this evidence is very recent, and not all investigators agree over the details of its interpretation. One thing is abundantly clear, however, and this is that in the period from about 1602 to 1609 Galileo's empirical bent asserted itself much more strongly than is indicated in the treatises already reviewed, or indeed in any of the works he was to produce thereafter.

a. CORRESPONDENCE WITH GUIDOBALDO

The key problem Galileo faced in this venture, as should be clear from the foregoing, was one of formulating *definitiones* and *suppositiones* that

[83] Sec. 5.2.c.

would be capable of furnishing a true mathematical description of the (at times defective) operations of nature. This would be no easy task, particularly when Commandino and Guidobaldo del Monte had already rejected the simplest suppositions that would make a mathematical physics conceived along such lines possible. It is somewhat ironic, therefore, that Galileo should signal the program on which he was embarking in a letter to Guidobaldo dated November 29, 1602. Apparently their differences over technical matters had not destroyed their friendship, and he and Guidobaldo had continued to correspond. In fact there are indications that Guidobaldo had performed some primitive experiments before 1600 attempting to trace the paths followed by projectiles, and had informed Galileo of these.[84] Somewhat later Galileo had apprised Guidobaldo of an interesting result to which he himself had come, namely, that an object descending along a circular arc (say, along the lower quadrant of a circle in the vertical plane) would always reach the bottom in the same time interval regardless of the point on the arc from which its motion started—a result possibly suggested by experiments with pendulums. Guidobaldo's reaction to this is lost, but from Galileo's reply (the letter just referred to) it would appear that he objected on two counts: it seemed impossible that one body would move many miles in the same time interval it would take another body to move several inches, and, moreover, Guidobaldo's own experiments with balls rolling inside a vertical hoop did not substantiate Galileo's findings. The latter's answer to these difficulties is illuminating for the view it provides of experimental method and the way Galileo thought it could solve problems of this type.

In the face of Guidobaldo's rejection of the truth (*verità*) of his proposition, Galileo begins by asserting that a particular experiment (*esperienza*) has made him certain (*certa*) of it, and that he will take the occasion to explain this clearly (*apertamente*) so that his correspondent, by repeating the experiment, can become convinced (*accertarsi*) of it also (10:98.10-14). He then gives instructions explaining how to make two pendulums with lead bobs and to set them swinging in unison, the one with a large arc and the other with a very small one, so as to ascertain that they complete their swings over precisely the same time interval. This quite obviously agrees with the result Galileo had calculated. With regard to the experiment (*esperienza*) Guidobaldo had performed, on the other hand, Galileo is not surprised at its failure. Its results could be somewhat lacking in certitude (*incerta*), he writes, perhaps because the surface he used was not well polished (*ben pulita*), or because it was not perfectly

[84] The experiments are described in Ronald H. Naylor, "The Evolution of an Experiment: Guidobaldo del Monte and Galileo's *Discorsi* Demonstration of the Parabolic Trajectory," *Physis* 16 (1974): 323-346.

circular (*perfettamente circolare*), or even because in a single passage one cannot well observe (*bene osservare*) the precise moment of the motion's beginning (10:99.48-52). In other words, these were all so many impediments that would obstruct nature's operation or prevent anyone from observing its proper course. Galileo's experiment was superior to Guidobaldo's precisely because it got around these impediments: the swing of a pendulum is as smooth and as circular as one could wish, and its built-in ability to accumulate the results of innumerable swings frees one from the uncertainties that attended the observation of a single, short motion.[85] Plainly stated, the ingenuity of his own test had unmasked the ratios of motions that were here at work, whereas Guidobaldo's had failed to do so.

With regard to his correspondent's claim that his result was impossible (*impossibile*)—whereas Galileo preferred to see it as remarkable (*mirabile*, 10:97.3-5)—he goes on to point out that the slope of a river could be so slight that an object would float only a few inches along it during the same time interval that a body dropped from a great height or impelled by a great force (*grandissimo impeto*) would cover a hundred miles (10:99.60-65). The example is clearly a physical one, and in Galileo's opinion it is no more lacking in verisimilitude than the proposition that two triangles with equal base and altitude will have equal areas, even though the sides of one would be very short and those of the other a thousand miles long—an obvious mathematical parallel (10:99.65-69). The conclusion he has demonstrated (*dimostrato*), therefore, is no more untenable (*inopinabile*) than the mathematical example (10:99.69-72).

Galileo goes on from this to discuss other propositions relating to motions along inclined planes that he has succeeded in demonstrating (*ho dimostrato*) without transgressing the limits of mechanics (*i termini mecanici*), although he also mentions a result he has been trying to demonstrate but has been unsuccessful up to that point (10:100.82-88). Such statements reveal that by late 1602 he had already come to a fairly sophisticated notion as to how a science of mechanics that included moving objects would have to be constructed. Even more revealing is the way in which he responds to a query contained in Guidobaldo's letter relating to percussive phenomena, replying that he has been speculating about how to measure the force of percussion (*per misurare il momento della percossa*, 10:100.91). Guidobaldo's own observations, he writes, are well put, for when one begins to be concerned with matter (*materia*), its very contingency (*contingenza*) alters propositions that are considered in the

[85] It is difficult to trace the source of Galileo's inspiration here; some possibilities are canvassed in Bert S. Hall, "The Scholastic Pendulum," *Annals of Science* 35 (1978): 441-462.

abstract by the geometer; when dealing with phenomena perturbed in this way one can no longer claim to have certain science (*certa scienza*)—a problem from whose consideration the pure mathematician is absolved (*è assoluto il matematico*, 10:100.92-96). It seems quite clear from this that Galileo had no illusions about the difficulty of the enterprise in which he had become involved. In some cases the hindrances that sensible matter presented to the physicist should be removed by proper experimental techniques; in others their removal would not be so easy, and one would have to be cautious before claiming the certitude of scientific results.

b. MANUSCRIPT EVIDENCE OF EXPERIMENTS

The letter of Guidobaldo indicates that by 1602 Galileo was already experimenting with pendulums and with motions along inclined planes, and that the degree of accuracy attainable in these experiments was crucial for him in accepting a result as demonstrated or not. It has long been known that Galileo performed such experiments, if only because of his references to them in the *Two New Sciences*. The recent discovery of manuscript evidence already alluded to furnishes much more detail about these and other experiments that he performed before his work with the telescope, and this new evidence now merits brief evaluation.

In addition to the codices of Galileo manuscripts previously discussed, another codex, MS 72, contains a large number of folios on which are recorded diagrams, calculations, lists of numerals, and drafts of various texts in Latin and in Italian explaining proofs or solutions of problems, most of which pertain to the science of motion. Since this material obviously was relevant to the *Two New Sciences*, the editor of the National Edition, Antonio Favaro, transcribed a representative portion of it and published it in volume 8 along with the text of that work. He left a substantial amount unpublished, however, including a number of folios containing diagrams and calculations that are now recognized to be vestigial records of experiments Galileo performed but never reported in his published writings. Favaro recognized that some of this material, moreover, dated from Galileo's Paduan period, but having no precise way of ordering in a chronological sequence the parts he selected, he merely grouped them according to various propositions they resembled in the *Two New Sciences*. In the past decade, owing mainly to the researches of Stillman Drake, a new technique of dating the folios contained in MS 72 has been worked out, and this now permits a different ordering of the folios that better reflects the temporal order of their composition.[86]

[86] Drake describes this technique in his *Galileo at Work*, pp. 76-78; see also his "On the Probable Order of Galileo's Notes on Motion," *Physis* 14 (1972): 55-68.

The technique is based on a study of watermarks, and involves a comparison of the watermarked paper of the various folios with letters written and dated by Galileo on paper with similar watermarks. The method is not foolproof, but taken along with other clues it provides a chronological record of Galileo's experimentation that has found fairly uniform acceptance among scholars.

Not so uniform is the agreement over the precise experiments Galileo performed and the results he intended to confirm by them. This is readily understandable, for although some folios contain rather complete diagrams and measurements obviously related to them, others are quite cryptic, bearing only lists of numerals and a few lines that would suggest one figure or another depending on how they were interpreted. Working from such clues, however, a number of investigators have attempted to reconstruct, and then actually perform, the experiments that are recorded on the various (dated) folios.[87] Such reconstructions have a twofold objective: not only do they furnish a good idea of Galileo's investigative procedures at a particular time, but they also permit one to judge the degree of accuracy attainable in the various experimental arrangements he was employing. Some of the disagreement just mentioned is occasioned by different views as to how accurately one could have measured various parameters, especially time, in such reconstructed tests if they were performed in the early 1600s. Other disagreement comes from the way various researchers view all of this experimentation as being related to the propositions of the *Two New Sciences* in their attempts to differentiate the order of exposition in that work from the order of discovery they see as leading to its composition.

For purposes of this study much of this disagreement is immaterial and so can be disregarded. It seems sufficient to note, therefore, that Galileo's experiments around 1602 seem to have been concentrated on the type of evidence one could obtain by swinging pendulums and by rolling balls down planes inclined at various angles to the horizontal. Shortly after that time, say from 1603 to 1605, he seems to have gotten interested in tracing the paths of projectiles, for a number of his diagrams contain parabola-like figures that would be associated with such paths.[88]

[87] Stillman Drake and James MacLachlen have pioneered in this research; other significant results have been obtained by R. H. Naylor and T. B. Settle, among others. See the bibliography at the end of this volume.

[88] Apart from Drake's "Galileo's Experimental Confirmation of Horizontal Inertia: Unpublished Manuscripts (Galileo Gleanings XXII)," *Isis* 64 (1973): 291-305, and Drake and MacLachlen's "Galileo's Discovery of the Parabolic Trajectory," *Scientific American*, vol. 232, no. 3 (March 1975): 102-110, the more important articles are R. H. Naylor, "Galileo: The Search for the Parabolic Trajectory," *Annals of Science* 33 (1976): 153-172; "The Role of Experiment in Galileo's Early Work on the Law of Fall," *Annals of Science* 37 (1980): 363-378; and "Galileo's Theory of Projectile Motion," *Isis* 71 (1981): 550-570.

One folio, dating from about 1605, shows parabolic paths that intersect a horizontal at various levels, together with measurements of distances along these levels, suggesting experiments with balls projected horizontally from surfaces at different heights above a floor and then impacting on the floor at corresponding distances along the horizontal. Later still, at about 1608 or 1609, Galileo seems to have combined inclined plane and parabolic-path investigations in a more sophisticated type of experiment wherein he rolled balls down an inclined plane set on a table at or near its edge, then allowed the balls to fall freely to the floor. The ball's path in its free descent was then regulated by him in a variety of ways. In some experiments, using a small angle of incline, he apparently rolled balls down the plane for various lengths so that they would follow the path of the incline when leaving the table and then curve downward to the floor, hitting it at different distances from the base of the table depending on their length of roll. In other tests he apparently used a steeper incline and moved the inclined plane back from the edge of the table so that the ball, on leaving the incline, would be directed horizontally along the line of the table top; then it too would follow various trajectories in its fall to the floor, depending on the height above the table from which it was allowed to drop. In yet other tests he seems to have used an incline of very steep pitch, approaching the vertical, and then arranged a curved deflector at the base of the incline that would give the ball a slight rise as it was projected from the table top; in such tests it would follow yet different paths to the floor. Since many of the diagrams recording such experiments also provide figures, and occasional indications of results that were calculated and then measured, with the differences between the two also noted, they provide a rich source for evaluating Galileo's experimental program during these years at Padua in terms of accuracy and other objectives he probably had in mind.

Apart from such diagrams and associated calculations, a few folios of MS 72 contain portions of texts that are directly relevant to Galileo's attempts to formulate laws of motion for free fall and for descent along inclined planes. One such (fol. 128) bears the same watermark as the cover sheet of a letter Galileo had written to Paolo Sarpi on October 16, 1604, and is obviously connected with the contents of that letter. Toward the end of the passage written on the folio, Galileo makes explicit mention that the distances traversed by a naturally falling body are as the squares of the times of fall, and as a consequence that the spaces passed over in successive equal time intervals are as the odd numbers from unity; he corroborates these statements by saying that they agree with what he has said all along and has observed in experiments (*che risponde a quello che ho sempre detto e con esperienze osservato*, 8:374.6-7). This is an important indication that by October of 1604 Galileo felt he had experimental confirmation of the times-squared law and the odd-number property of naturally accelerated motion. At an earlier date he apparently had given up the explanation of the increase of speed experienced by falling bodies that he had proposed in the *De motu antiquiora*, wherein acceleration was viewed merely as a temporary phenomenon that lasted only until a *virtus impressa* or extraneous lightness had been used up.[89] Thus he had at least implicitly adopted the scholastic view that the falling body varied in its velocity continuously and uniformly throughout the length of its fall, or, in other words, that its motion was *uniformiter difformis*.[90] Possibly because of his preoccupation with measuring distances, Galileo initially speculated that the speed of fall would increase uniformly with distance of fall. He knew, however, that this was merely an assumption on his part, for in his letter to Sarpi he acknowledged that he did not have a principle that was unquestionable and could serve as an axiom for his reasonings (*mi mancava principio totalmente indubitabile da poter porlo per assioma*, 10:115.3-4). Lacking such a principle, he went on, he was forced to employ one that had much of the natural and evident about it (*la quale ha molto del naturale et dell'evidente*, 10:115.5), namely, that the falling body goes increasing its speed in proportion to the distances it traverses during its fall (10:115.5-15). This principle being supposed (*questa supposta*)—a clear reference to the doctrine on supposition—Galileo feels that it will be sufficient for demonstrating the various *accidenti* or properties of falling motion that have been observed by him (*per dimostrare li accidenti da me osservati*, 10:115.2-3), presumably in his experiments.

This passage, when taken in conjunction with the notations on fol. 128, has important methodological significance. It shows that at this stage of his investigations Galileo was not employing a hypothetico-deductive methodology wherein the truth of his principles would be judged by the truth of the consequences he could derive from them. Although he claimed knowledge of the truth of the times-squared law and of the odd-number rule (cf. 8:374.7), and believed that he could demonstrate these from the principle he had assumed, such confirmation in his view was not sufficient to establish the truth of the assumed principle. To serve as a principle for a demonstrative science, there would have to be independent evidence of its truth, either as *per se nota* in its own right or as demonstrated on other grounds.[91] The way Galileo begins the passage on fol. 128 gives clear indication of his thinking in this regard. "I suppose," he writes, "and perhaps I shall be able to demonstrate this"

[89] Sec. 5.2.

[90] Sec. 4.2.d

[91] Cf. secs. 1.3.b, 3.1.a, and 3.2.a.

(*Io suppongo, e forse potrò dimonstrarlo*, 8:373.1), "that the naturally falling body goes continually increasing its velocity according as the distance increases from the point from which it departed" (8:373.1-3). A supposition is not enough to ground a *scientia*; to do this, it would have to be either true or demonstrated. In 1604 Galileo was optimistic that he could produce such a demonstration, but later, as will be seen, he found he could not do so and had to substitute a different, more accurate principle in its place.

Other folios in the same codex record changes in terminology that appear to be speculatively significant for Galileo's understanding of motion at that time. For example, on fol. 172, dating from 1603-1604, he noted that the changes in the velocity of motion (*motus velocitas*) down an inclined plane would be proportional to the changes in the moments of gravity (*gravitatis momenta*), thus correlating speed with weight and using the term *momentum* to designate a particular value of weight at a given place and time.[92] Again, on fol. 85v, written in 1604, Galileo described velocity as increasing with distance of fall in such a way that its value changes at every point along the line of descent (*velocitas augetur consequenter in omnibus punctis lineae*, 8:383.10-11). Now the usual way of referring to the velocity of a motion, generally adopted in the lectures at the Collegio Romano, was to designate this by a *gradus velocitatis*, or degree of velocity—a terminology Galileo himself consistently used before 1605.[93] But on fol. 179, which can be dated around 1605 or 1606, Galileo started to substitute the expression *momentum velocitatis* for the more usual *gradus velocitatis*, and thenceforth used the former expression to designate a particular value of instantaneous velocity associated with an accelerated motion. Such a change, wherein the Latin term *momentum*, which can have a wide variety of meanings, came to be used as a measure of instantaneous velocity, on the basis that the values of such measures were changing continuously throughout the duration of a motion, indicate that Galileo was doing more than experimenting throughout this period. He was thinking deeply about motion as a continuous phenomenon which, while capable of being measured, also presented problems in its philosophical understanding that would have to be worked out before a *scientia de motu locali* could be properly formulated.

[92] For an exhaustive study of Galileo's use of the term *momentum* in his various writings, see Paolo Galluzzi, *Momento: Studi galileiani* (Rome: Edizioni dell'Ateneo & Bizzarri, 1979); Galluzzi's thesis, which is not without bearing on the overall argument of this volume, is summarized by the author in his review of the work in *Renaissance Quarterly* 34 (1981): 400-402.

[93] For example, this expression is used by Vitelleschi in his teachings surveyed in sec. 4.2.d.

c. CONTACTS WITH THE JESUITS

During this period of Galileo's experimental activity, aside from a few manuscript fragments of the type just mentioned, there is little documentary evidence available to connect his thought with that of the Jesuits of the Collegio Romano. Contacts were not completely lacking, however, for Christopher Clavius still retained his interest in Galileo, and in fact wrote to him on December 18, 1604, bringing him up to date on his work and his publications. Among other things he mentioned his observations on the nova of 1604 and inquired whether Galileo had any data to add to his (10:121.22). This could have been Clavius's main reason for writing, but it is also possible that Clavius was reminded of Galileo by conversations with another Jesuit, Andreas Eudaemon-Ioannis, who had been at Padua and apparently had engaged in discussions with Galileo about the latter's experimentation.[94] That Galileo was on friendly terms with the Jesuits at this time is clear from a letter written by Blancanus to Christopher Grienberger on June 14, 1611, in reply to a request from the Collegio for information about an attack made on Galileo at Mantua for his teaching about mountains on the moon. Blancanus hastened to point out that he had attempted, without complete success, to head off the attack, and that he was unhappy that it had given Galileo offense, particularly because the latter had good reasons behind his position. He went on to add that he liked and admired Galileo very much, "not only for his exceptional teaching and inventions, but also for the friendship that I formed with him at Padua," whose courtesy and warmth were such that he still felt tied to him (11:126.13-16). Later Galileo, having seen a copy of Blancanus's letter to Grienberger, wrote to the latter on September 1, 1611, acknowledging his indebtedness to the Jesuit mathematician for the sentiments there expressed (*io resto infinitamente obligato al P. Biancano*, 11:180.47-48). The favorable treatment of Galileo and his father in Blancanus's publications of 1615 and 1620, together with the similarity of their ideas on the nature and importance of mathematics, gives no reason to doubt that this friendship, which probably started in the early 1600s, had continued throughout the years.[95]

The evidence for Galileo's contacts with Eudaemon-Ioannis is more indirect, being found in a letter written to Galileo from Rome by Mario Guiducci on September 13, 1624. In it Guiducci reports a visit to the Collegio Romano during which a Jesuit recounted a story about a distinguished colleague, "Father Andrea, a Greek" (13:205.9). The latter, apparently some twenty years before, had been told by Galileo while at

[94] Sec. 1.1.g.

[95] Cf. secs. 3.4.c and 4.4.b, and note 156 to chap. 3.

Padua of an experiment he had performed in which a body was let drop perpendicularly from the mast of a ship, and was found to fall at the foot of the mast whether the ship was in motion or at rest (13:205.11-13; cf. 6:545.30-34). It is known that Eudaemon-Ioannis was a Greek, and there is no doubt that he is the Father Andrea referred to by Guiducci. Now Eudaemon-Ioannis had entered the Society of Jesus in 1581 (eleven years before Blancanus), and assuming a two-year novitiate, would have begun logic at the Collegio under Lorinus in 1583 and then had Mutius de Angelis for natural philosophy in 1584 and Valla for metaphysics in 1585.[96] In 1596 Eudaemon-Ioannis was back again at the Collegio, where he taught the philosophy cycle himself, beginning with logic in the fall of that year and ending with metaphysics in the fall of 1599. At that time he apparently moved to Padua, where he was engaged in the teaching of theology. How long he remained there is somewhat problematical, but since he published a long series of apologetical works beginning in 1605 and ending in 1617, it is safe to assume that he left Padua around 1604. In any event, neither he nor Blancanus would have been there after 1606, for the Jesuits were expelled from the Venetian republic that year (cf. 10:158.14-21).

The lecture notes of Eudaemon-Ioannis for his course on the *Physics, De caelo*, and *De generatione* are still preserved in the Collegio archives, and these show that he covered systematically the same materials as were taught in the 1580s, although in less detail and with greater emphasis on philosophical issues.[97] Appended to a set of his lectures on the *De caelo*, however, is an interesting tractate in which he attempts a more or less mathematical analysis of action and reaction and of the motion of projectiles.[98] The exposition follows a pattern of first giving *definitiones*, then a number of *suppositiones*, and lastly deriving from these several series of *propositiones* relating to the matter at hand. Included among the definitions is one for *uniformiter difformis*, and many of the propositions explain how qualities are quantitatively distributed in action and reaction, usually illustrated with geometrical diagrams in the margins of the manuscript. The treatment of the motion of projectiles is less mathematical, and a negative attitude is maintained toward the existence of a *virtus impressa*

[96] For particulars concerning Eudaemon-Ioannis, see Sommervogel, *Bibliothèque*, vol. 3, cols. 482-486.

[97] The manuscripts pertaining to natural philosophy are: Cod. APUG-FC 1006, Quaestiones in libros Aristotelis De physico auditu, De caelo et mundo, anno 1598; Cod. APUG-FC 713, In libros De caelo and In libros De generatione; and Cod. APUG-FC 555, Quaestiones in libros Aristotelis De generatione et corruptione.

[98] Cod. APUF-FC 713 (lacks complete foliation, but following fol. 177r, at which the exposition of the *De generatione* concludes), Tractatus primus . . . ; De iis quae in actione et passione physica contingunt, Liber secundus . . . ; Quaestio de motu proiectorum. . . .

in the projectile. Undoubtedly it was Eudaemon-Ioannis's work on this tractate that stimulated his interest in Galileo's experiment, and this perhaps also explains why his report on it was recalled twenty years later at the Collegio. It would be very difficult, of course, to explain the result claimed by Galileo without some doctrine of impetus or other proto-inertial principle, and this must have posed a problem for Eudaemon-Ioannis.

In any event, this Greek Jesuit was apparently well acquainted with the tradition of the medieval *calculatores*, and thus he, along with Blancanus, could have been additional sources through which scholastic terminology continued to enter into Galileo's writings of the period. In his surviving notes Eudaemon-Ioannis has no treatment of motive qualities or of the motion of heavy and light bodies, but from the fact that his teachers were Mutius de Angelis and Valla one would expect that his ideas were quite similar to those contained in the set of notes closest to those used by Galileo when composing his physical questions.[99]

With respect to the speculative treatment of motion as a type of continuous entity, it should be observed that all of the Jesuit lectures on the *Physics* devoted considerable space to problems relating to the structure of the continuum.[100] Eudaemon-Ioannis's notes are somewhat distinctive in that he has no treatment of time when giving his commentary on the fourth book, as was the usual custom, but deferred this to his commentary on the sixth book, where he took up problems relating to time only under the formality of its being a species of continuous quantity.[101]

Regarding Galileo's change of terminology from *gradus velocitatis* to *momentum velocitatis*, it is no simple matter to locate the source from which this might have come. It is noteworthy, however, that the sixth book of the *Physics* provides the context in which such a change can be understood. Both time and motion are examples of "flowing" or successive continua, different from static or permanent continua such as a line segment in that all of their parts do not exist at once, but come into existence and pass out of existence in sequential fashion. All that exists of time is the instant known as the present, which terminates the past and begins the future. Similarly, all that exists of any motion is the "indivisible part" of it, to use Rugerius's expression, that completes the distance already traversed and begins that about to be traversed, analogous to the point that connects the parts of a line segment. However, although there is no terminological difficulty in identifying the point that

[99] Secs. 2.3 and 2.4

[100] Sec. 4.1.e.

[101] Cod. APUG-FC 1006, In librum sextum physicorum, Quaes. 2, De speciebus quantitatis continuae, pp. 819-854.

connects parts of a line, or the instant that connects parts of time, there is such a difficulty in naming this indivisible part of motion. Now Thomas Aquinas was one of the few Aristotelian commentators who suggested that the term *momentum* be used to designate the indivisible of motion, just as the point or the "now" designates the indivisibles of the line and of time, respectively. Thus, in a Thomistic context, *momentum velocitatis* would clearly refer to the velocity at the moment or instant the motion had just achieved, which is to say its instantaneous velocity at that particular moment. Thus it is quite appropriate to use this expression to indicate the degree or value of velocity attained at any instant, which is clearly the meaning intended for it by Galileo at the time he began to use it.[102]

d. UNIFORMLY ACCELERATED MOTION

Precisely in what year Galileo discovered that the correct principle on which to base a science of falling motion was to regard that motion as *uniformiter difformis*, not with respect to distance of fall but rather with respect of time of fall, is difficult to ascertain. All scholars are agreed that this could not have been before October 1604, because of the letter to Sarpi. Since the incorrect principle was not presented in that letter as completely unquestionable and Galileo was actively attempting to demonstrate its truth, he could have revised it shortly after the time. Indeed, if his review of the Collegio notes or his discussions with Blancanus and Eudaemon-Ioannis had uncovered the views of Domingo de Soto on falling motion, he would have seen that the scholastic teaching applied the expression *uniformiter difformis* to such motion in a temporal, rather than a spatial, sense.[103] This would have immediately given the clue to the correct principle, although it would not have supplied the demonstration for which Galileo was seeking.

However this may be, the earliest incontrovertible written evidence of Galileo's having a principle of uniform velocity increase with time of fall is found on manuscript fragments that appear to date from 1609. The first of these, on fol. 91v of MS 72, states the principle explicitly, namely, that in motion from rest the instantaneous velocity increases in the same ratio as does the time of the motion (*In motu ex quiete eadem ratione intenditur velocitatis momentum et tempus ipsius motus*, 8:281.12n), and then offers a proof of the principle. Since this proof was later used almost verbatim by Galileo in one of his theorems on projectile motion in the *Two New Sciences*, there is reason to suspect that it was his experimen-

[102] For Aquinas's usage, see Galluzzi, *Momento*, pp. 136-137.

[103] Sec. 4.2.d; see also Wallace, *Prelude to Galileo*, pp. 91-109.

tation with projectiles, especially those projected from a table top using the techniques already described, that led him to the demonstrations.[104]

The other evidence is contained in the celebrated *De motu accelerato* that is now bound in MS 71 along with all the other materials of the *De motu antiquiora*, probably inserted there by Galileo himself to furnish the correct answer to the problem he had worked on at Pisa. This document is not as easy to date as is the first fragment, since arguments can be adduced for locating it as early as 1604, as did Favaro in the National Edition, or as late as 1630, as do Drake and Wisan on the basis of its relationship to the *Two New Sciences*, to be discussed in the following chapter.[105] Both Wohlwill and Koyré, however, date it in 1609, and the terminology Galileo employs together with a few stylistic features of his writing are compatible with that dating. For example, on fol. 91v just discussed and in his manuscript draft of the *Sidereus nuncius*, the latter surely written in late 1609, Galileo indicates transpositions by underlining successive passages and then numbering them in the order in which they are finally to be read; this practice is also followed in the *De motu accelerato*.[106] The fragment obviously summarizes important experimental work done by Galileo, presumably at Padua, and is probably a first (or very early) draft of a new treatise on motion he mentions as requiring completion in May of 1610. Thus it probably belongs to the end of his Paduan period—a fitting recapitulation of his efforts there to develop his *scientia de motu locali*.

Galileo begins the *De motu accelerato* with the notation that this is the second book (*Liber secundus, in quo agitur de motu accelerato*), following

[104] The proof is sketched by Wisan, "The New Science of Motion," pp. 227-228. Her paraphrase shows that Galileo erected his demonstration (concluding with the clause *Quod erat demonstrandum*) on the basis of the double-distance rule. The argument is essentially mathematical, but if Galileo already had experimental support for this rule, he would also have physical grounds for holding that in free fall the velocities acquired are proportional to the times of fall. See the passage referenced as note 78 of chap. 6, *infra*.

[105] Drake presents his evidence in *Galileo at Work*, p. 315, and Wisan, who acknowledges that "an early date is not impossible," gives her preference on pp. 277-278, notes 2-4, of "The New Sciences of Motion." Apart from Favaro's dating of 1604 (*Opere* 2:259), Wohlwill dates the fragment in 1609 and Caverni in 1622-1623; Koyré, surveying the work of his predecessors, argues for 1609. Much of the fragment is translated in Alexandre Koyré, *Galileo Studies*, trans. John Mepham (Atlantic Highlands, N.J.: Humanities Press, 1978), pp. 95 ff. The problem of dating is discussed on p. 124, n. 137. See sec. 6.3.d for further implications relating to the thesis of this volume.

[106] Wisan reports such underlining in the fragment on fol. 91v of MS 72, but misinterprets its meaning ("The New Science of Motion," p. 227, note 22); the word order as given in Favaro, *Opere* 8:281-282, with notes, is therefore correct. The *De motu accelerato* fragment exhibiting the same underlining is on fol. 39r of MS 71; the passage underlined is correctly ordered by Favaro in *Opere* 2:261.18.

another that treated of the properties (*accidentia*) that are found in uniform motion (*in motu aequabili*, 2:261.1-3). The first thing to be done, he observes, is to seek out and explain a definition of accelerated motion that best agrees with what nature itself employs (*ei quo utitur natura*, 2:261.5). For, although it is not improper to make up any kind of local motion (*aliquam lationis speciem*) and consider the properties (*passiones*) that follow from it—as some have imagined for themselves helical and conchoidal lines arising from certain motions, granted that nature does not employ these, and have commendably demonstrated *ex suppositione* their properties (*earum symptomata ex suppositione demonstrarunt cum laude*)—nonetheless nature does employ a definite kind of acceleration for certain motions such as those of descending heavy objects (2:261.6-12). He has therefore decided to look into their properties (*passiones*) to see if the definition of accelerated motion he is about to adduce agrees with the essence (*essentia*) of naturally accelerated motion (2:261.12-15). And finally, he writes, after lengthy agitation of mind, he is confident that he has found this, chiefly for the very powerful reason that the properties thereafter demonstrated by him (*symptomatis deinceps a nobis demonstratis*) correspond to, and are seen to be in agreement with, the data that physical experiments present to the senses (*ea quae naturalia experimenta sensui repraesentant*, 2:261.15-19).

Following this introduction, to which he adds an observation about the simplicity of nature's motions in the cases of birds and fishes, Galileo inquires into the way increments of velocity are acquired in falling motion. In a passage reminiscent of the *De motu antiquiora* he notes that the moving body remains the same and so does the motor principle (*idem est mobile, idem principium movens*, 2:262.6), so why should not everything else be constant also. Not the velocity, he goes on, for this would make it uniform; where the identity or uniformity or simplicity needs to be found is rather in the additions to the velocity, or, in other words, in the acceleration (*non in velocitate, sed in velocitatis additamentis, hoc est in acceleratione*, 2:262.7-11). The key to the simplicity of nature's operation lies in seeing the greatest possible affinity between time and motion (*maximam temporis atque motus affinitatem*, 2:262.14)—a rather constant theme in Aristotle's *Physics*. Galileo exploits this affinity by developing an obvious parallel between uniform motion and accelerated motion: just as we call a motion uniform or equable if it traverses equal spaces in equal time intervals, so we may conceive of equal increments of speed (*celeritatis incrementa*) taking place in equal parts of time, and then the motion will be uniformly and continually accelerated (*motum illum uniformiter atque eodem modo continuo acceleratum esse*, 2:262.15-22). One will not be far from right reason, he concludes, if one holds that the intension of velocity takes place according to the extension of the time (*si accipiamus intensionem velocitatis fieri iuxta temporis extensionem*, 2:263.1-2)—a quite precise and scholastic way of formulating the principle of uniform acceleration in cases of free fall.

The remainder of the fragment is concerned with the resolution of philosophical difficulties that seem to arise from this formulation. One consequence of it, he observes, is that in an extremely short time interval it would not be possible to have such a small velocity that an even shorter one might not be assigned; indeed a velocity could be infinitely small, so small in fact that an object moving at its rate would traverse only a few inches in an entire year (2:263.4-16). Consistent with his earlier teachings,[107] Galileo maintains that there is nothing false or absurd in this, even though it *is* remarkable (*mirum*), and that it can be shown to be possible by an experiment (*experientia*) that is hardly inferior to a demonstration (*qualibet demonstratione haud infirmior*, 2:263.16-19). The experiment consists in dropping a weight from various heights on top of a stake driven into the earth; here the cause of the stake's penetration into the ground is the velocity with which the weight strikes it (2:263.19-26). Since the height from which the weight is dropped can be made as small as one likes, even to the thickness of a sheet of paper, there is no limit to the smallness of the velocity that can be imparted to the stake (2:263.26-264.3). The conclusion to be drawn from this is that any object, even the heaviest, when descending naturally from rest will pass through all the degrees of slowness (*per omnes tarditatis gradus*) and will not delay at any; rather, following the succession of the instants of time it will always acquire new and larger degrees of swiftness (*novos maioresque semper acquirere celeritatis gradus*, 2:264.3-7). Many other experiments (*plures alias experientias*), he adds, he could bring forth to confirm this conclusion, but he will put these in his *Mechanicae quaestiones* as in a more convenient place (2:264.7-9). The reference to experiments here is quite significant, for apparently these are not the same experiments as those he had earlier referred to, but rather were made explicitly to confirm the truth of the principle on which the first had been based.

The concluding portion of the *De motu accelerato* continues to explore the infinitesimal implications, if one may use the modern term, of the new principle to which Galileo has come, and has special interest because of the way in which it rejoins topics previously discussed by him in the *De motu antiquiora* and in *Le meccaniche*.[108] Galileo draws attention to the fact that a variety of causes (*pluribus ob causis*) can affect the way in which a particular degree of velocity (*velocitatis gradus*) is acquired over larger

[107] Sec. 5.4.a.

[108] Secs. 5.2 and 5.3.b.

or smaller time intervals, and that one of the most important of these, for the present investigation, is the path followed by the descending body (2:264.10–13). Not only do heavy objects descend perpendicularly as they tend to the center, but also along planes inclined toward the horizontal, and the more slowly they descend the greater is this inclination; therefore, they will travel most slowly along planes only slightly tilted above the horizontal, and finally will come to infinite slowness or rest (*infinita demum tarditas, hoc est quies*) on the horizontal plane itself (2:264.13–18). Thus one can see, he observes, the great difference in the degrees of swiftness acquired over various time intervals, for the particular degree acquired by a heavy body falling perpendicularly for one minute might only be acquired by another object descending along an inclined plane for an hour, a day, a month, or a year, even though its motion was being continually accelerated (2:264.18–22). Galileo then proceeds to illustrate this by a geometrical example in which both velocities and times are represented by line segments, and wherein he shows that the properties of the continuum apply equally to motions and to lines (2:264.25–266.18)—a procedure very similar to that employed by Aristotle in the sixth book of the *Physics*. Galileo's reasoning here is particularly instructive, for it shows how convinced he was at this time that nature could be geometrized, as it were, and that a strict mathematical physics could be formulated through the use of appropriate principles.

e. PROSPECTS FOR A MATHEMATICAL PHYSICS

In a letter already referred to, written to Belisario Vinta on May 7, 1610, and wherein he contrasts his lengthy studies in philosophy with his brief instruction in mathematics, Galileo describes the works he must bring to completion should he make the move from Padua to Florence and enter the employ of the Grand Duke of Tuscany.[109] Among these he lists three books on local motion, of which the second is undoubtedly that on accelerated motion—the draft of whose beginning has just been reviewed. Galileo says that these books represent an entirely new investigation in a field where no previous worker, ancient or modern, has discovered any of the remarkable properties that he demonstrates (*i moltissimi sintomi ammirandi che io dimostro*) to exist in both natural and violent motions, and thus he may call this a new science (*scienza nuova*) that he has discovered from its first principles (*da i suoi primi principii*, 10:351.107–352.112). He adds that he must also complete three books on mechanics, two concerned with demonstrating its principles and foundations (*due*

[109] Sec. 5.1.c.

attenenti alle demostrazioni de i principii et fondamenti) and another with its problems (*uno de i problemi*, 10:352.112-113)—the last possibly the same work referenced as the *Problemi meccanici* in *Le meccaniche* and as the *Mechanicae quaestiones* in the *De motu accelerato*. If Galileo is to be taken at his word, and there is no reason why he should not be, these statements mean that prior to the discoveries with the telescope in late 1609 he was well along with the development of the sciences of motion and mechanics that have been a major concern in this chapter. A brief evaluation of his progress to that date would therefore seem to be indicated.

The terminology of the letter leaves little doubt that the ideals of the *Posterior Analytics* as set out in his logical questions remained Galileo's guide for the two decades spanning his investigations.[110] Practically every treatise he wrote in that time speaks of science, demonstration, and the principles on which both are based, and this whether its major concern was mathematics or physics in the sense of an applied branch of mathematics. The teaching on supposition likewise manifests itself with great frequency, whether in the simple sense of an hypothesis that functions in an astronomical theory or in the more difficult-to-verify sense of a principle or definition that will guarantee the truth and reality of conclusions deduced from it.[111] The existential strictures placed on certain subjects of investigation because of spatio-temporal factors or other physical impediments that would have to be removed to permit demonstration seem also to be recognized over the years. Most importantly, Galileo seems well aware that the truth of the principles on which he would base his demonstrations could be certified in a variety of ways: his early preference was for principles that are obvious in sense experience, but he recognized that physical causes are frequently hidden and require the work of intellect to be uncovered; the important thing, nonetheless, was that they be demonstrable and true, even if not immediately evident. The abundance of reflective statements supporting this methodological framework makes it the unquestionable standard against which his work will therefore have to be judged.

Similar considerations in the realm of natural philosophy suggest that Aristotelian concepts as set forth in the physical questions and their supporting works provided an enduring background for Galileo's studies of motion. The anti-Aristotelian invective of the *De motu antiquiora*, much of which was suggested by and was compatible with developments within scholasticism, became more attenuated after he had left Pisa; even before that, however, nature and violence were the twin concepts dominating his thought.[112] He was continually aware of the difficulties of working

[110] Secs. 1.3, 3.1, and 3.2.

[111] Secs. 3.1 and 3.2.a.

[112] Secs. 4.1, 4.2, and 4.3.

with sensible matter, the accidents and impediments it presented to the intellect, and the variability and contingency of nature's operations. Local motion, itself an Aristotelian specification, was his predominant interest, and its acknowledged causes, motive powers such as *gravitas* and impressed forces, were his chief concern.[113] The impetus theory with which he started, and the possibility of preternatural motions that are neutral with respect to nature and so could endure forever in the absence of impediments, continued to exert its influence over a lengthy period.[114] Above all he seems to have had a clear knowledge of the category of quantity, not only in the discrete sense of number but also in the sense of extensive magnitude as found in the continuum. The different ways in which these could be studied by the mathematician and the physicist, with particular advantage to the latter, remained a dominant characteristic of his thought.

Against this more or less permanent logical and physical backdrop Galileo set out, during his Pisan period, to develop a mathematical physics that would be more rigorous than the other sciences with which he had come in contact. Under the tutelage of Ricci, and with assistance from Clavius and Guidobaldo del Monte, he developed considerable competence in mathematics; this he always regarded as a *scientia* to which his studies could be subalternated, even though it meant rejecting an alternative view of mathematics (e.g., that of Pererius) popular in his day.[115] Another option on which he made an early decision had to do with simplifying assumptions or suppositions that would make physical problems tractable by mathematical methods. In this choice he decided against Guidobaldo, Benedetti, and Commandino and sided instead with Tartaglia and a medieval-scholastic tradition deriving from Jordanus de Nemore.[116] Possibly because of this, and with encouragement from Mazzoni, he adopted a conviction that was to perdure during the entire period under review, namely, that although the abstract matter of the geometer is not identical with the sensible matter of the physicist, it is possible to have a close enough fit between geometry and nature to generate scientific certitude in conclusions about their joint subject matters.[117] He was particularly adroit at furnishing examples that show, through a proper choice of suppositions and limiting cases, how one can approach quantities as closely as one wishes in both spheres. And yet Galileo also manifested a sophisticated awareness that the gap between the physical and the mathematical could not always be bridged, and especially of how important it would be to exercise experimental ingenuity to circumvent the

[113] Secs. 4.2.a and 4.3.
[114] Sec. 4.1.d.
[115] Sec. 3.4.a-b.
[116] Sec. 4.4.a-b.
[117] Secs. 5.1.c and 5.2.b.

impediments that inevitably seemed to arise. Even with such ingenuity, moreover, he seems to have recognized that error-free measurements would not be possible; what he apparently aimed for was a degree of accuracy wherein such errors would be acceptable as negligible, or as *insensibile* (to use his expression), compared to the magnitude he was attempting to verify.[118]

Undoubtedly this attitude toward the development of a mathematical physics was largely Galileo's own, compatible with the position pioneered by Clavius at the Collegio Romano, and yet having no explicit counterpart in the Jesuit lecture notes. The view of science that it entailed was probably refined and clarified over the two decades at Pisa and Padua, but it had its start at Pisa and there is no indication of any abrupt departure from the program there initiated. Yet it was at Padua that Galileo's skill as an instrument maker and designer of experiments clearly manifested itself, and it was here that he first seriously tackled the problem of impediments and their removal. The pendulum obviously eliminated surface irregularities and friction, and the inclined plane was an ideal instrument to slow speeds of descent and permit measurements of times and distances of travel. Crude though such measurements may have been, they permitted Galileo to reject the explanation of velocity change in natural motion that was proposed in the *De motu antiquiora* and to recognize that a type of uniform acceleration should be substituted instead. Unfortunately his choice for an alternate principle was incorrect, even though he thought he could deduce from it some of the properties of falling motion, such as the times-squared law. More experiments were needed, as was deep thought over the continuous properties of time and motion and the possibilities of an infinite range of instantaneous velocities, before he could arrive at the proper principle, namely, that in natural motion velocity increases uniformly with time of fall. It does not seem that Galileo ever achieved a direct experimental confirmation of that principle, mainly because of the difficulties in measuring time with the instruments he had available. But refined experiments with parabolic trajectories, wherein accurate distance measurements could be substituted for those of time, gave him the indirect proof he sought, and this was all he needed, by his own canons, to found a new science of motion.

Confirmatory evidence that Galileo was actively pursuing a *scientia media* of this type, whose *suppositiones* would be established by effect-to-cause reasoning based on experiment, is contained in a response from Luca Valerio dated July 18, 1609, to a letter of Galileo written on June

[118] See sec. 5.4.a; also *Opere* 2:217.7, 2:181.6–12, 2:183.21, M273.1 (cited on p. 250 *supra*) and 7:256.35–37. Note also *Opere* 17:91.85–91, cited on p. 337 *infra*.

5, which unfortunately is no longer extant. Valerio admits that the principles on which a *scienza di mezo* (10:248.4) is built cannot be seen by the light of intellect alone, but their truth can hardly be doubted if one is able to account for quantitative variations in an effect in terms of proportional variations in its cause. The only restriction Valerio places on such a proportionality, for him equivalent to a self-evident truth (*verità nota per sè*, 10:248.9), is that every kind of impediment (*ogni sorte d'impedimento*, 10:248.11-12) be removed in the course of the experiment. The precise nature of the experiment being discussed in the letters is not clear, though obviously it had to do with the velocity of a body moving down an inclined plane, as shown on a diagram. For purposes here it suffices to note that Galileo had presented two suppositions for Valerio's consideration (as witnessed by the expression, *la seconda suppositione*, 10:249.31), and that Valerio felt that the experimental evidence Galileo had adduced was sufficient to establish their truth—if not according to the canons of pure geometry, at least according to those acceptable "in such middle sciences" (*in queste scienze medie*, 10:248.19).

By the end of 1609, before the telescope would dramatically change his life, Galileo therefore stood on the verge of the remarkable achievements he would outline to Vinta in his letter of May 1610. Then in his forty-sixth year, no longer a juvenile by any standard, he had persisted in a task he had set for himself some twenty years earlier. Clavius was still alive, his relationships with the Jesuits were still friendly, and there is no indication that he had changed the conceptual framework—from the viewpoint of both logic and natural philosophy—in which he had situated his investigations from the start. The basic physical definitions and suppositions were under control, a draft of the most difficult principle and its justification had probably been written, and sketches for the proofs of key propositions were already at hand. Galileo had an intuitive conviction, with good basis in fact, that the elements were available from which he could quickly put the sciences of motion and mechanics in proper mathematical form, with definitions, axioms or postulates, theorems, corollaries, lemmas, problems, and scholia, just as in the classical works of the Greek mathematicians. He had grasped the essential demonstrations; all he needed was the time to work out their rigorous formulation.

CHAPTER 6

Galileo's Later Science (After 1610)

As 1609 drew to a close Galileo was already past mid-life and could look back on a series of respectable accomplishments: his competence as a mathematician was recognized, he was valued as a teacher, and he had had considerable success as an instrument maker. He had not published anything significant, however, and he was hardly known outside the confines of Tuscany and the Venetian Republic. His work with the telescope changed all that, for shortly after the publication of the *Sidereus nuncius* in March of 1610 his name was known and his fame resounded all over Europe. Such prominence gained for him immediate advantages. For one, it helped to free him from his teaching duties at the University of Padua and to obtain the coveted post in the Florentine court as mathematician and philosopher to the Grand Duke of Tuscany. Along with this came a position of prominence from which he could go about propagating his ideas about the system of the world—now radically changed on the basis of the celestial observations he had made. But it also took its price, for as a public figure Galileo could no longer pursue his researches in mechanics and astronomy in the ways he had at Pisa and Padua. He could henceforth make his new views known, to be sure, but it also became incumbent on him to defend such views against those (and there are many) who were intent on preserving the status quo. From what has been seen of his earlier writings, Galileo was never afraid of taking on adversaries, but with his move to Florence such a task became almost an obligation. As the sequel shows, being placed in such a role encouraged him to develop his considerable skills as a controversialist and polemicist, skills second only to his abilities as a mathematician and natural philosopher.

This new development complicates enormously the task of tracing Galileo's understanding of science and its methods throughout his later period. Not one of the writings or publications that came after the *Sidereus nuncius*—and these are the works that are thoroughly combed by historians and philosophers of science for hints on this subject—was written as a disinterested scientific treatise that would stand or fall on its own merits. Every one was written as a discourse, letter, or dialogue whose literary form permitted rhetorical elements to enter into the presentation. This being so, any statement he makes must be subjected to a twofold critique, one based on its scientific claims and their justification and the other based on the immediate purposes for which he employs it. The first type of critique is hard enough, but the second makes it extremely difficult to discern his true position (assuming he had one) on the scientific value of the arguments he was elaborating.

By the end of 1609, as argued in the previous chapter, Galileo's notion of the science of nature based on experimentation and mathematical reasoning was already quite established. There were defects in it, of course, mostly traceable to its largely intuitive character and the fact that he had not yet had the opportunity to organize his thoughts reflectively and present them in a systematic treatise. Of the two subjects he had been teaching, mechanics and astronomy, the first had received almost his entire attention and so he was quite confident that he had within his grasp a "new science" of motion. Now, with the availability of unexpected evidence against which the *ippotesi* and *supposizioni* of astronomy could be checked, he was prepared to move into this field also and show how this could be put on the same basis as the science of mechanics. Instead of restricting itself to "saving the appearances," as he had characterized it in the *Trattato della Sfera*, the science of the heavens would henceforth be able to employ "necessary demonstrations" and so establish with apodictic certitude the true system of the world. And this system would not be the Ptolemaic system he had expounded in the *Trattato* but rather that of Copernicus. His observations of the moons of Jupiter (and later of the phases of Venus) furnished convincing evidence that the universe could not be completely geocentric, and thus the other alternative, heliocentrism, clearly had to be embraced. So Galileo set out, fatefully, on a crusade that would absorb most of his energies over the next two decades in a courageous attempt to prove the truth of the Copernican system.

1. The Copernican Debates

The first attacks against the *Sidereus nuncius* came from those who were not prepared to accept the data Galileo alleged and so rejected the observational base on which he hoped to construct his science. Among the first to come to his defense was Father Clavius and his associates at the Collegio Romano, who had constructed a small telescope and attempted to verify the findings reported in the *Sidereus*. Galileo wrote to Clavius on September 17, 1610, reporting some more of his observations, and on December 17 the latter, then in his seventies, informed his astronomer friend that he had finally been able to confirm his statements about the satellites of Jupiter and the general shape of Saturn, and awaited his coming to Rome for a fuller account. A fortnight later, on December 30, 1610, Galileo expressed his delight at having such a fine testimonial of the truth of his new observations (*un tanto testimonio alla verità delle mie nuove osservazioni*, 10:499.5-6), and promised to come as soon as he could. Further exchanges between Galileo and Grienberger followed, giving evidence of the amicable relations with the Collegio at that time.

a. GALILEO AT THE COLLEGIO ROMANO

By early 1611 two books had already appeared against Galileo, one by a Florentine mathematician, Francesco Sizzi, who accused him of corrupting the purity of the mathematical sciences, which had always provided the most certain demonstrations, through the use of an optical instrument that yielded spurious and fictitious data (3:211). When Galileo arrived in Rome in late March of that year, he immediately visited Clavius and had a lengthy discussion with him and other mathematicians on the faculty. On April 1 he was able to write Belisario Vinta that these professors and their students were reading Sizzi's book, "not without considerable laughter" (11:79.14), for they had already been observing Jupiter's moons for two months and were completely convinced of their reality. Galileo made a copy, in fact, of the positions they had recorded for the various satellites between November 28, 1610, and April 6, 1611, which is still preserved among his manuscripts (3:863-864). Later in that Spring, on May 18, 1611, a solemn convocation was held at the Collegio honoring Galileo and his discoveries, during which addresses were given by a Belgian professor, Father Odo van Maelcote (3:293-298), and some students explaining the import of these new findings. Many years later another Belgian Jesuit, Gregory of St. Vincent, wrote to Christian Huygens that he had partaken in the festivities, wherein they had clearly proven—though not without some dissension from the philosophers (*non absque murmure philosophorum*)—that Venus revolves around the sun.[1] Gregory's report attests to the fact that the oppositions between the philosophers and the mathematicians at the Collegio, which Clavius had striven to mollify for over twenty years, were still exerting their influence.[2]

[1] Villoslada, *Storia del Collegio Romano*, p. 198.

[2] Sec. 3.4.a-b.

During his visit to Rome, Galileo gave demonstrations of his telescope and invited a number of influential people, including Cardinal Robert Bellarmine, to use it to examine the celestial phenomena for themselves. Shortly after having done this, and possibly prompted by inquiries into the orthodoxy of Galileo's views in the papal curia, Bellarmine wrote to Clavius and his colleagues on April 19, 1611, to seek their opinion on the validity of the astronomer's claims (11:87-88). The points on which he desired information were these: whether there are multitudes of invisible stars in the skies, and whether the Milky Way and nebulae are composed of these; whether Saturn is not a simple star but three stars joined together; whether Venus changes its shape, waxing and waning like the moon; whether the moon has a rough, uneven surface; and whether four movable stars revolve around the planet Jupiter, each with a different movement but all with great speed. A few days later came the reply from the Collegio, signed by Clavius, Grienberger, Maelcote, and John Paul Lembo (the last the priest who built the telescope) wherein they generally answered Bellarmine's queries in the affirmative (11:92-93). They did express doubts, however, as to Saturn's real shape, merely remarking that it was not spherical as were Jupiter and Mars, and they admitted some disagreement about the unevenness of the moon's surface, noting that Clavius thought the irregularities were merely apparent whereas others thought them real. This expert opinion became known to Galileo, who cherished it as his favorite testimonial from "his great friends, the most distinguished men" in the Jesuit Order (11:102.9-10).

b. *DISCOURSE ON BODIES ON OR IN WATER*

Upon his return to Florence, Galileo became involved in another controversy with a staunch Aristotelian there, Ludovico delle Colombe, over a problem in hydrostatics. Colombe and his associates disagreed with Galileo over the role of shape in determining whether a body would float or sink in water, and proposed to challenge the latter's position by experimental test. Rather than enter into a public debate Galileo decided to write a discourse on the subject, which was published in 1612 as the *Discourse on Bodies on or in Water* (4:57-151).[3] Although he included in

[3] Drake, *Galileo at Work*, pp. 169-170. Drake has recently published a new English translation of this work, around which he has interwoven a dialogue of his own creation that is intended to serve as a commentary on it: *Cause, Experiment and Science.* A Galilean dialogue incorporating a new English translation of Galileo's "Bodies That Stay atop Water, or Move in It" (Chicago-London: The University of Chicago Press, 1981). Regrettably, Drake takes no account of Galileo's extensive use of causality in his other writings, and his commentary lacks historical as well as systematic value.

the essay some materials pertaining to heavenly bodies, it was not directly relevant to the Copernican debates; moreover, much of the experimental evidence was based on surface-tension effects, which were not well understood either by Galileo or by his adversaries. This publication did gain for Galileo some acclaim from contemporary mathematicians, however, and his methodological comments are of significance for seeing how he viewed the science of hydrostatics in the aftermath of his discoveries with the telescope.

Addressed to the Grand Duke and written in the form of an expository essay, the *Discourse* proposes to settle a dispute that had recently arisen at Florence. Predictably it begins by rejecting what Aristotle had taught about the behavior of bodies in water and aims to clarify the true, intrinsic, and total cause (*la vera, intrinseca e total cagione*, 4:67.5) of such phenomena. This has been treated by Archimedes, Galileo acknowledges, but he seeks to find a better demonstration that will employ a different method and other means to reach the same conclusion, namely, by reducing the causes of such effects to more intrinsic and immediate principles (*riducendo le cagioni di tali effetti a'principii più intrinsechi e immediati*, 4:67.18-19). Galileo notes that in his reduction he will follow "the demonstrative progression" (*la progressione dimostrativa*, 4:67.23).[4] This somewhat unusual expression ties his methodology here to the *progressiones* he had written about in his logic notes (LQ7.3)—effectively the resolution and composition involved in the demonstrative regress, which he saw there as necessary for proofs in the physical sciences.[5] First he will define his terms and then he will explain his basic propositions, so that he can subsequently use these as true and manifest (*come di cose vere e note*, 4:67.22-25). Among his definitions is one for specific gravity (*grave in ispecie*, 4:67.26), which he henceforth uses as a technical term. His two principles are taken from the science of mechanics (*dalla scienza meccanica*), the first stating that equal weights moving at the same *velocità* will exercise the same *forza* and *momento*, and the second that the *forza* and *momento* of a moving weight increases with the *velocità* of its motion (4:68.9-33). The latter principle, he explains, is behind the operation of all machines and was basic to the Aristotelian *Quaestiones mechanicae* (4:69.20-22)—an implicit admission that he is supplementing Archimedean statics with Aristotelian dynamics to reach a correct explanation. Another overtone from the *Quaestiones mechanicae* had been sounded earlier by Galileo when

[4] Drake translates this expression as "demonstrative advance," and in his comment has Salviati describe it as a type of hypothetico-deductive methodology that would "successfully predict by calculation"; see his *Cause, Experiment and Science*, pp. 26-27. This characterization, though ingenious, actually obscures the technical meaning of *progessione dimostrativa* as described in Galileo's logical questions.

[5] Sec. 3.3.b and note 75 of chap. 3.

he spoke of the "marvelous and almost unbelievable effects" (*qualche accidente ammirando e quasi incredibile*, 4:67.20) that could be explained through these principles.[6] Among these and other paradoxes is the fact that a very small amount of water can support a weight a hundred or a thousand times heavier than itself—the principle behind the hydraulic lift—which Galileo effectively explains through an application of the principle of the balance.

In discussing the role of a body's shape in determining whether it floats or sinks in water, Galileo admits that many causes can be operative in such cases, but that the most important thing is to determine which among these is the proximate and immediate cause (*la causa prossima e immediata*, 4:87.8). Here, as in his earlier writings, he proposes to isolate this by experimental means, using a method that is more refined (*modo più esquisito*, 4:91.20) than that of his adversaries. He describes a ball of wax that can be molded into a variety of shapes and shows how none of these causes it to sink; the addition of a small piece of lead to the wax, however, will cause it to do so regardless of its shape. Why a thin plate of ebony floats on the surface of water whereas a ball of ebony sinks (the counterexperiment proposed by Colombe) offers Galileo more difficulty. His reply to this is that one must consider not only the volume displaced by the plate but also the volume of air in the depression below the water's surface, assuming that the top of the plate is not wetted by the water, when calculating the forces acting on the plate and keeping it afloat. Additionally, one must realize that a cause is that which, being present, the effect is there, and being removed, the effect is taken away (4:112.21-23).[7] Taking these considerations into account, one will see that the true, natural, and primary cause (*la stressa vera naturale e primaria cagione*) of a body's floating or sinking is the excess or defect of the *gravità* of the water compared to the *gravità* of the total volume (*mole corporea*) of the materials displacing it (4:120.23-28).

Apart from such simple experiments, which Galileo alleges can fully demonstrate the truth of his proposition (*sopra tale esperienza fondato*,

[6] Sec. 5.3.a-b.

[7] Drake seizes on this expression as a new way of characterizing a cause that would make it specially apt for scientific inquiry, and sees it as one of Galileo's significant contributions; see his Introduction to *Cause, Experiment and Science*, pp. xxv-xxvii. Actually there is nothing novel about it, since Galileo's aim is to find the intrinsic cause (4:67.5) of flotation, and intrinsic causes are always coexistent (*simul*) with their effects; extrinsic causes, on the other hand, may coexist with their effects or they may precede or follow them. Galileo touches on this teaching in his physical questions, PQ2.4.4-5; see Wallace, *Galileo's Early Notebooks*, pars. F4 and F5. For further details of Galileo's views on causality in his various writings, see the author's "The Problem of Causality in Galileo's Science."

[8] Secs. 5.2.b and 5.3.a-b.

affermo d'aver pienamente dimostrata la verità della mia proposizione, 4:122.8-9), two other points should be mentioned in conjunction with the *Discourse*. The first is Galileo's concern with the resistance encountered by objects moving through water, wherein he rejoins topics discussed in his Paduan compositions.[8] Even the smallest particle will make its way to the bottom in a vessel of water, he observes, and a large ship can be drawn from place to place on the water's surface by the pull of a single hair (*un solo capello di donna*, 4:104.26). Considering such examples, he cannot see how one can imagine any power or force so small (*minima virtù e forza*) that the resistance offered by the water would not be smaller, and thus he is forced to conclude that it is actually zero (*per necessità si conclude che ella sia nulla*, 4:104.14-18). Water of itself offers no resistance to division; it is only when one tries to move bodies through it quickly that it does so, and this in proportion to the velocity of the motion (4:104.33-36). This being the case, one obviously need not worry about *impedimenti* in developing a science of hydrodynamics, for if velocities are kept negligibly small there can be an exact fit between physics and mathematics.

The other point is related to this, since it concerns the propriety of introducing mathematics into the study of natural pheonomena, a procedure which conservative Aristotelians, while willing to consider experiments, would not generally countenance. Shortly after the publication of the *Discourse* it was attacked in print by a number of Galileo's adversaries. The first of these, who identified himself only as the Unknown Academician (*Accademico incognito*), argued that mathematical propositions and proofs are incapable of demonstrating the force and the true causes (*la forza e le vere cagioni*) behind the operations of nature (4:165.10-11). He was followed in 1613 by Vincenzio di Grazia, who attempted a more extensive refutation based on texts in Aristotle's *Physics* and *Posterior Analytics*. Each science, he wrote, has its own proper principles and causes, and from these it demonstrates properties of its proper subject; such being the case, it is improper to use the principles of one science to prove properties in another. This is particularly inappropriate where mathematics and physics are involved, since the proper concern of the physicist is motion and the subject of mathematics abstracts from all motion (4:385.21-28).[9] Interpretations of texts such as these were apparently much discussed in the Florentine court, and it is worth noting that Galileo gave his own opinion of them in one of the draft versions of the *Discourse*. Those who argue in this way are apparently unaware that the truth is one, he states, as if geometry would prevent one from

[9] Cf. sec. 3.4.

developing a true philosophy (*vera filosofia*). One can be a geometer and a philosopher as well, since the knowledge of geometry does not preclude that of physics, nor does being a geometer prevent one from treating physical matters physically (*trattar delle materie fisiche fisicamente*, 4:49.24-33). What is noteworthy about this statement is that it shows Galileo regarding himself as both a mathematician and a philosopher who is perfectly capable of making judgments about sensible matter and knowing when quantitative considerations are appropriate or not. Such a view was consistent with what Clavius had been fostering at the Collegio Romano, and it is noteworthy that both Blancanus (as already mentioned) and Grienberger (11:477) gave enthusiastic approval to the *Discourse* when it appeared.[10] The Aristotelianism, if one may call it that, being developed at the Collegio was thus far different from that professed by Galileo's enemies at Florence and their predecessors at Padua and Pisa.

c. *LETTERS ON SUNSPOTS*

While the debates about floating bodies were in progress, another development took place that was to challenge more seriously Galileo's concept of a mathematical physics, this time as it could be applied to yield certain knowledge of the heavens. The occasion for the dispute was a series of three letters on sunspots published pseudonymously at Augsburg in 1612 and actually the work of a German Jesuit, Christopher Scheiner, who had been forbidden by his superiors to reveal his identity lest he be in error. Scheiner's observations of the spots, possibly suggested by a communication from the Collegio Romano after Galileo's mentioning them during his visit there—though this point is arguable—led him to believe that there were only two realistic explanations for them.[11] Either they were imperfections of the sun itself, moving along its surface, or they were actually small stars rotating around the sun's sphere and continually eclipsing it (5:26.20-29). Scheiner's preference was for the latter explanation, which allowed one to preserve the unalterability of the heavens. He was convinced, however, that Venus rotated around the sun (5:28.21), and at the end of his disquisition called attention to Clavius's warning, in the last edition of the *Sphaera* before his death, to the need for working out a new system of the heavens (5:69.4-8). In particular, he was certain that the common teaching of astronomers on the hardness and constitution of the heavens, especially the spheres of the sun and of Jupiter, could no longer be maintained (5:69.2-4).

[10] Sec. 4.4.b

[11] See Drake, *Galileo at Work*, p. 175; also M. L. Righini Bonelli, "Le Posizioni Relative di Galileo e dello Scheiner nelle Scoperte delle Macchie Solari nelle Pubblicazioni edite entro il 1612," *Physis* 12 (1970): 405-410.

Stimulated by the receipt of this work, Galileo renewed his own observation of the spots and in turn wrote three letters wherein he contested the findings and conclusions of Scheiner, unknown to him at first, but whose identity he eventually discovered. These were published at Rome in 1613 with a title indicating that they were an account and demonstrations concerning sunspots and their properties (*Istoria e dimostrazioni intorno alle macchie solari e loro accidenti*, 5:71-249). The work is significant not only for the fact that it marks the beginning of the deterioration of Galileo's relationships with the Jesuits, but also for the claims he would make for the possibility of attaining true and certain knowledge of the heavenly bodies.

In his first letter Galileo is restrained in his claims, clearly aware, as he states, that it is easier to tell what the sunspots are not than what they are, and that it is much more difficult to discover the truth than it is to disprove the false (5:95.11-13). In his second letter, however, after continued daily observation, he has more confidence and is prepared to maintain that of the two *posizioni* proposed by Scheiner, namely, that the spots are bodies rotating at some distance from the sun or that they are contiguous with the sun's surface, the second is true and the first false (5:118.10-11). He has come to this conclusion, he says, by reasoning from particular properties furnished by sense observation (*da certi particolari accidenti che le sensate osservazioni ci somministrano*, 5:117.34-35). He goes on to enumerate three *accidenti*, the last of which provides such strong evidence that it is sufficient of itself to demonstrate (*a dimostrar*) his position (5:121.4). All of the appearances observable in the spots correspond exactly (*puntualmente rispondono*) with their being on the sun's surface and are opposed to any other position (*ad ogni altra posizione*) one might attempt to propose (5:127.13-18). After listing other necessary demonstrations that admit of no reply whatever (*le dimostrazioni necessarie e che non ammettono riposta veruna*, 5:128.16-17), Galileo invokes the authority of Aristotle himself, and say that Aristotle would have concluded similarly if he had had Galileo's evidence, since he not only based his reasoning on sense observations and obvious experiences (*sensate osservazioni . . . manifeste esperienze*) but actually accorded them primacy in his investigations (5:138.27-139.3).

Despite the confidence of these assertions, Galileo had already pointed out in his first letter the great difficulty of making any judgment about the sunspots' essence and substance (*essenza e sustanza*), noting that the properties they manifest are so common (*gli accidenti . . . comunissimi*) that they can provide little reliable information (5:105.18-106.3). In his third letter he elaborates on this point further and argues that the same difficulties attend man's efforts to attain to the true and intrinsic essence of any natural substance (*l'essenza vera ed intrinseca delle sustanze naturali*), whether it be an earthly substance close at hand or one in the celestial regions, and that one should content oneself with a knowledge of their properties (*affezioni*, 5:187.8–188.4). The properties he apparently has in mind are those relating to the shape, position, and motion of bodies, all of which are quantitative characteristics that can be established by *a posteriori* demonstration using mathematical middle terms. Such statements need not be interpreted as indications of skepticism or agnosticism on Galileo's part, since it was commonly accepted in his day that even Aristotle had used similar demonstrations to establish the shape of the earth and the moon. Blancanus likewise makes much of the valuable information that can be provided by a study of quantitative *accidenti*, and scholastic philosophers were quite agreed that knowledge of the true natures of material substances is indeed beyond human grasp.[12]

These brief selections give evidence that by this time Galileo was convinced that astronomy need no longer be a *scientia secundum quid* but could now, through the use of telescopic observations, achieve the status of a strict demonstrative science. An important passage in the first letter signals this attitude. There, after admitting that Scheiner had made some progress toward good and true philosophy, he chided him for the uncritical stand he had taken on the reality of Ptolemaic eccentrics, deferents, equants, epicycles, etc. The latter, he went on, are posited by pure (i.e., mathematical) astronomers to facilitate their calculations, but they are no longer retained by philosophical astronomers who, going beyond the requirement of saving the appearances in some way, seek to investigate the true constitution of the universe (*la vera costituzione dell'universo*), the most important and wonderful problem there is (5:102.15–19). Such a constitution exists, and it is unique, true, real, and cannot be otherwise (*solo, vero, reale ed impossibile ad esser altramente*), and should on account of its greatness and dignity be considered foremost among the questions of speculative interest (5:102.20–23). There can be no doubt that, by this time, Galileo had abandoned the instrumentalist view of astronomy entertained by Mazzoni and was convinced that the *immaginaria supposizione* (5:103.7) of the Ptolemaists could be replaced by others that would reflect the true system of the univserse.

Not only this, but by the time Galileo was writing the second letter on sunspots, he had also proceeded to the stage where he felt that the newly discovered principles of his science of motion could be applied to the heavenly regions, where they might serve to explain the movements

[12] Blancanus makes the point in his *Dissertatio*, p. 26, that a true knowledge of material substance can never be attained, and the same point is hinted at in Galileo's physical questions, PQ5.4.8, where he attributes to St. Thomas Aquinas the view that the substantial forms of the elements are hidden from us. See secs. 2.3.c and 4 4.b.

of the spots. In a much-cited passage in this letter he reiterates his earlier teaching on the three types of motion seen in natural bodies: one to which they are naturally inclined by some type of intrinsic principle (*per intrinseco principio*), as in the movement of heavy bodies downward; another to which they have a natural repugnance and can only be forced by an external mover (*da motore esterno*), as in the movement of heavy bodies upward; and a third to which they are indifferent (*si trovano indifferenti*), having neither an inclination nor a repugnance, as in the horizontal movement of the same (5:134.5-15).[13] This being so, he continues, if all external impediments are removed (*rimossi tutti gl'impedimenti esterni*), a heavy body on a spherical surface concentric with the earth will be indifferent to rest and to movement (5:134.16-18). Thus a ship, having received an impulse through the tranquil sea, would move continually around the globe without ever stopping, and placed at rest would remain at rest forever, if in the first case all external impediments could be removed (*si potessero rimuovere tutti gl'impedimenti estrinseci*) and in the second no external motive cause were applied (*qualche causa motrice esterna non gli sopraggiugnesse*, 5:134.21-135.5). In a similar way the sun, following the motion of its ambience, would have no intrinsic repugnance or external impediment (*intrinseca repugnanza ne impedimento esteriore*) to its rotation, and this could serve to explain its continual motion as well as that of the spots on its surface (5:135.10-33). Clearly, the principles of a celestial mechanics were already taking form in Galileo's mind, as he was attempting to forge a unity of the two sciences he had been studying since his early days at Pisa.

d. REPLY TO THE THEOLOGIANS

Thus far attention has been focused mainly on Galileo's reflections concerning the scientific validity of his conclusions. Both treatises discussed, the *Discourse on Bodies on or in Water* and the *Letters on Sunspots*, also present a rhetorical or polemical dimension, the first for the obvious reason that it was directed to the Grand Duke to discredit Galileo's adversaries at Florence, and the second for reasons that are less clear but that seem to be aimed at discrediting Scheiner. In any event both works contain considerably more than scientific reasoning, being embellished with ridicule, invective, and critical or ironical statements, not without humor, which were calculated to gain an advantage for Galileo, in the eyes of the reader, over those who opposed his views. They were written in Italian, moreover, a language wherein Galileo was skilled at enhancing this effect. Such a style of writing, particularly when directed against

[13] Sec. 5.2.b.

Scheiner, who had published his disquisitions in Latin, could only elicit an unfavorable reaction, and may help to explain (apart from their dispute over the priority of the sunspot discoveries) the intense animosity the German Jesuit would later develop toward Galileo.

In the wake of the sunspot controversy and undoubtedly prompted by Galileo's increasing advocacy of the Copernican system, opposition rose against him from yet another quarter, namely, that of the theologians. A number of statements in the Scriptures, taken in a literal sense, were seen to conflict with the claims of the Copernicans that the sun was immobile at the center of the universe and that the earth was endowed with a twofold movement, one of diurnal rotation and the other of annual revolution around the sun. The possible conflict was discussed at Florence, and led to two other compositions of Galileo, neither of which was published at that time, but which circulated widely in manuscript form and came to be well known. The first was a letter to a Benedictine monk, Benedetto Castelli, who was Galileo's former student and had succeeded him in the post at Pisa, and the second an enlargement of this letter written in 1615 to the Grand Duchess Christina, mother of Galileo's patron. Both were concerned with the relationships between science and religion and made numerous assertions about the validity of scientific conclusions vis-à-vis the Scriptural interpretations of theologians. The *Letter to Christina* (5:307-348) being the more polished of the two, a brief review of its contents will prove helpful at this point.

Galileo begins his letter by observing that there are many who use the Bible to disprove arguments they do not understand, whereas those who are expert in astronomical and natural science (*scienza astronomica e . . . naturale*, 5:310.14) have accepted his views as soon as they were presented. He is forthright in admitting that he holds, on the basis of his studies in astronomy and philosophy, that the sun is motionless and at the center of revolution of the heavenly spheres, whereas the earth revolves on its axis and rotates around the sun (5:311.1-4). His bases are not merely the refutation of Ptolemy's and Aristotle's arguments but also positive reasoning based on natural effects (*effeti naturali*), whose cause can perhaps not be assigned in any other way (*le cause de'quali forse in altro modo non si possono assegnare*); in addition he has astronomical evidence that clearly refutes the Ptolemaic system, while it admirably accords with and confirms the alternate *posizione* (5:311.4-11). Since he possesses this proof, and since both nature and Scripture proceed from the same source and cannot be in conflict, it is necessary to reinterpret any statements in the Bible that seem to controvert the Copernican system. Galileo is explicit that Copernicus's account of the motion of the heavens is well founded on manifest experience and necessary demonstrations (*ben fondata sopra manifeste esperienze e necessarie dimostrazioni*, 5:312.27–28), and repeats this or equivalent expressions more than twenty times throughout the letter. Yet, oddly enough, although he does mention such phenomena as the phases of Venus (5:328.22–24), he nowhere gives a single argument or scientific demonstration that establishes his case. Indeed, in a rhetorical flourish, he turns the tables on the theologians and says that it is they who have the obligation of showing that the Copernican system has not been rigorously demonstrated (*ch'ella non sia dimostrata necessariamente*, 5:327.23).[14] One cannot condemn as heretical a proposition that might turn out to be true, he concludes, and thus people are wasting their time when they seek the condemnation of a system if they have not yet demonstrated it to be impossible and false (*se prima non la dimostrano essere impossibile e false*, 5:343.11–15).

From the viewpoint of scientific methodology it is difficult to know what to make of this *Letter to Christina*. At about the same time as Galileo wrote it, he had begun to work on a proof for the earth's motion based on the tides (5:371–395), later to appear in the *Two Chief World Systems*. This will be discussed presently, for it apparently developed the argument from effect to cause mentioned in the letter. Otherwise he seems to act as if only two *posizioni* were possible in this matter, analogous to the two explanations of sunspots debated with Scheiner, and thus if he had succeeded in proving the one false the other must necessarily be true. But this assumption seems simplistic, for Galileo knew that Tycho Brahe had offered a *tertium quid* between the Ptolemaic and the Copernican systems that would explain the phenomena Galileo had observed and at the same time be concordant with the Scriptures. It seems more likely that at this time Galileo had an intuitive grasp of the necessary demonstrations he was about to offer, somewhat similar to the demonstrations on which he was to base his new science of motion, but had not yet had a chance to work them out in definitive form. So sure was he that he could do so, given the time and effort, that he made the strong claims in the letter—even though they would soon be contested by Grienberger and other scientists in the Society of Jesus.

Support for this interpretation comes from a sketch of a reply Galileo had prepared to a letter from Bellarmine which is preserved under the title, *Considerazioni circa l'opinione Copernicana* (5:349–370). On April 12, 1615, Bellarmine had written to a Carmelite priest who had attempted to reconcile Copernicianism with the Scriptures, Paolo Foscarini, that he commended him and Galileo because they had spoken only *ex sup-*

[14] For a rhetorical analysis of the letter that brings out some of these points, see Jean Dietz Moss, "Galileo's *Letter to Christina*: Some Rhetorical Considerations," *Renaissance Quarterly* 36 (1983): 547–576.

positione on this matter and not absolutely (12:171-172). Bellarmine's view was that the Copernican system was still only an hypothesis, and that it would have to be given demonstrative proof (which he did not believe existed) before one would have to reinterpret the Scriptures on its basis. Galileo's reflections on the letter are instructive, for they center on the different meanings the term *suppositio* can take on in scientific discourse. He maintains, against those who hold that Copernicus used the expression *ex hypothesi* in the traditional sense of an astronomer who was merely showing that his system better satisfied and saved the apparent heavenly motions (5:354.28-31), that the Polish astronomer took the *posizione* he did not as a pure (i.e., mathematical) astronomer but to satisfy the necessity of nature (*per satisfare alla necessità della natura*, 5:355.33-34). With regard to the heavens, he goes on, two kinds of *supposizioni* have been made by astronomers: some are primary and with regard to the absolute truth in nature, whereas others are secondary and are posited imaginatively to account for the apparent movements of the stars, whose appearances they show are somehow not in agreement with the primary and true *supposizioni* (5:357.10-14). The first kind Galileo characterizes as *supposizioni naturali*, which are established (*stabilite*) and show what truly takes place; the *ipotesi* of Copernicus, he has no doubts, if properly considered will be placed among *posizioni* that are primary and necessary in nature (*prime e necessarie in natura*, 5:357.21-34). The second kind, on the other hand, are the chimerical and fabulous (*chimerica e favolosa*) hypotheses of Ptolemy, which are false in nature (*falsa in natura*) and are introduced into the science merely for the sake of astronomical computation (5:358.35-359.2). He himself is willing, for those who wish to employ *ex supposizione* in Bellarmine's sense, to put the earth's motion on the same plane as eccentrics and epicycles, provided that his proof of the nonexistence of the latter will force his adversaries to admit the truth of the former (5:360.10-25). As to Copernicus, Galileo believes that this was his evaluation of the matter, although one might gain the opposite impression from reading the preface to the *De revolutionibus*, which was not signed by him and clearly was the work of another (5:360.33-361.16). Finally, to remove all doubts (*per levar ogn'ombra di dubitare*), his telescopic observations of Venus correspond exactly (*puntualmente rispondere*) with all the effects one should expect, and therefore the Copernican *posizione* must be regarded as true and real (*vera e reale*, 5:362.24-37).

Galileo's main effect-to-cause proof of the earth's motion, hinted at in the *Letter to Christina*, was then sketched in his letter to Cardinal Orsini of January 8, 1616. The treatise it presented, entitled *Discorso del flusso e reflusso del mare*, in fact lays the groundwork for the causal analysis of the tides that would bring to conclusion the *Two Chief World Systems* of

1632.[15] The marvelous problem he is addressing, Galileo writes to Orsini, is that of finding the true cause (*la vera cagione*) of the ebb and flow of the sea, hidden and difficult to discover, but now fortunately laid bare by him (5:377.11-13). This true cause (*vera causa*) readily and clearly explains all the effects and properties (*tutti i particolari sintomi e accidenti*) of the ocean's motion (5:377.17-18). Because of the complexity of these movements, however, it is necessary to assign a primary cause (*cagion primaria*) for them and then to add other secondary and concomitant causes (*cause secondarie e . . . concomitanti*) to account for their diversity (5:378.18-23). The principal cause, on further consideration, turns out to be itself twofold. The first and more simple is the alternate acceleration and deceleration of the earth produced by the composition of its two motions, diurnal and annual. These add to and subtract from each other and so cause the waters of the ocean to slosh back and forth in a 24-hour cycle. The other cause depends on the *propria gravità* of sea water (5:391.8-9), which alters the primary motion in various ways depending on the dimensions of the sea bed in which the water moves. This has the additional effect of producing periods or cycles of various durations in different parts of the world (5:390.35-319.15). Such seems to Galileo to be the *causa adequata* (cf. 5:381.9) of tidal effects, and on its basis he is proposing the Copernican system not as a fictitious hypothesis but rather as based on principles that reflect the true structure of the universe.

Despite Galileo's conviction in this matter, a group of theological consultors to the Holy Office rendered the judgment on February 23, 1616, that it was "formally heretical" to hold that the sun was immobile at the center of the universe and "erroneous in the faith" to hold that the earth was in motion (19:321). Three days later Galileo was informed by Bellarmine of this decision, and apparently told to conform to it in his teaching. On March 5, 1616, Foscarini's book defending his interpretations of Scripture was placed on the Index, and Copernicus's *De revolutionibus* was suspended until corrected. These events, it goes without saying, could not help but have a profound effect on everything Galileo would henceforth say about the science of astronomy.

e. THE DISPUTE OVER COMETS

Late in 1618 a series of comets appeared in the skies and stimulated a number of publications on these fascinating, and seemingly portentous, phenomena. Among these was a public lecture given at the Collegio Romano by Orazio Grassi, one of Clavius's successors (6:21-35). The

[15] Sec. 6.2.d.

tone of the work was unexceptional, and there is no evidence that it was directed against Galileo, although it advocated the Tychonian system of the universe and could be construed as an implicit attack on the Copernican view. At the time of its publication one of Galileo's disciples, Mario Guiducci, who had also studied at the Collegio Romano and was friendly with the Jesuits, was preparing a series of lectures to be given at the Florentine Academy. Undoubtedly he solicited Galileo's help, and this was offered in superabundance, for drafts of the discourses reveal them to be largely the work of Galileo. The lectures were published at Florence in 1619 under Guiducci's name as *Discorso delle comete* (6:37-108), and turned out to be a critique and refutation of the analysis given by Grassi. The latter quickly discerned that Galileo was the real author behind the composition, and by the end of 1619 he had prepared and published, under the pseudonym of Lothario Sarsi, a detailed reply to Galileo with the title, *Libra astronomica ac philosophica* (6:109-180). The resulting challenge was too much for Galileo to ignore, so he set to work on a devastating counterreply to Grassi. Long in preparation, this became a materpiece of polemical literature, and when it was finally published in 1623 as *Il Saggiatore* (6:197-372) Galileo dedicated it to Matteo Barberini, recently elected to the papacy as Urban VIII, a fellow Florentine and literary patron. Since a number of philosophical positions are developed throughout these debates, though their methodological significance is not as great as one might expect, they merit summary consideration as Galileo's last public controversy before the appearance of the *Two Chief World Systems*.

The criticism directed in the *Discorso delle comete* against the Aristotelian theory of the formation of comets is that Aristotle's account in the *Meteorologica* is completely full of *supposizioni*, which, if not obviously false, at least stand much in need of proof (*se non manifestamente false, almeno molto bisognose di prova*), whereas what is supposed in the sciences should be completely evident (*che si suppone nelle scienze doverrebbe esser manifestissimo*, 6:53.19-22). This, of course, is the theme of the *Posterior Analytics* that most scholastics and Aristotelians of the time would accept without question.[16] In the *Discorso* Galileo-Guiducci use it not only to discredit Aristotle but also to call into question the arguments Grassi had used to establish the positions and paths of the comets. Aristotle had thought that comets could be located under the orb of the moon, they argue, but astronomers have conclusively demonstrated that they are far above the moon (*necessariamente dimostrato gli astronomi altissimo sopra la luna*, 6:64.23). The proofs the astronomers use are based on measurements of parallax,

[16] Secs. 3.1 and 3.2.

but these are valid only when one is certain of the existence of the objects one is measuring. Now there are two kinds of visible objects (*oggetti visibili*): some are true, real, unique, and unchangeable (*veri, reali, uni ed immobili*); others are mere appearances, light reflections, images, and errant counterfeits (*solo apparenze, reflessioni di lumi, immagini e simulacri vaganti*, 6:65.12-15). Grassi has not proved that the comets of 1618 are of the first type, since they might be mere optical illusions. Thus, using a technique he had already employed in the *Letter to Christina*, Galileo passes the burden of proof on to his adversary, requiring him first to prove that the comets are real before invoking parallax measurements to establish their position. Other defects in Grassi's arguments are found in a statement he makes about the different magnifying powers of a telescope with respect to near and distant objects (6:73.23-34). The position taken by Galileo-Guiducci, against that of Tycho Brahe who assigned comets an orbit around the sun (6:88.11-13), is that comets actually move in a path perpendicular to the earth's surface, which they try to establish by mathematical arguments (6:95-97). These do not square with the appearances, as their authors recognize (6:98.16-24), and thus one is left with the impression that the intent of the *Discorso* was to discredit Grassi's analysis without having anything constructive to put in its place.

Grassi's reaction was predictable, and his *Libra astronomica ac philosophica* was not without its polemical overtones, although it also contained some telling criticisms of the *Discorso*, namely, that it disparaged measurements of parallax while at the same time invoking them as a refutation of Aristotle, and that it perversely tried to maintain a theory of cometary motion that was in manifest opposition with the facts (6:118-120). Galileo used his reply, however, to show further weaknesses in Grassi's position and to use the meager knowledge provided by sense observation of the comets to exploit a number of philosophical speculations of his own. He thus begins *Il Saggiatore* with an autobiographical description of his commitment to mathematics and geometrical demonstrations as the key to a true knowledge of nature. His discourse on floating bodies, he complains, was immediately attacked by those who did not understand that his views were confirmed and proved with geometrical demonstrations (*confermato e concluso con geometriche dimostrazioni*), not realizing that to contradict geometry is openly to deny the truth (*contradire alla geometria è un negare scopertamente la verità*, 6:214.6-9).[17] Later, accusing Grassi of finding his philosophy in books rather than in nature, Galileo makes his oft-quoted manifesto that philosophy is written in that greatest of all books standing continually open before our eyes, that is, the universe,

[17] Sec. 6.1.b.

and this book is written in the language of mathematics (*è scritto in lingua matematica*), without whose means it is humanly impossible to understand it (6:322.11-18). Such a statement does not indicate, as some have interpreted it, that Galileo was a naive Platonist or Pythagorean, for elsewhere in the same treatise he acknowledges that his philosophy and his mathematics (*la nostra filosofia e le nostre matematiche*) are both necessary for certain knowledge (*le cose che noi sappiamo indubitatamente*, 6:279.2-3), and that the rigor of mathematical demonstration is too dangerous (*la severità di geometriche dimostrazioni è troppo pericoloso*) to employ if one does not know how to use it, since it offers no middle ground between truth and falsity (6:296.19-22). In the same vein, and in an attempt to cover up some of the extravagant claims of the *Discorso*, Galileo reminds Grassi that Guiducci proposed whatever new he had to offer tentatively and conjecturally (*dubitativamente e conghietturalmente*), not forcing others to accept his conclusions as certain when both he and Galileo regarded them as probable at best (6:303.23-30). On his part, moreover, Galileo was not adverse to employing his own conjectures, for in the latter parts of *Il Saggiatore* he discusses the nature of heat and alleges the existence of invisible particles to account for thermal phenomena, and then, in passages that are much studied by historians of philosophy, argues that sensations are merely subjective impressions and so are not to be taken as qualities really existing in the objects of experience. The incipient skepticism does not enter into his scientific reasoning, however, and thus is not pertinent to the theme being developed in this study.

Although it is commonly conceded that Galileo got the better of Grassi in the overall exchange, and in so doing made an important contribution to polemical literature, one important fact about the controversy should not be overlooked. This is that *both* Grassi and Galileo were competent mathematicians and experimentalists who shared a large common ground as to what constituted convincing proof in matters relating to astronomy and natural philosophy.[18] The same could be said, though to a lesser extent, of Scheiner and Galileo. Galileo could enter into prolonged debate with these Jesuits precisely because he understood very well the commitments and the methods that lay behind their arguments. And this would hardly have been possible had he not pursued the course he had from his first contacts with Clavius all the way down to the accession of Pope Urban VIII.

[18] This point is elucidated by W. R. Shea in his *Galileo's Intellectual Revolution: Middle Period, 1610-1632* (New York: Science History Publications, 1972), pp. 75-108. Some of Grassi's experiments are described, in English translation, in *The Controversy on the Comets of 1618*, trans. Stillman Drake and C. D. O'Malley (Philadelphia. University of Pennsylvania Press, 1960), pp. 86-132.

2. The *Two Chief World Systems*

In April of 1624 Galileo made a visit to Rome, and before his return to Florence at the beginning of June had managed to have six audiences with the newly elected pope. Unfortunately for historians, the details of his discussions with Urban VIII are not known, but it seems clear that by the time of his departure Galileo was assured that he had the pope's permission to continue writing on astronomical matters. Meanwhile the Jesuits were continuing to explore the ramifications of the Copernican and Tychonian hypotheses. In the same year, 1624, Scheiner was transferred from Ingolstadt to the Collegio Romano, and Guiducci had visited Grassi at the Collegio and entered into a friendly discussion with him—in fact, it was Grassi who told him about Father Andrea and the experiments involving the ship's mast.[19] Scheiner was not so benign, however, and it may have been he who informed the Holy Office that *Il Saggiatore* should be prohibited or at least corrected. In any event, Guiducci wrote to Galileo on April 18, 1625, that "a pious person" had done this, and that the book had been given by the Holy Office to Giovanni di Guevara, the priest whose views on the science of mechanics were discussed in chapter 4, for review. Guevara reported back that he was rather pleased with Galileo's teaching and saw no reason why the book should be condemned; on this basis, wrote Guiducci, the Holy Office had decided to let the matter rest (13:265.9-19).

So encouraged, Galileo continued work on the manuscript that was to become the *Two Chief World Systems*. He did not complete this until January 1630, and in May of that year traveled to Rome to obtain the necessary permissions to have the book printed there under the auspices of the Accademia dei Lincei. The pope's theologian, Niccolò Riccardi, whose official title was Master of the Sacred Palace, seemed favorable to the enterprise, although he was also aware of difficulties that would be raised by the Jesuits and others.[20] When the decision was made to have the book printed in Florence rather than in Rome, Riccardi wrote to the inquisitor of Florence explaining that the manuscript, which had earlier been entitled *De fluxu et refluxu maris*, was ready for printing, that he had seen and approved it, and that it was agreed that Galileo would make some changes in it and return it to him for final approbation before printing. For good reasons, Riccardi continued, Galileo was unable to do this, and so he was willing to have the Florentine inquisitor give the final permission. The pope's mind (*esser mente di Nostro Signore*), however,

[19] Sec. 5.4.c.

[20] For further details on Riccardi and his involvement with Galileo, see Ambrosius Eszer, "Niccolò Riccardi, O.P., il 'Padre Mostro' (1585-1639)," *Angelicum* 60 (1983): 428-457.

was that the tides were not to be in the title or to be the subject of the book, which was to deal only with a mathematical consideration of the Copernican position (*della mattematica considerazione della posizione Copernicana*), its aim being to show how, Church doctrine aside, the appearances can be saved in this position (*potrebbono salvare le apparenze in questa posizione*), but that the position was not to be regarded as having absolute, but only hypothetical truth, and this apart from the Scriptures (*non mai si conceda la verità assoluta, ma solamente la hipothetica e senza le Scritture*, 19:327.124-130). If these precautions were observed, Riccardi concluded, the book would encounter no obstacle in Rome. The fact that he called attention to the pope's instructions regarding the book is a fairly good indication that this represented the substance of Urban's conversations with Galileo in 1624, when Galileo sought his first approval for the composition.[21]

In apparent compliance with these instructions, and pursuant to further explicit suggestions from the censor at Florence, Galileo composed an appropriate preface and conclusion for the book, and it was printed in February 1632 with the simple title of *Dialogo* but with a lengthy subtitle indicating that it contains discourses held over meetings of four days on the two chief world systems, the Ptolemaic and the Copernican, proposing philosophical and natural arguments indifferently (*indeterminatamente*) for the one as for the other (7:25). Galileo acknowledges in the preface that he has taken the Copernican side in the discourse, while treating it as a pure mathematical hypothesis (*pura ipotesi matematica*), and this mainly to show that Italians were not ignorant of reasons being offered north of the Alps in support of the system, though they themselves rejected it for religious reasons, being aware of the divine omnipotence and the weakness of the human mind (7:29.24-30.30).

a. THE FIRST DAY

The discussions of the first day are preparatory to the book's thesis as a whole, for they are designed to provide a background against which the arguments for the earth's rest and the sun's motion can be critically evaluated. The main thrust is toward the unity of the world, showing that bodies move naturally in both the heavens and on earth and do not require the Aristotelian distinction between curvilinear and rectilinear

[21] In his commentary on the *Two Chief World Systems*, Giorgio de Santillana maintains that the pope actually dictated to Galileo the wording he wished to have used in a conversation with him in 1630. See Galileo Galilei, *Dialogue on the Great World Systems in the Salusbury Translation*, revised, annotated, and with an introduction by Giorgio de Santillana (Chicago: The University of Chicago Press, 1953), p. 471, note 31.

motion in order to do so, that the heavens are alterable just as is the earth and thus there is no substantial difference between the celestial and the terrestrial, and finally and more particularly, that the moon and the earth share a common nature. The methodological statements made in the course of the dialogue are not numerous and they are all intelligible in light of the purpose Galileo has in mind.[22]

Aristotle had invoked mathematical reasoning to establish his preliminary distinctions. Galileo does not wish to exclude such reasoning from the study of nature, since so much of his own argumentation depends upon it, and thus he simply makes the point that arguments such as "*three is a perfect number*" *should be left to the rhetoricians* (*a i rhetori*) and that proofs should be established with necessary demonstrations, as those that are properly made in the demonstrative sciences (*con dimostrazione necessaria, chè così convien fare nelle scienze dimostrative*, 7:35.10-11).[23] He does not deny, moreover, that Aristotle had reasoned in the latter way, and offers an emendation of his proof that bodies do not have, and cannot have, more than three dimensions (*le dimensioni non esser, nè poter esser, più de tre*) as evidence of this (7:36.1-2). Against the objection that the necessity of mathematical demonstration is not always to be sought in physical matters (*nelle cose naturali non si deve sempre ricercare una necessità di dimostrazion matematica*), Galileo readily concedes this point, merely insisting that if such a demonstration is at hand one should not be unwilling to use it (7:38.17-20).

The alterability of the heavens is a topic that enables Galileo to stress the importance of sense evidence in scientific reasoning, for he argues that had Aristotle seen the new effects and observations (*accidenti ed osservazioni nuove*) now available, he would have changed his view that the heavens are inalterable; indeed such a change would have followed from his method of philosophizing (*dal suo stesso modo di filosofare*), wherein he properly gave priority to sense experience over natural discourse (*anteposto, come conviene, la sensata esperienza al natural discorso*, 7:75.9-21). Here Galileo differentiates between the method that Aristotle used when presenting his doctrine (*la sua dottrina*) and that he used in discovering it (*la investigò*)—the obvious scholastic distinction between the *via doctrinae*

[22] A full identification and analysis of all these statements are to be found in M. A. Finocchiaro, *Galileo and the Art of Reasoning: Rhetorical Foundations of Logic and Scientific Method*. Boston Studies in the Philosophy of Science, vol. 61 (Dordrecht-Boston: D. Reidel Publishing Company, 1980), especially pp. 67-150. In general this is a penetrating analysis of the *Two Chief World Systems*; a few points on which Finocchiaro departs from the thesis being argued in this volume are indicated in the author's review of Finnochiaro's work in the *Journal of the History of Philosophy* 20 (1982): 307-309.

[23] Secs. 1.3.b and 3.2 a.

and the *via inventionis*—noting that in the latter Aristotle started with the senses, experiments, and observations (*per via de'sensi, dell'esperienze e delle osservazioni*) to gain the greatest possible assurance for his conclusion (7:75.26-30). Once having the conclusion, he goes on, he sought the middle terms to prove it (*ricercando i mezi de poterla dimostrare*), as one does generally in the demonstrative sciences (*nelle scienze dimostrative*, 7:75.30.32). If the conclusion is true, using the resolutive method (*servendosi del metodo resolutivo*) one may come to a proposition already demonstrated (*già dimostrata*) or to one *per se nota* (*per sè noto*); if it is false, one may go on to infinity (*in infinito*) without coming to a known truth, unless he ends with an impossibility or obvious absurdity (*alcun impossibile o assurdo manifesto*, 7:75.32-37). It should be noted that this description of resolutive method is precisely that given by Blancanus in his account of how resolution is used in the mathematical sciences, and is not quite the same as its employment in the natural sciences.[24] Galileo here has conflated the two usages. It is perhaps noteworthy that his student, Benedetto Castelli, had given almost an identical account of resolution (and its opposite, composition) in a defense of Galileo's work on floating bodies published in 1615 (4:521.19-23); Galileo's hand is detectable in Castelli's composition, and one should not be surprised, considering the fact that both were mathematicians, that they should almost automatically describe how resolution is used in their discipline.[25]

The comparisons of the earth with the moon present Galileo with the opportunity of showing how experiments and the construction of models can aid in establishing conclusions concerning natural phenomena. The experiments he describes relate to the ways in which one can show how light is reflected from one surface to another, and how these are relevant to understanding the illumination of the moon and its surface. One of the discussants is persuaded, from the very beginning of an experiment (*da questo poco principio di esperienza*), that the moon does not have a polished surface (7:96.37). A second proposes to remove all doubt with another experiment (*con altra esperienza rimouovere ogni scrupolo*, 7:100.10), and adduces yet another to make certain (*assicuriamo con l'esperienza*, 7:108.22-23) that an objection has no validity against his conclusion. The appearances the moon presents, moreover, can be duplicated by constructing a model with prominences and cavities on it; when properly

[24] Secs. 3.3.a and 3.4.c.

[25] Nicholas Jardine surveys some of these texts in his interesting article, "Galileo's Road to Truth and the Demonstrative Regress," *Studies in History and Philosophy of Science* 7 (1976): 277-318; although he mentions Galileo's logical questions in the article, he does not go into the teachings contained in them relating to the *regressus*, nor is he aware of the sources on which they are based. This vitiates the conclusions to which he comes.

illuminated this will show the same views and changes (*rappresenteranno l'istesse viste e mutazioni*) as can be seen in the moon (7:111.35-112.5). But one should not be disappointed that man's knowledge of the moon is still limited, for man's mind is not the measure of nature and there is not a single effect in nature (*non è effetto alcuno in natura*) that he can completely understand (7:127.1). Human knowledge is therefore to be contrasted with divine knowledge, and yet there are some propositions that man can grasp with absolute certainty (*assoluta certezza*), namely, those of the pure mathematical sciences (*le scienze matematiche pure*, 7:128.36–129.1). In such matters the human intellect has the same objective certainty (*certezza obiettiva*) as the divine (7:129.4). The only difference is that man arrives at his conclusions discursively, in a step-by-step process, whereas God knows his at a simple glance (*di un semplice intuito*, 7:129.19).

b. THE SECOND DAY

With this as a background, Galileo is able to move into the discussions of the second day, where he takes up the arguments on the Copernican side to show that the earth has a diurnal motion, that is, one of daily revolution on its own axis. Here he has to invoke some experimental knowledge to support this position, but he also has to be wary of sense knowledge, since there is very little sense evidence of the earth's rotation, and thus he is forced to advance the claims of reason over those of the senses, especially when reason has the powerful assistance of mathematical methods.

The argument from simplicity is favorable to the Copernican hypothesis, Galileo notes, but one cannot regard this as a necessary demonstration since it confers only a greater probability (*una maggior probabilità*) on the conclusion (7:144.25). He is well aware that a single experiment or a conclusive demonstration (*una sola esperienza o concludente dimostrazione*) can knock down his case (7:148.21-22). The path he therefore adopts is to examine all of the objections to the earth's rotation that have been raised by classical and contemporary authors to show how they are completely lacking in force and thus cannot be used to reject the Copernican thesis.

Basic to Galileo's rejection of Aristotle's view in this matter is Aristotle's contention that only two types of motion are possible, natural or violent, and since a rotary motion could not be natural for the earth it would have to be violent, and so could not be perpetual (7:159.31-34). Along with this, Aristotle did not have a proper understanding of projectile motion, and thus was incapable of seeing how a previously impressed impetus can explain the continued motion of objects. (Both of these teachings, as has been seen, had earlier been rejected by scholastic commentators on Aristotle, although they continued to be propagated by those who took their inspiration from Aristotle's text.)[26] In the projectile context, Galileo brings up the very appropriate experiment (*l'esperienza tanto propria*) of the stone dropped from the top of a ship's mast to expose the complexity of issues involved in experimental proofs (7:167.29-30)—his point being essentially that the evidence brought against the earth's rotation yields the same result whether the earth is moving or at rest. The Aristotelians are then ridiculed for alleging an unverifiable conclusion without having performed the experiment (7:170.19), while Galileo, who had already done so, stresses that he did not have to bring in experimental evidence because he already knew (presumably on the basis of his knowledge of mechanics) that the effect would happen the way it did (7:171.7-8).[27] This should not be interpreted as a rejection of experimentation or of its devaluation, but rather as an emphasis on the care required to extract from it a true conclusion. After discussing another experiment, that of the crossbow fired from a moving carriage (7:194-197), Galileo indeed points out that fallacies arise from continually supposing as true the precise matter that is being questioned (*nel suppor sempre per vero quello che è in quisitione*, 7:200.14-15)—a typical example of circular reasoning. The same thing will happen in all experiments (*tutte l'altre esperienze*) attempting to disprove the earth's rotation, whereas those who understand that the earth communicates its own motion to all objects on its surface will have no difficulty interpreting their results (7:209.3-13).

As to the method to be used to discover the proper principles behind a science of motion, Galileo observes that it is a lack of knowledge of geometry and not of any other discipline (*solo per difetto non di logica o di fisica o di metafisica, ma di geometria*, 7:244.1-2) that has impeded their discovery, and again, that to seek to treat physical questions without geometry (*voler trattar le quisitioni naturali senza geometria*) is to attempt the impossible (7:299.18-20). On the more difficult problem of how to apply abstract mathematical knowledge to physical reality, Galileo maintains that this can be done conditionally (*condizionatamente*), in the sense that if material objects satisfy the conditions of being spheres and planes, to take a simple instance, then they truly touch in only one point (7:233.14-18)[28]—a clear example of *ex suppositione* reasoning. It is difficult, he goes on, to discern in the concrete effects that have been demonstrated in the abstract, but if the geometrical philosopher wishes to do this, he must deduct the impediments of matter (*quando il filosofo geometra vuol riconoscere*

[26] Secs. 4 1.d and 4.3.

[27] Sec. 5.4.c.

[28] Cf. sec. 4.1.e.

in concreto gli effetti dimostrati in astratto, bisogna che difalchi gli impedimenti della materia); should he be able to do this, his calculations will agree with the reality (7:234.5-9). Errors therefore do not lie in the abstract or in the concrete, nor do they lie in the geometry or the physics, but in the calculator who does not know how to take these considerations into account (7:234.9-11). Later on Galileo gives examples of how mathematical demonstration (*dimostrazion matematica*) can yield knowledge of beautiful properties belonging to natural motion and to projectiles (*bellissime passioni attenenti a i moti naturali e a i proietti*, 7:248.28-31), drawing on his discoveries made at Padua. That he does not expect a perfect fit between mathematics and nature, however, is clear from the statement he makes about the results obtained in his experiments with pendulums: if the periods are not exactly equal, granted that they should be, the difference is insensible, as one can see by performing the experiment (*se pur non sono del tutto equali, son elleno insensibilmente differenti, come l'esperienza vi può mostrare*, 7:256.35-37).

Galileo then replies to two of the more contemporary objections to the earth's rotary motion that have methodological import. The first is that the earth, like other heavy and light bodies, can have neither an internal nor an external principle of such motion, and thus it cannot be rotating. Galileo denies that this is true, and affirms instead that it must have either the one or the other, but that he does not know which it has. He speculates that the earth's rotation is caused by the same motive power as moves Mars, Jupiter, and the heavenly spheres, and that it is probably no more unknowable than the cause of falling motion (7:260.19-31).[29] To the further objection that the latter cause is well known and that it is gravity, Galileo makes a statement that is often cited for its skeptical overtones: he says that we merely name this cause *gravità*, but we do not know its essence (*essenza*), as indeed we do not know the essence of any other motive forces, whether these be a *virtù impressa*, an *intelligenza* (understanding this as a form that is either *assistente* or *informante*), or even *natura* as this is the cause of all movement (7:260.32-261.10). His point is not to deny the validity of causal investigations, for he has been concerned with such problems from the time he wrote his physical questions, wherein he opted for the position that intelligences are assisting forms rather than informing forms,[30] and the *De motu antiquiora*, wherein he provided his own analyses of *gravitas* and *virtus impressa*.[31] Instead, Galileo's point is simply to reaffirm that in natural

[29] Cf. sec. 4.2.b.

[30] PQ3.6.29-31, Latin text in *Opere* 1:108.13-16; see Wallace, *Galileo's Early Notebooks*, pars. L29-L31, and commentary on the same.

[31] Sec. 5.2; cf. also secs. 4.2.a and 4.3.

science it is no simple matter to identify the essence of a cause from the study of its effects, and this being so, one cannot validly argue from the relative ignorance of a cause to the nonexistence of an effect that is traceable of it.

The second objection is that the senses can deceive us in attempting to judge whether a motion is truly straight or only apparently so, and similar matters, and so they are useless in trying to decide the question of the earth's rotation. Indeed, in Copernicus's view, one must deny one's own senses (*in via del Copernico bisogna negar le sensazioni proprie*, 7:279.17-18) in order to accept his position. Galileo, fairly enough, does not deny this, but rather urges that an appearance on which all agree should be set aside, and that instead one should have recourse to reason in order to confirm its reality or to discover that it is a mere illusion (*per confermar la realtà di quella, o per iscoprir la sua fallacia*, 7:281.6–8). This is straightforward scholastic epistemology, wherein the ability of the intellect to correct sense impressions was commonly admitted, and which therefore provided the obvious way to meet such anti-Copernican objections.[32]

c. THE THIRD DAY

The dialogue on the third day focuses on a yet more debatable feature of the Coperican system, namely, that in addition to its rotary motion the earth is making an annual revolution around the sun. In attempting to gain further acceptance for the hypothesis he is arguing, Galileo rejoins some of the arguments he has offered in earlier Copernican debates, and thus casts fuller light on the intention that probably lay behind them.

The first serious problem Galileo must face in arguing for the earth's motion around the sun is the absence of stellar parallax, an evidence that the Greeks had sought but could not find, and therefore had held for the earth's immobility. Galileo's tactic is to focus on the extreme difficulty of making quantitative determinations in astronomy, and thus to maintain that measurement is an unreliable guide in a matter such as this. At very great distances, he observes, the smallest errors made by an observer with his instrument will change a location from finite and possible to infinite and impossible (*minimissimi errori fatti dall'osservatore sopra lo stru-*

[32] Grassi acknowledges as much in his controversy with Galileo over the comets: ". . . nisi ratio aliud suadeat, sensu ad scientiam ducimur. . . ." See his *Ratio ponderum librae et simbellae*, reprinted in *Opere* 6:480.3. Note, in this context, the discussion of the sense-reason relationship in Luigi Olivieri, "La problematica del rapporto senso-discorso, tra Aristotele, aristotelismo e Galilei," in *Aristotelismo Veneto e Scienza Moderna*, Luigi Olivieri, ed., vol. 2, pp. 769-785.

mento rendono il sito di terminato e possibile, infinito ed impossibile, 7:317.30-33). Thus he is forced to discard in this situation the means that he would ordinarily depend upon (cf. 7:413.29-414.10).

The telescope, however, has furnished a different type of evidence, not available to Copernicus, that can supplement the simplicity and insight that enabled him to see the truth of his system. Observations of the planets, which are most evident and offer conclusive proof (*concludesi da evidentissime, e perciò necessariamente concludenti, osservazioni*), exclude the earth from the center of the universe and put the sun there in its place (7:349.24-26). Of these the most certain evidences attest that Venus and Mercury revolve around the sun, particularly the changes in Venus's shape, which concludes necessarily (*la mutazione di figure in Venere conclude necessariamente*, 7:350.2-6). The apparent anomaly of the moon going around the earth every month and the earth in turn circling the sun every year is also removed when one sees Jupiter, in its twelve-year orbit, being accompanied not by one moon alone but by four moons (7:367.35-368.11). Not only, therefore, do the planets move around the sun, but it seems likewise very probable and perhaps necessary (*probabilissimo e forse necessario*) to concede that the earth circles it also (7:368.25-31).

The next difficulty to be disposed of is the contention that Copernicus had based his system on mere *supposizioni* and thus had no way to be assured of their truth. Here Galileo resorts to his previously made distinction between mathematical astronomers (*i puri astronomi*) who are only concerned to give an account of the appearances of the heavenly bodies (*delle apparenze ne i corpi celesti*), and thus are content to work with assumptions that are false in nature (*con assunti falsi in natura*), and the philosophical astronomer (*astronomo filosofo*) who has sought to obtain true suppositions (*vere supposizioni*) on which to base his reasoning (7:369.1-28).[33] Ptolemaic astronomy is a good example of the work of an astronomer who is a pure calculator (*astronomo puro calcolatore*), whereas Copernican astronomy, with its two new suppositions (*con queste due nuove supposizioni*) of the earth's diurnal and annual motion, can lay claim to being a true physical account of the structure of the universe (ibid.).

Further evidence for the earth's annual circuit of the sun is given by the study of sunspots, for if the earth does so move, some strange changes (*strane mutazioni*) should be expected in their apparent movements (*i movimenti apparenti*, 7:374.25-26). After having explained the geometry behind this expectation and the observations he had made during the controversy with Scheiner,[34] Galileo announces that the events he has observed agree exactly with his predictions (*si trovano gli eventi puntual-*

[33] Sec. 6.1.d.

[34] Sec. 6.1.c.

mente rispondere alle predizioni, 7:379.16-17). This, in his view, is apparently analogous to the confirmatory experiments he had performed to establish the truth of his principle of uniform acceleration in free fall, and thus he regards it as having strong probative value. He admits that, from the point of view of logical form, his argument may not be conclusive (7:379.26-28), and yet it serves to explain so many divergent phenomena, such as the motions of the planets and those of the sunspots, that it easily and clearly reveals the true cause (*facilmente e lucidamente rendan la vera cagione*) that accounts for them (7:383.15-16).

Toward the end of the third day's conversation Galileo enters into a discussion of William Gilbert's *De magnete*; in this he is somewhat critical of the English scientist for not being more of a mathematician (*un poco maggior matematico*), and especially for his deficiency in geometry, which would have made him more rigorous in searching out the *verae causae* of magnetic phenomena (7:432.20-23). In this context Galileo gives another account of the method to be used in investigating the unknown causes of a conclusion (*nell'investigar le ragioni delle conclusioni a noi ignote*), and again describes the resolutive method as it is used in the mathematical sciences, as he did on the first day, apparently still unaware that this differs from the regressive methodology used in the study of nature (7:434.36-435.5).[35] The digression is illuminating, however, for its confirmation of Galileo's commitment to mathematical method and his confidence in this method as a way of uncovering the causes of physical phenomena.[36]

d. THE FOURTH DAY

Had Galileo ended the dialogue at this point, notwithstanding his occasional lapses into physical argument in the preceding day, there is a good chance he would not have run afoul of the Inquisition and so would not have suffered the indignities of 1633. Instead, however, Galileo decided to add a fourth day during which he would discuss an ingenious speculation (*una fantasia ingegnosa*) that he had mentioned in his preface (7:30.14). This was the proof of the earth's motion from the ebb and flow of the tides, already sketched in his letter of 1616 to Cardinal Orsini.[37] It was this topic that Riccardi, on the pope's instruction, had told

[35] Secs. 3.3, 3.4.c, 6.2.a, and 6.4.a.

[36] Donald W. Mertz identifies this method as one of "causal proportionality" that has a structure similar to that used by Galileo in his search for the cause of the tides. See his "On Galileo's Method of Causal Proportionality," *Studies in History and Philosophy of Science* 11 (1980): 229-242, especially pp. 240-241. For further details on this method and its relationship to the demonstrative *regressus*, see sec. 6.4.a.

[37] Sec. 6.1.d.

him to avoid, for it would bring his treatise out of the mathematical realm of saving the appearances and into the physical sphere of the causes and effects of natural motions. Why Galileo embarked on this dangerous course is difficult to ascertain. The best explanation would be that he saw, after the discussion of the first two days, that he had simply countered the traditional objections against the earth's motion but had given no convincing evidence in its support, and further, after those of the third day, that he had shown the untenability of the Ptolemaic hypothesis but had given no proof for the Copernican system vis-à-vis that proposed by Tycho Brahe. Since Catholic astronomers who had acquiesced to the decree of 1616, and particularly the Jesuits, had used Galileo's own telescopic evidence as support for the Tychonian system, he probably felt that his own work would come to naught if this alternative were not eliminated. Physical proof of the earth's rotation would, of course, settle the issue in favor of the Copernican opinion, and it was probably for this reason that Galileo decided to attempt it. It is generally admitted that the argument he developed based on the tides, which he had been working on for many years, is inconclusive and so did not achieve the objective he had in mind.[38] Yet the discussions of the fourth day are important methodologically, for they show Galileo's understanding of the terminology of cause and effect and how this would be involved in a scientific proof.

Having rejected as unsatisfactory earlier explanations of the tides, Galileo observes that in natural questions (*questioni naturali*) of this type it is a knowledge of the effects (*la cognizione de gli effecti*) that leads to the investigation and discovery of causes (*conduce all'investigazione e ritrovamento delle cause*, 7:443.18-20).[39] It is thus necessary to have a full knowledge of the effects, for if among these one is able to identify those that are principal (*le principali*), from these it will be possible to discover the true and primary causes (*poter pervenire al ritrovamento delle vere cause e primarie*, 7:443.25-444.1). Furthermore, although an identification of all the proper and sufficient (*proprie ed adequate*) causes of these effects may not be possible, if one carefully studies effects that are similar in kind (*gli effetti che son del medesimo genere*) one will be led by this ultimately to a single true and primary cause (*finalmente una sola ha da esser la vera e primaria causa*), since there can be only one true and primary cause for any one effect (7:444.2-14). Galileo embarks on this program, and so comes to the conclusion that the composition of the earth's two motions

[38] See Shea, *Galileo's Intellectual Revolution*, pp. 172-189, and the fuller analysis in Mario G. Galli, "L'argomentazione di Galileo in favore del sistema copernicano dedotta dal fenomeno delle maree," *Angelicum* 60 (1983): 386-427.

[39] Secs. 3.2.b and 3.3.b.

(*dalla composizione di questi due movimenti*, 7:452.18-20), that is, the uneven result than comes when the diurnal component is added to and subtracted from the annual component, is the most efficacious and primary cause of the tides' ebb and flow (*la potissima e primaria causa del flusso e reflusso*, 7:454.1). In addition to this, there are many and various particular accidents (*multiplici e varii sono gli accidenti particolari*) that must be accounted for, but these can be shown to be dependent on different other accompanying causes (*da altre diverse cause concomitanti dependano*), though they are also connected with the primary cause (7:454.2–7). These secondary causes Galileo identifies as various actions associated with liquids, such as their oscillations and other hydrodynamic behavior, which can account for all the particular accidents he is able to enumerate (7:454.7-456.23). To illustrate these he recounts his experiments (*possiamo esperimentare*) with a model (*una machina*), namely, a barge filled with water, wherein one can observe in detail the effects of these marvelous compositions of motions (*nella quale particolarmente si può scorgere l'effetto di queste meravigliose composizioni di movimenti*, 7:456.23-37). The results are hardly confirmatory, and are not proposed as such in any quantitative way, but they attest to Galileo's experimental ingenuity in attempting to establish his proof.

An objection that can be raised against this entire procedure, which is recognized explicitly in the dialogue, is that the reasoning is *ex suppositione* and thus is entirely dependent on what has been assumed, namely, the two motions attributed to the earth (7:462.16-19). Not only is the assumption gratuitous, moreover, but it can be used to deduce consequences that are contrary to fact, such as motions of the air on the earth's surface analogous to those of the water, which should be perceived as winds 7:462.19-463.5). This is Galileo's only explicit mention of reasoning *ex suppositione* in the *Dialogo*, although he had hinted at this procedure indirectly during the second day's discussions when treating of the application of mathematics to the study of nature. There he showed how it could be used to generate certain proof, whereas in this context the expression is used by an objector to his teaching, and thus has the same sense as when employed by Bellarmine against the Copernican hypothesis generally. Here Galileo does not resort to his normal reply, namely, that some *supposizioni* are true and others false (his being the former),[40] but rather spends all his time showing how winds and like phenomena would not actually result from the primary and secondary causes he has already established (7:463.6-470.20).

Another objection would introduce as an alternative explanation of the

[40] Sec. 6.1.d.

tides the possible influence of the moon and the sun on the oceans' waters (7:470.24-25). Whereas in his early notebooks Galileo had mentioned such *influenciae* and had left the question of their existence an open one,[41] here he finds them completely repugnant to his thought (*totalmente repugna al mio intelletto*) and would much prefer to take an agnostic position on the cause of the tides than allow them to enter into his speculation (7:470.26-36). It is regrettable, to be sure, that his earlier professed unwillingness to investigate the nature of gravity, coupled with his present rejection of occult qualities and similar vain imaginings (*qualità occulte ed a simili vane immaginazioni*, 7:470.29), should have precluded his discovery of the *vera causa* of tidal phenomena as later revealed by Kepler and Newton. Galileo is confident, nonetheless, that he can explain all the secondary effects that are alleged in terms of concomitant variations in his principal cause. The reasoning he invokes is that one effect can have only one primary cause (*di un effetto una sola sia la cagion primaria*), and since there is a fixed and constant connection (*una ferma e costante connessione*) between the cause and the effect, any fixed and constant alteration in the effect must be traceable to a fixed and constant alteration in the cause (7:471.7-11).[42] This allows him to speculate about possible variations in the diurnal and annual movements of the sun that can serve as so many accidental causes (*cause accidentarie*, 7:485.13) of particular tidal variations, which he leaves to others to investigate in all their detail (7:485.12-15).

Having thus presented his case, Galileo concludes by making a bow in the direction of Pope Urban VIII by having the conservative Aristotelian in the dialogue, Simplicio, admit that Galileo's thoughts are ingenious (*ingegnoso*) but that he does not regard them as true and conclusive (*verace e concludente*, 7:488.25-27). The reason now is the pope's own, namely, that God could have effected all of these movements in ways completely beyond human comprehension, and so it would be excessively bold to limit and restrict the divine power and wisdom (*limitare e coartare la divina potenza e sapienza*) to some particular fancy (*fantasia particolare*) of one's own (7:488.27-37). This is the so-called "medicine of the end" (*medicina del fine*, 19:326.91) that so infuriated Urban VIII, and, coupled with Galileo's betrayal of the confidence he had placed in him, led to the ignominious trial and condemnation of 1633.[43]

[41] PQ6.1.7; *Opere* 1:159.16-17. See Wallace, *Galileo's Early Notebooks*, par. V7, and its commentary.

[42] For fuller particulars, see the author's "The Problem of Causality in Galileo's Science."

[43] For the author's evaluation of the trial in light of the background here supplied, see his "Galileo's Science and the Trial of 1633," *The Wilson Quarterly* 7 (1983): 154-164, and a more technical version of the same, "Galileo and Aristotle in the *Dialogo*," *Angelicum* 60 (1983): 311-332.

3. The *Two New Sciences*

In the penultimate sentence of the *Dialogo* the expectation is voiced that a similar discussion can be arranged to lay bare the elements of a work mentioned during the dialogue, namely, Galileo's *nuova scienza* dealing with local motion, both natural and violent (*intorno a i moti locali, naturale e violento*, 7:489.18). The state in which Galileo was left after the trial, together with the prohibition of any future publication on his part, did not augur well for the fulfillment of that expectation. Yet the archbishop of Siena, Ascanio Piccolomini, in whose residence Galileo was detained as part of his sentence, proved to be a gracious and stimulating host, and with his help Galileo was able to resume work on what was to prove to be his masterpiece. All of his manuscripts were preserved intact during his ordeal in Rome, as he was to find when he returned to Arcetri at the end of 1633, and there, with the help of his daughter, Sister Maria Celeste, he regained his composure and began to bring his science of motion to its final form.

a. PRELIMINARY CONSIDERATIONS

During his stay in Siena in the latter part of 1633 Galileo engaged in amicable discussions with a professor of philosophy at the university there, Alessandro Marsili, who was quite competent in Aristotle. This circumstance may help to explain the better treatment accorded to the Aristotelians in the *Two New Sciences*, especially in the person of Simplicio, when compared to that in the *Two Chief World Systems*. Another factor that is noteworthy was the publication of a book in late 1633 and dedicated to Pope Urban VIII that defended Aristotle's teaching against the attacks made by Galileo in the *Dialogo*. The author of the work entitled *Esercitationi Filosofiche*, was Antonio Rocco, and it is to Galileo's credit that he read and annotated Rocco's critique and even wrote out a series of replies to him, some of which later appeared in the *Two New Sciences*. Finally, Bishop Guevara, who by this time had sent two copies of his *Mechanica* to Galileo, continued to importune him about the solution he had developed to the paradox of the continuum posed by the "wheel of Aristotle," and stimulated the astronomer-physicist to devote more attention to his own solution, which he likewise presented in his last work.[44]

One section of Galileo's reply to Rocco's critique, written at about the same time as Galileo was previewing portions of Bonaventura Cavalieri's

[44] Cf. sec. 4.4.b.

Geometria indivisibilibus continuorum nova quadam ratione promota, to appear in print in 1635, is of special interest because of its remarks on the structure of the continuum. Rocco was not completely satisfied with Galileo's discussion of the perfect sphere touching the perfect plane at a point, and the way in which Galileo used this to analyze the application of mathematics to the world of nature. His own conviction was that the mathematical sciences were sciences of the real, and thus they could be concerned with reality and applied to existent things (*alle cose esistenti*), without one's having to be troubled with such problems as perfect planes and figures that are most perfectly spherical (7:683.23-26). Galileo does not address this remark, but he uses Rocco's mention of the statement, *Sphaera tangit planum in puncto*, to enter into the problem of the structure and composition of the continuum (7:745.13-14)—a tactic, as has been seen, that was also used by the Jesuits of the Collegio Romano in a similar context.[45]

The commonly accepted teaching of Aristotle was that the continuum could not be composed of indivisibles, and thus a line could not be composed of points, and Galileo admits that philosophers and mathematicians regard the contrary as patently false, since every continuous quantity consists of parts that are always divisible (7:745.14-18). His personal view, as opposed to this, is that it is not false to hold that the continuum is composed of indivisibles; rather, since truth is one, it is the same thing to say that the continuum is composed of parts that are always divisible and to say that it is composed of indivisibles. Galileo's argument in support of this contention is that the continuum is divisible into parts that are always further divisible precisely because it is composed of indivisibles (7:746.1-2). If division and subdivision go on forever, the parts must be infinite in number, and if they are such, they must be *parti non quante*, for otherwise they would go to make up an infinite quantity (7:746.2-9). The process of division, therefore, will never arrive at the primary components (*a'primi componenti*), which are the non-further-divisibles (*i non più divisibili*), or, in other words, the indivisibles (*gl'indivisibili*, 7:746.9-12). Galileo goes on to criticize, and reject, the way in which his adversaries employ act and potency to solve problems of this type. Whether actual or potential, the *parti quante* of the continuum can never be infinite in number, and thus the act-potency distinction is of no help in this matter. What is interesting about Galileo's treatment is that he speaks of indivisibles as parts of the continuum, that is, as *parti non quante* or unquantified parts, while admitting that there are *parti quante* or quantified parts in it also, though the latter can never be infinite in

[45] Sec. 4.1.e.

number (7:746.13-31). His solution may have been difficult for Rocco to accept, but actually it is not far different from that proposed more than forty years earlier by a progressive Aristotelian such as Rugerius, if one may call him that, who insisted that both *partes indivisibiles* and *partes divisibiles* enter necessarily into the composition of the continuum.[46]

By mid-1635 Galileo had completed the first half of the new work, which would go into the first two days of its dialogue and would be devoted to a new science of mechanics, patterned somewhat on the Aristotelian *Quaestiones mechanicae* but concerned with a more subtle inquisition into the strength of the materials of which machines are composed. There seems little doubt that Galileo incorporated materials he had earlier written at Padua on these mechanical problems, much of which he put into the second day of the dialogue, to become part of the first of the new sciences in the treatise. The second new science, which would occupy the third and fourth days of the discussions, would then be the science of local motion in both its natural and forced aspects. This also would include a substantial amount of previously written materials, some of which has already been discussed.[47] Possibly because the latter drafts were already in Latin, Galileo decided to present the discussions of the last two days in two languages, having the interlocutors speaking in Italian and reflecting on the content of a Latin manuscript written by the Academician—the term Galileo used to describe himself. All of the discussions of the first two days, on the other hand, are in Italian, and it seems likely that the drafts of his more developed treatise on mechanics, like his earlier *Le meccaniche*, were also composed in that language.[48] The dialogue form of the *Two Chief World Systems* was preserved and the same three interlocutors employed, this time without the strong polemical overtones, but nonetheless with a good admixture of rhetoric designed to gain support for Galileo's novel ideas.

With the aid of sympathetic friends Galileo was able to send his manuscript to the Elzevirs, a publishing firm in Leiden and thus out of reach of the Inquisition, who began the printing in 1637, even before the text of the fourth day's discussions was completed. The work came off the press in June of 1638 with a simple title indicating that it contained discourses and mathematical demonstrations concerning two new sciences pertaining to mechanics and local motion (*Discorsi e dimostrazioni matematiche intorno a due nuove scienze attenenti alla mecanica e i movimenti locali*, 8:41), now usually abbreviated as the *Two New Sciences*. The title was not Galileo's own, his having apparently been lost, and he was not satisfied with it (17:370,373). It accurately portrays the contents of the

[46] Ibid.

[47] Secs. 5.2 and 5.4.

[48] Sec. 5.3.a-b

work, however, and indeed emphasizes the mathematical demonstrations that enabled him to make the new scientific claims for which the treatise would best be remembered.

b. THE FIRST DAY

The setting for the first day's conversations is the arsenal of Venice, where the participants are examining ships and weapons under construction, and, in the spirit of earlier writings on mechanics, are intent on the investigation of the causes of effects that are not only striking, but are also hidden and almost unknowable (*nell'investigazione della ragione di effetti non solo maravigliosi, ma reconditi ancora e quasi inopinabili*, 8:49.16-17).[49] Their starting point is the commonly accepted principle that mechanics has its foundation in geometry (*tutte le ragione della mecanica hanno i fondamenti loro nella geometria*, 8:50.15), for this seems to be at variance with the experience of artisans in the arsenal who note that, though the geometry of large and small structures is similar, the materials of which they are constructed seem to make the small structure stronger than the large. The reason for this disparity would seem to lie in the forces that hold materials together, and thus the discussion turns to finding out the binding force that holds the parts of solids united (*d'intender qual sia quel glutine che sì tenacemente ritien congiunte le parti di i solidi*, 8:56.4-5). They are not to be tied down to a tight and concise method (*non obbligati a un metodo serrato e conciso*), 8:55.32-33), but they are aware that the moment the cause is discovered their surprise will vanish (*la riconosciuta cagion dell'effetto leva la maravigilia*, 8:53.25-26). And although Aristotle has made a start in this type of investigation, he was not able to prove his results with necessary demonstrations from their primary and certain foundations (*nè . . . da i loro primarii e indubitati fondamenti con necessarie dimostrazioni provate*, 8:54.24-26); the Academician, on the other hand, has done this, demonstrating his results geometrically (*tutte . . . geometricamente dimostrate*), and on this account his may truly be called a new science (*una nuova scienza*) of mechanics (8:54.19-22).

Despite this auspicious beginning, by modern standards Galileo is not very successful in tracking down the causes he is seeking. Setting aside materials that have a fibrous structure, he speculates that the coherence of the parts of substances can be explained in terms of two types of causes, the first being the repugnance nature exhibits toward a vacuum (*quella repugnanza che ha la natura all'ammettere il vacuo*) and the second some viscous or pasty glue that firmly binds together the particles of which the body is composed (*qualche glutine, visco o colla, che tenacemente*

[49] Ibid.

colleghi le particole delle quali esso corpo è composto, 8:59.6-11). The existence of the first type of cause is demonstrated by a convincing experiment (*con chiare esperienze*)—otherwise well known in the sixteenth century—namely, the placing together of two flat plates that slide freely over each other but strongly resist any effort to pull them apart (8:59.11-19).[50] The second type proves more difficult to investigate, though Galileo is convinced that it exists in liquids as well as in solids, and in fact will serve to explain why siphons and suction pumps are unable to raise water to a height beyond eighteen Florentine cubits (8:63.34-64.25). Guided, however, by the principle that for one effect there must be a single true and optimal cause (*d'un effetto una sola è la vera e potissima causa*), he surmises that the gluey effect itself must be caused by minute vacua that hold together the smallest particles (*delle minime ultime*) of material substances (8:66.1-10).[51] This suggests the further speculation concerning how many such vacua can exist in a finite extent of matter, for obviously the cumulative effect of a very large number would be required to generate any sensible cohesive force (8:67.21-36). The problem will be solved, of course, if one can show that within a finite continuous magnitude it is possible to discover an infinite number of vacua (*in una continua estensione finita non repugni il potersi ritrovar infiniti vacui*). In the same stroke, moreover, one will have the solution to the paradox presented by the "wheel of Aristotle," first discussed in the *Quaestiones mechanicae* and more recently by Bishop Guevara in his learned treatise (8:68.1-10). Galileo therefore hopefully embarks on this mathematical problem, using a method different from Guevara's, but which he is confident will lead to the discovery of the cause of the cohesiveness of solids.

The paradox of the wheel, as formulated in the *Quaestiones mechanicae*, arises from the fact that two concentric wheels on the same axle but of different radii will roll over circumferences of different lengths in the course of one complete revolution. The question then is this: how can a point on the circumference of the smaller wheel traverse the same distance as a point on the circumference of the larger without skipping or slipping over the length difference between the two circumferences? Guevara's solution is that there is an infinitely infinite number of points (*numerus infinities infinitus punctorum*) in both the perimeters of the wheels and the plane they traverse, and in view of the [one-to-one] correspondence that can be seen between these infinite points and the parts of the plane (*correspondentiam quam praestare debet infinitis punctis ac partibus plani*), there is no difficulty (*nullum relinquitur inconveniens*) if the smaller cir-

[50] For a discussion of these and similar experiments, see Schmitt, *Studies in Renaissance Philosophy and Science*, Essay 7.

[51] Again see the author's "The Problem of Causality in Galileo's Science."

cumference traverses a larger space due to the unequal displacement.[52] Galileo, interestingly enough, modifies Guevara's solution by first approximating the circles of the wheels with regular polygons of large numbers of sides, on the basis that circles are polygons of infinitely many sides (*i cerchi . . . son poligoni di lati infiniti*, 8:71.29), and then introducing the distinction he had previously made (in the reply to Rocco) between the *parti quante* and the *parti non quante* of the continuum (8:71.35-72.1). The length difference between the two circumferences can obviously be made up by admitting a sufficient number of vacua into the composition of the longer length, he observes, assuming that these are quantified void spaces (*spazii quanti vacui*); however, if the ultimate resolution is made to the primary components of the continuum, which are unquantified and infinite (*i primi componenti non quanti ed infiniti*), then one need not employ such quantified void spaces but can instead use infinitely many unquantified vacua (*vacui infiniti non quanti*) to achieve the same result (8:72.1-19). This is clear evidence in his mind, apparently, that it is possible to discover an infinite number of vacua (granted, unquantified) within a finite continuous magnitude, and these can therefore be invoked to explain the cohesiveness of solids. Galileo goes on to explain how they can also be used to explain the contraction and expansion of bodies (*la condensazione e rarefazzione de i corpi*) without admitting the compenetrability of matter or the introduction of quantified vacua to explain their size variation (8:96.6-8). Neither argument, it might be noted, is accepted with great enthusiasm by the interlocutors (cf. 8:96.9-31), a possible indication that Galileo himself was not convinced by his deft proposal.

This lengthy discussion of the vacuum permits a natural transition to Aristotle's disproof of the existence of vacua in Book 4 of the *Physics*, where he introduces two quasi-mathematical arguments to make this point. Basic to Aristotle's reasoning is the principle, probably advanced by his opponents, that the velocity of fall of a heavy object is directly proportional to its weight and inversely proportional to the resistance it encounters in passing through a medium. The inverse proportionality part of this principle provides Aristotle's ground for rejecting the existence of vacua, for if, in accordance with the accepted premise, the medium offers no resistance whatever, as would be true of a vacuum, falling motion would be instantaneous, and this is contrary to fact. The ensuing discussion does not turn on the inverse proportionality part of the premise, however, but concentrates on Aristotle's seeming admission that, in any one medium, two bodies of different weights will move with different

[52] Guevara, *Mechanica*, p. 219.

speeds that stand to one another in the same ratio as their weights (*si muovano nell'istesso mezzo con diseguali velocità, le quali mantengano tra di loro la medesima proporzione che le gravità*, 8:106.1-3). The example is given of a body ten times heavier than another falling ten times more rapidly through the same medium, to which the oft-analyzed objection is made that Aristotle had not experimented to see whether this was true or not (*io grandemente dubito che Aristotele non sperimentasse mai quanto sia vero* . . . , 8:106.25-26). One might expect that Galileo would introduce his own experimental evidence at this place, but curiously he does not, preferring to give rather an intuitive refutation of Aristotle's so-called dynamic law. He asserts that, even without experiment, he can clearly prove with a concise and conclusive demonstration (*senz'altre esperimente, con breve e concludente dimostrazione possiamo chiaramente provare*) that a heavier body does not fall more rapidly than a lighter one, provided they are made of the same material and all other conditions are the same (8:107.5-9). The argument is similar to one offered by Benedetti in 1554, and invokes the premise that if two stones of the same weight fall at a given speed, tying them together will not cause that speed to double—a result Galileo establishes by a *reductio ad absurdum* line of reasoning (8:107.17-109.5). On its basis he is able to conclude that large and small bodies move with the same velocity provided they are of the same specific gravity (*i mobili grandi e i piccoli ancora, essendo della medesima gravità in spezie, si muovono con pari velocità*, 8:109.5-7). Further discussion reveals that, although appreciable velocity differences are seen in media that are more resistant (*ne i mezzi più e più resistenti*), such variations in air are practically insensible (*quasi del tutto insensibile*), and therefore in a medium totally devoid of resistance all material objects will fall with the same speed (*se si levasse totalmente la resistenza del mezzo, tutte le materie descenderebbero con eguali velocità*, 8:116.28-29).

The problem of resistance to motion leads Galileo back to one of his favorite themes, namely, the accidents or impediments that lead to a disparity between the propositions he is able to establish and the way these will be able to be verified in practice.[53] The velocity variations one observes in falling bodies, he now says, are not caused by the different specific gravities (*non ne sia altramente causa la diversa gravità*); rather they depend on external *accidenti* and especially the resistance of the medium (*dependa da accidenti esteriori ed in particolare dalla resistenza del mezzo*, 8:118.4-8). Therefore the truth of his proposition must be understood as applying whenever all external and accidental impediments have been removed (*si deve intender verificarsi tutta volta che si rimovessero tutti gl'impedimenti acci-*

[53] Secs. 1.3.a, 3.1.c, 5.2.b, and 6.1.b.

dentarii ed esterni, 8:118.32-33). This is almost impossible to do, he admits, when dealing with natural fall because of the omnipresence of surrounding media (8:118.34), and yet he is confident that one will find a much closer experimental agreement with his computation than with that proposed by Aristotle (*credo che troveremo, l'esperienze molto più aggiustatamente risponder a cotal computo che a quello d'Aristotele*, 8:120.24-26).

The remainder of the dialogue of the first day is spent largely discussing experimental techniques that have been developed by Galileo to manifest the truth of his principles. Some of these, as has been seen in chapter 4, are experiments with a long history, such as that with the inflated bladder to prove that air has weight in air (8:122.33-123.4).[54] Others required more ingenuity, such as those designed to prove that a difference in *gravità* is without effect in changing the *velocità* of a falling body—a conclusion so novel and apparently remote from the truth that he must neglect no experiment or argument (*esperienza o ragione*) to establish it (8:127.3-11). So as to employ the slowest speeds possible and thus reduce the effect of the resisting medium, he first allowed bodies to descend along a plane only slightly inclined to the horizontal (*fare scendere i mobili sopra un piano declive, non molto elevato sopra l'orizontale*, 8:128.18-22). Then he further wished to free himself from any *impedimento* that might arise from the contact of the heavy object with the inclined plane, and so began his pendulum experiments (8:128.22-129.9). The latter offered another distinct advantage: they enabled him to repeat many times the fall through a small height, so that the sum of the intervals of time that elapsed would not only be observable, but patently observable (*cosi congiunte facessero un tempo non solo osservabile, ma grandemente osservabile*, 8:128.5-13). These and other experiments, including those with vibrating strings (8:138.19-150.14), were obviously performed many years previously, as is clear from Galileo's earliest writings.[55] They are all explained again here, however, in order to supply reasons, observations, and experiments that are common and familiar to everyone (*ragioni o osservazioni o esperienze tritissime e familiari ad ogn'uno*, 8:131.9-10), so that he can fulfill the ideal of the *scienze dimostrative*, namely, that they be based on principles that are well known, understood, and conceded by all (*principii notissimi, intesi e conceduti da tutti*, 8:131.13-15).[56] Even his Aristotelian adversary must admit the wisdom of Plato's advice to begin one's training with mathematics, a science that proceeds most cautiously (*molto scrupolosamente*) and admits nothing as established apart from what has been conclusively demonstrated (*fuor che quello che concludentemente dimostrano*, 8:134.28-31).

[54] Sec. 4.2.d.

[55] Ibid.

[56] Sec. 3.2.a.

c. THE SECOND DAY

The second day resumes the discussions of the strength of materials, only now in more systematic fashion than on the first day. As promised, one of the spokesmen brings with him some papers in which theorems and problems (*i teoremi e problemi*) that propose and demonstrate various properties of this subject matter (*le diverse passioni di tal soggetto*) are presented in a rigorous manner (*col necessario metodo*, 8:135.24-28). The discussion based on these papers is much briefer than that of the previous day, and it is evident that one of Galileo's earlier compositions is here being reworked in dialogue form. Whereas some of the materials pertaining to the first day seem to have been of recent composition, a study of the manuscript fragments for this day suggests that they were worked out by 1608, a full thirty years before they were finally published.[57]

The scope of the discussion is signalled at the outset: the previous dialogue was concerned mainly with the resistance bodies offer to fracture, and there the cause of their coherence (*la causa di tal coerenza*) was sought mainly in the vacuum (8:151.1-10). But bodies resist differently to a direct pull and to a bending force exerted on them, and since the first type of resistance has just been examined, the second type must now be considered (8:151.14-152.2). A beam that is mortised or cantilevered into a wall functions somewhat like a lever when forces are applied at the free end. Such are the cases Galileo proceeds to investigate, for their study enables him to take a variety of resistive phenomena and reduce them to problems that can be solved by a consistent application of the principle of the lever (5:152.3-7). And although the basis for such a study was laid by Aristotle, it was Archimedes in his *Equiponderanti* who gave it rigorous formulation and therefore will be Galileo's guide in its subsequent development (8:152.8-15).

From this it is apparent that the program of the second day is to reduce the new science of mechanics to a branch of statics, an undertaking that will reinforce Galileo's thesis that geometry is the most powerful tool available for the solution of physical problems. The propositions and demonstrations that make up this new discipline are not difficult to understand, once this aim is appreciated. Eight propositions are presented in systematic fashion, all of them concerned with the strength of cantilevered beams of various shapes and encompassing three different cases. In the first case an external force acts at the end of the beam, but the weight of the beam is neglected; in the second the beam's weight is taken into account; and in the third the external force is neglected and only the beam's weight considered (8:172.31-173.1). Following this a number of

[57] Drake, *Galileo at Work*, pp. 123, 134-135.

theorems are developed less systematically relating to beams supported at both ends and the ways in which these should be designed for maximum strength across a span. Galileo provides illustrations of such principles in the works of nature as well as in artifacts, including the way in which nature forms fish so that their bodies will be devoid of weight in their natural habitat (8:170.23-25), and the way in which nature employs hollow cyclinders, as in reeds and the bones of birds, to increase strength without the addition of weight (8:186.22-28). Whatever his difficulties may have been with discovering precisely how *natura* functions as a cause,[58] as voiced in the *Two Chief World Systems*, there seems little doubt that Galileo continued to view nature as a principle of operation, and indeed to acknowledge its wisdom in making correct use of mechanical principles.

Very few are the methodological observations in the course of the second day's conversations. One comment, however, has attracted some attention, and this is Galileo's comparison of geometry with logic as a tool of discovery in the sciences. After having worked through a goodly number of propositions, he has one of the discussants remark that geometry is the mind's most powerful instrument for sharpening its skills and for discoursing and theorizing (*discorrere e specolare*) effectively, thus reinforcing Plato's counsel to his students (8:175.16-20). Logic, on the other hand, while an excellent instrument for regulating discourse, cannot compare with geometry as a stimulus to discovery (*all'invenzione*, 8:175.25-27). To put the matter more clearly, logic enables one to evaluate whether or not arguments and demonstrations that have been already formulated are actually conclusive, but it does not teach one how to go about formulating them (8:175.28-31). Those acquainted with Aristotle's *Analytics* will recognize that this is not a pejorative evaluation of logic, for the description given by Galileo would accord with that of most of the Aristotelians of his day.[59] And although there are occasional lapses in his own logical argument, there is abundant evidence that Galileo was just as skilled at logic, in this understanding, as he was in the geometry that he sought to extol at every opportunity.

Several times during the discussions of this day Galileo mentions the parabola and its properties, and in one such case provides another clue to his experimental methods. The occasion is a question as to how an artisan might draw a parabola on a plane surface (8:185.13-15). Thereupon two methods are described, the first consisting in rolling a brass ball along the surface of a metallic mirror that is inclined somewhat to the horizontal; this method, says Galileo, offers clear and obvious evi-

[58] Sec. 6.2.a.

[59] Sec. 3.3.a.

dence (*chiara e sensata esperienza*) that the path of a projectile is a parabola (8:185.16-31). The second method is based on the supposition that a chain suspended from two nails at the same height on a wall will assume the form of a parabola (8:186.2-12)—a supposition that is incorrect, since the curve is actually a catenary. Both of these methods, it should be noted, are described in one of Guidobaldo de Monte's surviving notebooks dating from around the turn of the century; he is thus the probable source of Galileo's knowledge of them.[60]

d. THE THIRD DAY

The third day begins directly with the reading of a Latin manuscript entitled *De motu locali*, which purports to set forth a very new science dealing with a very ancient subject (*de subiecto vetustissimo novissimam promovemus scientiam*, 8:190.3). The aim of the Academician, that is, Galileo, in writing it is to report certain properties (*symptomata*) of local motion that are worth knowing but hitherto have not been observed or demonstrated (*adhuc inobservata necdum indemonstrata*, 8:190.5-6). Yet the author does remark that some things are known about the subject, for example, that the natural motion of falling objects is continually accelerated, but that the ratios according to which this occurs have not thus far been determined (*iuxta quam proportionem eius fiat acceleratio, proditum hucusque non est*, 8:190.7-9). The observation is of interest because of the fact that in his *De motu antiquiora* Galileo gave no indication of being aware that natural motion is continually accelerated. His admission of previous knowledge here, quite contrary to his normal custom of claiming every possible discovery as his own, could be a sign that he had learned about this feature of falling motion from the scholastic sources in which it was commonly taught.[61]

The treatise itself is divided into three books, the first concerned with uniform motion (*De motu aequabili*), the second with naturally accelerated motion (*De motu naturaliter accelerato*), and the third, whose discussion is postponed until the fourth day, with the motion of projectiles (*De motu proiectorum*). Of these the first is a simple propadeutic to the other two and on this account elicits no discussion from the participants. It begins with a definition of uniform motion, from this deduces four axioms, and then uses these to prove six different propositions, all of which are stated as theorems. The first three establish simple ratios between separate pairs of three quantities: distance or *spatium*, time or *tempus*, and speed or *velocitas*; and the last three specify compound ratios among all three

[60] See Naylor, "Galileo's Theory of Projectile Motion," p. 551; also sec. 5.4.a.

[61] Sec. 5.4.c-d.

quantities taken together. The reasoning is clear and straightforward, and it is concluded with the brief comment that this is what the author has written on uniform motion (*del moto equabile*), preparatory to a new and more subtle contemplation (*a più sottile e nuova contemplazione*) of naturally accelerated motion such as is generally executed by heavy falling objects (8:196.23-26).

The second book begins at this point, and does so with a passage that attempts to set forth, and justify, a definition of naturally accelerated motion that will form the basis for all the properties of such motion that are to be deduced. The extraordinary thing about the passage is that it duplicates, with only a few minor changes, the draft of the first three paragraphs of the *De motu accelerato* contained in MS 71, which was analyzed at the end of the last chapter.[62] In the previous discussion of this manuscript fragment it was pointed out that authorities date it as early as 1604 and as late as 1630, and reasons were given for placing it around 1609, after the completion of Galileo's experimental work on motion and before the discovery of the telescope. The terminology was there also indicated as similar to that employed by Galileo during his Pisan and Paduan periods, and the fragment itself was seen as terminating an inquiry begun by him in the *De motu antiquiora*, which his experimental work from 1602 to 1609 had enabled him to answer finally to his satisfaction. Now, it is entirely possible that the surviving fragment was not written until 1630, after Galileo had completed most of the *Two Chief World Systems* and was turning his thoughts away from the Copernican controversy and back to his science of motion.[63] If he did so, however, and even assuming that he worked from previous notes or drafts now lost, the important thing is that he resumed, after a period of more than twenty years, much the same terminology and thought context as had characterized his previous work. From the point of view of the continuity thesis being developed in this study, therefore, the later the date assigned to the fragment, the better indication this provides of a uniform methodology and conceptual framework extending from Galileo's early work to the beginning of his final period. As it happens, moreover, there is abundant evidence in the remainder of the *Two New Sciences* to confirm such continuity and uniformity, and when all this is taken into account the precise dating of the *De motu accelerato* becomes a matter of secondary interest.

Be this as it may, Galileo begins by noting that his search is for a

[62] Sec. 5.4.d. To be more specific, the Latin text contained between 8:197.1 and 8:198.15 is almost identical with that contained between 2:261.1 and 2:263.2, the only notable change being the omission of lines 2:262.6-11, which contains a rhetorical question and its answer, probably not deemed suitable by Galileo for his formal exposition.

[63] This is the view of Drake and Wisan; see chap. 5, note 105, *supra*.

definition that best agrees with the motion nature itself employs. One can make up many kinds of motion that nature does not employ, he observes, and from their definitions demonstrate *ex suppositione* the properties these might possess.[64] He himself is not content with this type of hypothetical procedure, for he is convinced that nature causes heavy bodies to descend in a determinate way, and he wishes to be sure that the definition he uses will agree with the essence of motions as these are actually found in nature. After much search, he is confident that he has found such a definition, for he has been able to demonstrate properties that are consequent on it and that agree with the results of experiments he has been able to perform. This, from a methodological point of view, represents an *a posteriori* form of reasoning, wherein effects are used to reason back to an underlying principle or cause that serves to explain them (8:197.4-16). Galileo is not content to rest with this, however, but proceeds to reinforce the definition with an *a priori* argument, namely, that nature always acts in the simplest way possible, and there is no simpler way for bodies to accelerate than to have equal increments of speed added in equal intervals of time (8:197.16-198.13). Thus he reasons to his definition: a uniformly accelerated motion is one which, starting from rest, during equal time intervals acquires equal moments of swiftness (*temporibus aequalibus aequalia celeritatis momenta sibi superaddit*, 8.198.17-18).

Galileo's way of presenting and justifying this definition has elicited criticism from some, who see him as employing a hypothetico-deductive method such as characterizes modern scientific investigations, and thus as falling into the fallacy of *affirmatio consequentis* when using the implied consequences of his definition to support it as the antecedent. It is true that the definition just proposed can be regarded as a *suppositio*, and therefore that the demonstrations to follow are made *ex suppositione*, as were those of Archimedes in his treatise *On Spiral Lines*, to which Galileo makes implicit reference in the passage. From a formal point of view, moreover, a *suppositio* has the character of an *ipotesi*, and thus its value might be judged by its ability to save the appearances, regardless of whether or not it describes a situation that is actually verified in the order of nature. It is for this reason that Galileo repeatedly makes the distinction between *supposizioni* that are true and absolute in nature, and those that are false and made purely for the sake of computation. There seems little doubt that Galileo, in the passage under discussion, regarded his definition of naturally accelerated motion as suppositional in the former rather than in the latter sense: his definition describes motions such as

[64] Secs. 3.1.a, 3.2.a, 5.1.a, 5.2.b, 5.3.b-c, 6.1.d, and the author's "Aristotle and Galileo: The Uses of *Hupothesis* (*Suppositio*) in Scientific Reasoning."

actually occur in nature, and not those that are mathematically elegant but that nature does not use. For this to be true, and for him therefore to have a *nuova scienza* that can demonstrate properties of natural motions in any meaningful sense, it is incumbent on him to manifest the adequacy of this definition. Most of the subsidiary discussion throughout the remaining two days, apart from the mathematical apparatus necessary to confer rigor on his deductions, goes to establish, by both *a priori* and *a posteriori* means, precisely this point. Galileo's principles, the definition of naturally accelerated motion included, must stand or fall on their own merits, and not merely on the basis of one or two consequences that happen to follow from them.

The first objection made by a discussant leaves little doubt in this matter; all definitions are arbitrary, he says, and thus one may doubt that any definition conceived and admitted in the abstract (*concepita ed ammessa in astratto*) would actually apply in nature (8:198.19-24). One difficulty is that a body falling from rest and increasing its speed with time of fall would have to be capable of moving at the smallest possible speed, and this seems contrary to experience, since the senses show that a falling body suddenly acquires great speed (8:198.34-199.19). Galileo's answer to this is in terms of the velocity of motion imparted to a stake by an object falling on it, as already analyzed in the *De motu accelerato* fragment, which shows that there is no limit to the slowness of a body's motion (8:199.20-200.9).[65] This is an experiment, he alleges, that shows the truth to be just the opposite of what one might ordinarily think (8.200.9-11). But even without experiment, reasoning alone can come to the same conclusion. Consider the case of the stone thrown upward that moves more and more slowly until it reaches the turning point, and then starts to fall downward in exactly the opposite sequence—the example he had previously made so much of in the *De motu antiquiora*.[66] Now Galileo uses it to show that, just as the ascending stone must pass through every possible degree of slowness (*passar per tutti i gradi di tardità*) before coming to rest, so it must pass through every possible degree of swiftness in its path of downward acceleration (8:200.11-30).Neither uniform deceleration nor uniform acceleration is inherently impossible, for each time interval, however small, may be divided into an infinite number of instants, and thus there will always be sufficient instants to correspond to the infinite degrees of diminished velocity (*son bastanti i rispondere a gl'infiniti gradi di velocità diminuita*, 8:201.1-2). The structure of the continuum, applied here to the infinite instants of time and corresponding infinite degrees of speed, supplies an *a priori* refutation of the objection brought against the definition.

[65] Sec. 5 4.d.

[66] Sec. 5.2.a-b.

A digression at this point suggests the value of Galileo's analysis for revealing the proper cause (*qual sia la causa*) of falling motion, and so the Hipparchian theory of the impressed impetus (*l'impeto impresso*) that had come to failure in the *De motu antiquiora* is again rehearsed (8:201.11-17).[67] This time Galileo will have nothing of it, diverting the discussion away from his former embarrassment with the remark that this is not the proper time for disputes of this type. Such investigations can and should be made, but they offer little profit to one who is only attempting to demonstrate some properties of an accelerated motion, whatever the cause of that acceleration might be (*dimostrare alcune passioni di un moto accelerato, qualunque si sia la causa della sua accelerazione*, 8:202.26-29). The important thing for Galileo at this stage is to establish his definition, and if he can show that the accidental features (*gli accidenti*) of the motion that follows from that definition agree with what is found in free fall, he will have the support he needs (8:202.34-36). This will not be a causal argument, but rather one made *a posteriori*; besides, he had already identified nature as the internal cause of the acceleration, and that is sufficient for his purposes. What must yet be established is the way nature operates in effecting such acceleration, and it is this with which the definition is concerned. Considering the context of this particular digression, it is odd that some have seen in it an attempt by Galileo to banish causes entirely from scientific investigation; nothing could be further from his intentions, or more foreign to the methodology he consistently employed in his work.[68]

Another proposal that sheds further light on the reasoning behind Galileo's acceptance of his definition is an alternate formulation of it wherein the speed would increase in the same ratio as the distance of fall (*lo spazio che si va passando*) rather than the time of fall, as just stated (8:203.5-7). This, it will be recognized, is precisely the principle he himself was entertaining in 1604 as the basis for deriving the properties of motion he had already observed at that time.[69] Galileo now sees the reason why that definition was not satisfactory, and he states it in a simple mathematical way. If the velocities acquired are in proportion to the spaces traversed, then all distances of fall will be covered in precisely

[67] Secs. 5.2.b and 4.2.d.

[68] Drake makes much of this point in the introduction and notes to his English translation of the *Two New Sciences* (Madison: The University of Wisconsin Press, 1974); see pp. xxvii-xxix and p. 159, note 12. In this note Drake states: "Rejection of causal inquiries was Galileo's most revolutionary proposal in physics. . . ." A fairer estimate would be that, had Galileo rejected causal inquiries, he would have no claim to being the Father of Modern Science. On this point, see the author's "The Problem of Causality in Galileo's Science."

[69] Sec. 5.4.a-b.

the same intervals of time; for example, one and the same body would take the same time to fall four feet as it would eight feet, a result that is clearly contradicted by experience (8:203.29-204.3). Thus the alternate definition is not only false, it is impossible (8:203.29). Galileo was extremely skilled in the use of the *reductio ad impossibile*, and thus if he could set up a dichotomous division of possibilities and then eliminate one alternative (as he seems to feel he had done with the Ptolemaic and Copernican systems), he regarded it as very strong proof. Yet this is still not a *demonstratio ostensiva* offering proof positive of his definition, but a negative argument showing why another way of stating it will not work.[70]

After the extensive discussion of his definition, Galileo proceeds with the task of systematically developing the second book of his science of motion on the pattern of a Euclidean-Archimedean mathematical treatise. The new definition is added to the definition and axioms of the first book, and then a principle is explicitly assumed; other principles are introduced informally in the subsequent discussion, and still others are used but never stated in the work. From these premises Galileo deduces thirty-eight propositions and intersperses among these a number of corollaries and scholia. The propositions are identified as either theorems or problems, and here Galileo follows the convention explained by Clavius and Blancanus: if the proposition indicates something to be proved, it is a theorem, whereas if it states something to be done (usually requiring a construction), it is a problem.[71] In erecting this formal structure Galileo draws on materials he had been developing over many decades, and when one considers the circumstances under which he assembled and made his final redaction of them, the rigor of his treatment is truly remarkable.

The next premise, which Galileo refers to as his *supposto* or *postulato*, states that the degrees of speed (*gradus velocitatis*) acquired by one and the same body moving down planes of different inclines are equal when the heights (*elevationes*) of the planes are equal (8:205.9-11). Of the truth of this statement he seems to have been convinced between 1600 and 1604, but he was unable to work out a satisfactory proof for it and so here presents it suppositionally. This does not prevent him, however, from hinting at its obvious and self-evident character and noting that it ought to be conceded without argument (*meriti di esse senza controversia conceduto*, 8:205.27). Apart from its verisimilitude, moreover, he notes that an experiment (*una esperienza*) can be adduced in its support that falls little short of a necessary demonstration (*necessaria dimostrazione*, 8:205.35-37). The experiment is one with which Galileo had long been

[70] Secs. 1.3.c and 3.2.b.

[71] Sec. 3.4.b-c.

fascinated, wherein a single pendulum is allowed to swing from a nail driven into the wall so that its plane of oscillation is parallel to the wall and is only two digits (*due dita*) in front of it. When a horizontal line is drawn on the wall from the point at which the bob is released, it is found that the bob practically reaches the same line on its upswing, with the very slight shortage (*piccolissimo intervallo*) that prevents it from getting there exactly (*toltogli il precisamente arrivarvi*) being attributable to the resistance of the air and the string (*impedimento dell'aria e del filo*, 8:206.1-20). Moreover, if the length of the pendulum is effectively shortened at the moment when it passes the vertical by having another nail driven in the wall below the point of suspension, and even below the horizontal line, the bob will always terminate its rise exactly at that line (*terminando sempre la sua salita precisamente nella linea*, 8:206.35-36). This experiment leaves little doubt of the truth of the supposition (*non lascia luogo di dubitare della verità del supposto*, 8:207.2-3), and yet some difficulty is posed by the pendulum traversing circular arcs rather than the straight lines of the inclined plane, and so Galileo agrees to take it as a postulate (*come postulato*). Its absolute truth, he observes, will be established when consequences derived from it will be seen to correspond with and agree exactly with experiment (*rispondere e puntualmente confrontarsi con l'esperienza*, 8:208.2-5). There can be no doubt, therefore, that in this case Galileo is asking for the acceptance of a *suppositio* that has some evidential support, but whose truth he feels will later be demonstrated by absolutely convincing experiments.[72]

On the basis of this assumption Galileo is now ready to proceed with his theorems. The first of these, sometimes referred to as the mean-speed theorem, states that the time in which any space is traversed by a uniformly accelerating body moving from rest is equal to the time in which the same body would traverse the same space when moving uniformly at a degree of speed (*velocitatis gradus*) one half that of the highest degree attained (8:208.9-13). This theorem was well known to the *calculatores* of the fourteenth century and one might suspect here a medieval influence on Galileo; the way in which he formulates its proof, however, is distinctively his own, and there are indications that he may not have worked it out until 1630 or later. The second theorem is the famous times-squared theorem (8:209.11-13), which Galileo already knew in 1604 but whose proof eluded him at that time. From this a number of interesting consequences follow, one of which Galileo presents as his first corollary, generally referred to as the odd-number rule (8:210.12-20) and also known to the medievals, but which he again proves in a distinctive way. Fol-

[72] Cf. secs. 3.1.a, 3.2.a, and 5.4.

lowing the proof he summarizes his results with the statement that a body starting from rest and acquiring velocity at a rate proportional to the time will, during equal intervals of time, traverse distances that are related to each other as the odd numbers beginning with unity, and in general, will stand in the same ratio as the squares of the times elapsed in their traversal (8:211.35-212.5).

This statement again evokes a challenge: such conclusions undoubtedly follow from the definition of accelerated motion that has been proposed, but one may still wonder if this is the acceleration nature employs in the descent of falling bodies (*l'accelerazione della quale si serve la natura nel moto de i suoi gravi descendenti*, 8:212.10-11). The request is made for some type of experimental proof, and Galileo sees this as most reasonable for a science such as he is developing, that is, one in which mathematical demonstrations are used to arrive at physical conclusions (*nelle scienze le quali alle conclusioni naturali applicano le dimostrazioni matematiche*, 8:212.17-18).[73] Thus he proceeds to explain his experiments with balls rolling down inclined planes, noting particularly how he went about measuring distances and times of travel, and concluding with the observation that the times of descent along different inclines maintained precisely (*esquisitamente*) the ratio he had assigned and demonstrated for them (8:212.32-213.30). It should be noted that the experiment is offered as direct proof of the times-squared law and not of the definition of naturally accelerated motion, but there seems little doubt that Galileo also regarded it as indirect proof of the latter (cf. 8:212.19-21).

Among the remaining propositions, two are especially worthy of comment. The first is presented as Theorem VI (8:211.8-10), commonly referred to as the law of chords, which was already stated without proof during the discussions of the first day (8:139.13-18). Galileo had hit upon this result by 1602 but apparently did not find a rigorous proof until later; it is possible that the discovery of this proof, together with the final formulation of the first two theorems, was what enabled him to synthesize his more fragmentary findings into the form in which they appear in the *Two New Sciences*. The second is Theorem XXII, the last theorem of the second book (8:261.30-262.4), which also was known to Galileo in 1602, when he was working on the classical problem of the brachistochrone, or the path of quickest descent from one point to another; this path he now indicates in a scholium to the theorem as not being a straight line but rather the arc of a circle (8:263.21-25).[74] These and other examples can be adduced to show that the order in which

[73] Cf. sec. 3.4.a.

[74] In fact the arc is not along a perfect circle but rather along a curve known as the cycloid.

Galileo presents his theorems, his *via doctrinae*, is very different from his *via inventionis*, the order in which he actually discovered them.

Other theorems investigate the properties of motions that combine vertical fall with descent along an incline, which suggest experiments made during the Paduan period but not described in this work, although related more directly to the theorems discussed on the fourth day. Another statement bearing on the same matter, and repeating a theme found in many of Galileo's earlier writings, is found in a scholium following Problem IX. This is that any degree of velocity will be by its nature indelibly impressed (*suapte natura indelebiliter impressus*) on a moving body provided external causes of acceleration or retardation are removed (*dum externae causae accelerationis aut retardationis tollantur*); this situation, however, occurs only on a horizontal plane (*in solo horizontali plano contingit*), and therefore motion along the horizontal is also eternal (*motum in horizontali esse quoque aeternum*, 8:243.17-22). The reason Galileo here alleges is that if the plane slopes downward a *causa accelerationis* is present, and if upward a *causa retardationis*; if neither, the motion must be uniform or equable (*aequabilis*), and thus not weakened or diminished, much less taken away (8:243.20-23). What is noteworthy here is not only the causal analysis but also the result to which it leads, similar to that of Vitelleschi and of Galileo in his *De motu antiquiora* and his *Letters on Sunspots*, except that a horizontal plane is now substituted for a surface that remains always at the same distance from the universal center of gravity.[75] Moreover, Galileo now maintains that any motion imparted to a body will be connatural to it, indelible, and eternal (*suapte natura indelebili atque aeterno*), and so will conserve itself permanently (*idem per se perpetuo . . . servari*) until acted upon by subsequent causes (8:243.23-244.1). This prepares the way for a mathematical analysis of motions that are projected upward after downward descent (8:243.33-34) and others with which he undoubtedly experimented in his Paduan period, and which will provide the basis for his final book.

e. THE FOURTH DAY

The fourth day of the *Discorsi*, as already indicated, is devoted to the contents of the third and last book of the *De motu locali*, concerned with the motion of projectiles. In it Galileo proposes to show how certain principal properties (*praecipua quaedam symptomata*), which he proposes to establish by solid demonstrations (*firmis demonstrationibus stabilire*), come to a body when its motion is compounded of two displacements, one

[75] Secs. 4.1.b, 4.1.d, 5.2, and 6.1.c.

uniform and the other naturally accelerated, such as is found in projectiles (8:268.7-11). With regard to the uniform component, Galileo illustrates this with a body projected along a horizontal plane, mentally conceived and so as lacking any impediments (*mente concipio, omni secluso impedimento*); such a body would have a motion that is uniform and perpetual if the plane were extended to infinity (8:268.13-16). If the plane is terminated, however, and set in an elevated position (and here Galileo seems to be thinking of a table top), a heavy body passing over its edge will acquire, in addition to its previous uniform motion, a downward propensity or component caused by its own weight (8:268.16-19). The motion that results, therefore, will be composed (*compositus*) of one that is horizontally uniform and another that is naturally accelerated downward, and it is this motion that may be referred to as projection (*quem proiectionem voco*, 8:268.20-21).

The first of the properties (*accidentia nonnulla*) of such a motion that Galileo proposes to demonstrate (*demonstrabimus*, 8:268.21-22) is stated as Theorem I of the third book, namely, that it follows the path of a semiparabola (*linem semiparabolicum*, 8:269.1-4). The proof requires some preliminary knowledge of conic sections, which is set forth in standard mathematical fashion; and, when such matters are presupposed (*suppongono*, 8:272.6), the demonstration is seen to follow in a straightforward manner (8:272.23-273.19). In the discussion that ensues it is noteworthy that these propositions taken from mathematics are not questioned at all; what is questioned, on the other hand, is the reasoning that enables them to be applied to the physical motion being investigated. So it is quickly pointed out that, although the argument is novel, ingenious, and conclusive, being made *ex suppositione*, it does suppose (*supponendo*) that transverse motion is always uniform, that downward motion is always accelerated according to the squared ratio of the times, that the resulting velocities can be added together without altering, disturbing, or impeding one another in any way, and that the path will not change ultimately into a different curve altogether (8:273.29-36). The last possibility seems inevitable on the basis of the projectile's tendency to seek the center of the earth during its fall, for if it begins its parabolic trajectory at some point directly above the earth's center, the farther it travels along the parabola the more it will depart from the earth's center, and this is contrary to its natural tendency (8:274.1-7). And the first supposition is questionable also, for to use it we must presuppose (*noi supponghiamo*) that every point on a horizontal plane is equidistant from the earth's center, which is not true, so that effectively motion along such a plane will eventually be uphill (*ascendendo sempre*); and, moreover, it is impossible to eliminate entirely the resistance of the medium (*impossibile lo schivar l'impedimento del mezo*, 8:274.8-18). All of these difficulties thus make it highly improbable that the results demonstrated (*le cose dimostrati*) from such unreliable suppositions (*con tali supposizioni inconstanti*) can ever be verified in actual experiments (8:274.19-21).

These criticisms, it should be noted, are the most pointed that can be made against the new science of motion, and thus it is instructive to see how they are handled by Galileo. He is quite ready to admit that conclusions demonstrated in the abstract are altered in the concrete, and in this sense can be falsified, but he notes that the same objection can be raised against Archimedes and other great men who employed suppositions (*hanno supposto*) of a like kind (8:274.22-31). In his demonstration of the law of the lever, for example, Archimedes supposed that the arm of a balance lies in a straight line equidistant at all points from the common center of gravity, and that the cords to which the weights are attached hang parallel to one another. Such license (*la qual licenza*) is permissible when one is dealing with distances that are very small (*piccole*) compared to the enormous distance to the earth's center (8:274.31-275.8). One should recall, moreover, that Archimedes and others based their demonstrations on the suppositions that the balance could be regarded as at an infinite distance from the center of the earth (*per infinita lontananza remoti dal centro*). Granted this supposition, his results are not falsified (*non erano falsi*) but rather drawn with absolute proof (*con assoluta dimostrazione*, 8:275.8-12). Finally, the objection from the resistance of the medium must be handled in an analogous way. Galileo is quick to admit that there are so many factors that affect the motions of bodies that it is almost impossible to have a firm science (*ferma scienza*) of them. But if one wishes to treat such matters scientifically (*scientificamente*), he needs to abstract from impediments of this type (*bisogna astrar da essi*). One must find and demonstrate conclusions in abstraction from such impediments (*astratte da gl'impedimenti*), and then be able to use them in practical situations under limitations that can be learned from experiment (*con quelle limitazioni che l'esperienza ci verrà insegnando*, 8:276.6-12). In this way the resistance of the medium can be minimized, and, if proper apparatus is employed, external and accidental impediments (*gli esterni ed accidentarii impedimenti*) can be reduced to the point where they are hardly noticeable (*pochissimo notabili*, 8:276.12-22).

In effect, what Galileo has done throughout this passage is distinguish various types of suppositions that will be employed in the process of developing a mathematical physics. First there are terms and definitions taken over from the abstract science of mathematics, which generally pose no difficulty. Then there are other suppositions that are made when applying such abstract definitions to the geometry actually found in physical situations, and here some adjustments must be made if simplifications are introduced; these are legitimate if one is aware of the orders of magnitude involved, and if the results demonstrated *ex suppositione* depart only in an insignificant way from those that would be deduced without the simplifications. Finally there are suppositions that look to the impediments that are omnipresent in nature and that seem to preclude the scientific treatment of phenomena as complex as motion. These are more difficult to formulate, but if one is sufficiently ingenious in techniques of experimentation, one will learn how to abstract even from these extrinsic and accidental factors and so arrive at the essences of the motions being investigated. In the last case, no less than in the second, one cannot expect a perfect fit between nature and mathematics. All that is required, however, is close enough agreement so that experimentally verified results do not depart significantly from those calculated on the basis of the suppositions.[76]

It is unfortunate that Galileo never saw fit to describe the techniques he actually employed to reduce these extrinsic and accidental impediments to the point where they would be negligible. As has been seen, in the discussions of the second day, he does allude to a type of experimental proof for the parabolic path of projectiles (8:185.28-29), but he gives no details that would enable one to judge the accuracy of his results. Since no other experiments are explicitly mentioned during the fourth day, it has heretofore generally been assumed that no further measurements were made by Galileo when investigating parabolic trajectories. Now, however, with the new knowledge of Galileo's experimental work at Padua before the discoveries with the telescope, discussed in the previous chapter,[77] sufficient clues are available to reconstruct the methods he used. These were capable of yielding a high degree of accuracy, and thus fully substantiate the statements he makes in the foregoing passage about the circumvention of impediments.

The remainder of Book Three consists of only thirteen propositions, some of which are theorems and others problems, the last nine of which describe the main mathematical properties of parabolic paths and directions for the computation of trajectories based on them. The second proposition, stated as a theorem, supplies the rule of vector addition for

[76] These various kinds of supposition are distinguished in Wallace, *Prelude to Galileo*, pp. 151-153; a fuller classification is provided in the author's "Aristotle and Galileo: The Uses of *Hupothesis* (*Suppositio*) in Scientific Reasoning." Galileo's verification of his suppositions in various experimental contexts is further discussed in the author's "Aristotelian Influences on Galileo's Thought," in *Aristotelismo Veneto e Scienza Moderna*, Luigi Olivieri, ed., vol. 1, pp. 221-277.

[77] Sec. 5.4.b.

the composition of velocities (8:280.1-14), which manuscript evidence shows was known to Galileo by 1609. The same can be said of the third proposition, also stated as a theorem (8:281.11-283.19), a draft of whose proof was written by 1609 and is contained on fol. 91v of MS 72, again mentioned in the previous chapter.[78] Since this draft, as noted, contains a proof for the definition of naturally accelerated motion based on parameters that could be measured accurately, there is good reason to believe that it was these experiments with parabolic paths, rather than the inclined-plane experiments recounted on the third day, that convinced Galileo of the physical truth of his definition. Such a proof would, of course, be *a posteriori*, for it would reason from the quantitative effects of such motion to its nature or essential characteristics, as these were set out early in the third day.[79]

Following the third theorem there is a brief digression wherein Galileo's thought is shown to be in agreement with a speculation of Plato regarding the origins of the universe (8:283.20-284.25). Apparently Plato proposed that God put the heavenly bodies in their proper orbits by first having them fall from a height and then deflecting them into circular orbits, where they will rotate uniformly forever—a conception that shows remarkable affinity with the experiments we now know Galileo himself had conducted. Earlier, in the discussions of the third day, after having stated and proved the law of chords, Galileo makes a similar allusion to a great mystery (*qualche gran misterio*) hidden in his fascinating conclusions, a mystery related to the creation of the universe (*alla creazione dell'universo*) and even to the residence of the first cause (*alla residenza della prima causa*, 8:229.29-32). These, plus other statements in the *Two Chief World Systems*, give indication that Galileo was not adverse to seeing profound metaphysical and theological implications in his science, though he never developed these at any length in his extant writings.

The balance of Book Three contains little of methodological interest. A corollary following Proposition VII, however, does make a statement about causes that is worthy of mention. The corollary notes the fact that a projectile shot from the earth's surface will attain maximum range when fired at an elevation of forty-five degrees. This result elicits a feeling of wonder and delight at the force of necessary demonstrations, such as are those of mathematics alone (*la forza della dimostrazioni necessarie, quali*

[78] Sec. 5.4.d.

[79] A possible line of reasoning used by Galileo is sketched in Wallace, *Prelude to Galileo*, pp. 150-156; another possibility would be the proof sketched by Wisan in "The New Science of Motion," pp. 227-228, to which attention has been called in chap. 5, note 104, *supra*.

sono le sole matematiche, 8:296.11-12). Already gunners had provided knowledge of the fact, but to know its cause (*la cagione*) infinitely surpasses what is learned from others or even from the results of repeated experiment (*molte replicate esperienze*, 8:296.12-19). The truth of these observations is then affirmed, and the statement is made even stronger: the knowledge of a single effect acquired through its causes (*la cognizione d'un solo effetto acquistata per le sue cause*) opens the mind to the understanding and certification of other effects without need of recourse to experiments (*senza bisogno di ricorrere alle esperienze*, 8:296.20-24). These, it seems obvious, could hardly be the convictions of anyone who had rejected causal reasoning from the sphere of scientific investigation. Indeed, it confirms the general attitude that has been seen to characterize Galileo's procedures throughout this study: physical causes are truly difficult to discover, but they can be uncovered through careful reasoning, and when known, particularly in their quantitative ramifications, they provide the highest type of knowledge attainable in the natural sphere.[80]

To the last day's discussions Galileo appends some propositions relating to the center of gravity of solids, which he acknowledges were written in his youth (*in sua gioventù*) and undertaken because the treatise of Commandino on that subject was not lacking in certain imperfections (*non mancasse in qualche imperfezzione*, 8:313.8-11). These are the *Theoremata* discussed at the very beginning of the previous chapter, Galileo's first technical essay whose favorable review by Clavius and Guidobaldo del Monte had launched him on his scientific career.[81] The inclusion of this work, written over fifty years earlier, as an appendix to the author's masterpiece, illustrates how Galileo valued everything he wrote. It also supplies mute testimony of the remarkable continuity of thought that evidences itself in his attempt to develop, over more than five decades, his *nuove scienze* of mechanics and local motion.

f. SUPPORTING LETTERS

Of the remainder of Galileo's writings, three letters are of interest for the reflections they contain on his methods and his use of *suppositiones* in formulating his demonstrations. The first was written to Pierre de Carcavy at Paris on June 5, 1637, while the *Two New Sciences* was still in press, and contains Galileo's answer to a query from the French math-

[80] Additional textual evidence is cited in the author's "The Problem of Causality in Galileo's Science."

[81] Secs. 2.4 and 5.1.b.

ematician, Pierre Fermat, concerning a passage in the *Two Chief World Systems* (7:190-193).[82] In this passage some materials that were later to appear in the *De motu locali* are being discussed, and Galileo's initial reasoning concerning the path that would be described by a heavy object falling from a tower if the earth were rotating in a direction away from the body's path of fall is detailed. The curve the object would follow toward the earth's center is said to be compounded of two motions, one straight and the other circular, on the analogy of Archimedes' treatment of spiral motion, and it is described, incorrectly, as a semicircle. Fermat noted the error and communicated his query, through Carcavy, to Galileo. The latter took the occasion not only to retract the earlier statement but to explain how he had derived his new parabolic curve, again noting the analogy with Archimedes' method. In his works, writes Galileo, Archimedes is supposing (*supponendo*), as do all engineers and architects, that heavy bodies descend along parallel lines, thereby leading one to wonder if he was unaware that such lines are not equidistant from each other but come together at the common center of gravity (17:90.59-65). From such an obviously false supposition (*falsa supposizione*) it is possible to raise objections of the type proposed by Fermat, continues Galileo, for at the center of the earth conditions are much different from those on its surface (17:90.65-72). He, however, is restricting himself now to motions of limited extent on the earth's surface, and his results are shown by experiment to be perfectly acceptable on the suppositions he employs.

As Carcavy and Fermat can soon see from his book that is already in the press, Galileo affirms that he argues *ex suppositione*, postulating a motion that departs from rest and increases its velocity in the same ratio as the time increases, and of this motion he conclusively demonstrates many properties (*io dimostro concludentemente molti accidenti*, 17:90.72-77). He notes further that if experiment were to show that these were the same properties as seen in the motions of bodies falling naturally, he could affirm that the latter was the motion he had defined and supposed (*fu definito e supposto*); even if not, however, his demonstrations would have been just as valid as those of Archimedes concerning motions along spiral lines, which do not occur in nature (17:90.78-91.85). But in the case of the motion he has supposed (*figurato da me*), writes Galileo, it has happened that all the properties he has demonstrated are verified in the motion of bodies falling naturally (*è accaduto che tutte le passioni, che io ne dimostro, si verificano nel moto dei gravi naturalmente descendenti*). They are verified, he emphasizes, in this way, that howsoever we perform experiments on the earth's surface (*che mentre noi ne facciamo esperienze sopra*

[82] Portions of this letter are translated into English in Wallace, *Prelude to Galileo*, p. 143.

la terra), at heights and distances that are practical there, we do not encounter a single observable difference (*non s'incontra niuna sensibile diversità*), even though he is aware that such a difference would not only be observable but large and immeasurable (*sensibile, grande et immensa*) if one would perform them much closer to the center of the earth (17:91.85-91). The reply is important not only for the information it provides about the limitations Galileo perceived in his experimentation, but also for the indication he gives there that it is the cumulative effect of all his experiments, and not merely one or another, that assures him of the truth of the definition he has supposed.

The second letter, written by Galileo to Giovanni Battista Baliani in Genoa on January 7, 1639, after the publication of the *Two New Sciences*, repeats in summary form the information contained in the letter to Carcavy.[83] Here too Galileo is clear that his argument is made suppositionally (*argomento ex suppositione*) and that, like Archimedes' demonstration, his would be valid whether or not the motion such as he had defined it actually corresponded with that found in natural fall. But in this matter, he goes on, he has been lucky, so to speak (*ma in questo sono io stato, dirò cosi, avventurato*), for the motion of heavy bodies, and the properties thereof, correspond exactly to the properties demonstrated by him of the motion as he defined it (*poichè il moto dei gravi e i suoi accidenti rispondono puntualmente alli accidenti dimostrati da me del moto da me definito*, 18:12.52-13.59). Here, and in the surrounding passages, Galileo makes the point that it is a large number of precise experimental confirmations that assure him that the definition of naturally accelerated motion he has supposed accurately portrays the one nature employs. And his acknowledgment to Baliani that it was by a stroke of luck (it was *avventurato*—note that in the letter to Carcavy it was by accident, *e accaduto*) that he was able to obtain these confirmations indicates that these experimental investigations caused him much trouble, and that it was not without considerable agitation of mind (as he put it in the *Two New Sciences*, 8:197.13, and even earlier in the *De motu accelerato*, 2:261.15-16) that he was able to assure himself that his supposition was ultimately justified.

Galileo's final reference to his *supposizioni*, it would appear, occurs in yet another letter he wrote, this time to Fortunio Liceti on September 14, 1640, where he sets out in more general terms what it means to be a true follower of Aristotle and then allows that he is really an Aristotelian in this sense himself. Galileo writes that to be truly a peripatetic one must philosophize as Aristotle did, using the methods (*methodi*), suppositions (*supposizioni*), and principles (*principii*) on which scientific dis-

[83] Portions of this are also translated in ibid., p. 144.

course (*lo scientifico discorso*) must be based, and supposing (*supponendo*) the kind of general knowledge from which one cannot deviate without serious error. Among these *supposizioni* is everything that Aristotle teaches in his logic (*nella sua Dialettica*),[84] which includes reasoning well so as to deduce necessary conclusions from the conceded premises (*dalla premesse concessioni*). In this matter, he goes on, he believes that he has learned skill in demonstration (*sicurezza nel dimostrare*) from the pure mathematicians (*dalla innumerabili progressi matematici puri*), with the result that he has almost never made mistakes in argumentation. This, therefore, he concludes, is the sense in which he himself is a peripatetic (*sin qui dunque io sono peripatetico*, 18:248.25-39).[85]

Considering Galileo's lifelong invectives against the Aristotelians, this, of course, is a surprising admission to make only sixteen months before his death. Yet, for one who has followed the course of his arguments and his many attempts to establish demonstrations and the *supposizioni* on which they would be based, it does sum up the methodological concerns that dominated his life and that are clearly implied in the documentation provided in this and the preceding chapters.[86]

4. The Novelty of Galileo's Contribution

Obviously Galileo had come a long way from the time he first penned MS 27 to his writing this letter to Liceti. Over the fifty years that had elapsed he was subjected to many influences not discussed in this volume but which undoubtedly shaped further his ideal of science and the ways he thought it could be attained. The thesis that has been argued throughout is that such an ideal received its preliminary formulation within a scholastic context provided by the Collegio Romano. Galileo's project,

[84] *Dialectica* (or, in the Italian, *Dialettica*) is, of course, Toletus's term for the whole of logic, see sec. 1.1.a, *supra*.

[85] An English translation of this statement is in the author's "Aristotle and Galileo: The Uses of *Hupothesis* (*Suppositio*) in Scientific Reasoning," p. 75.

[86] It is perhaps noteworthy that Galileo's student and successor as mathematician and philosopher to the Grand Duke of Tuscany, Evangelista Torricelli, continued to use Galileo's terminology relating to the *supposizioni* on which a mathematical physics would have to be based. On this, see Paolo Galluzzi, "Evangelista Torricelli: Concezione della matematica e segreto degli occhiali," *Annali dell'Istituto e Museo di Storia della Scienza* 1 (1976): 71-95, especially pp. 73-80. Although a passing statement made by Torricelli to Michelangelo Ricci in a letter of February 10, 1646, seems to indicate that mathematical suppositions are indifferent to the truth or falsity of propositions relating to natural motion, Galluzzi argues from other texts that Torricelli himself believed in an "effective correspondence" between Galileo's mathematical theory of motion and the way in which bodies fall in the order of nature (p. 80).

throughout the entire period, was the reduction of complex natural phenomena to obvious or certifiable principles through the use of quantitative techniques. Such a project was quite in accord with the goals and methodological canons outlined in the *Posterior Analytics*, particularly when interpreted along lines pioneered by the Jesuits, as opposed to the textual orthodoxy of the Aristotelians in the Italian universities. But Galileo did more than appropriate that interpretation. He also perfected it in distinctive ways, many of which have already been signalled in the passages cited from him in this and the previous chapter. These innovations, clearly more than the seedbed from which they sprang, are what earn for him the title of Father of Modern Science. It is only fitting that some concluding reflections be offered on the novelty of Galileo's overall contribution, to identify the elements that were essentially his own and not simply taken over from others.

a. A PROVISIONAL RECONSTRUCTION

Galileo's principal advance beyond the thought-context provided by the Jesuits lies in two related areas: in the mathematical techniques he perfected to make the new physics possible, and in the experimental methods he devised to make such techniques practicable in the study of local motion. With regard to the first contribution, Galileo was quite aware of, and in agreement with, the teaching of his day that mathematical concepts are more abstract than those concerned with sensible matter. His distinctive insight was that the abstractness of such concepts need not prevent their being applied to natural phenomena in order to yield a valid and scientific knowledge of nature. For him, they could do this only when employed *condizionatamente*, that is, when one knew precisely what suppositions to make so that any departures from mathematical rigor in their deduced consequences would be *insensibile* in observation and experiment. Such a program sheds light on the way in which mathematics could become, for Galileo, the impressive tool he claimed it to be for the natural sciences. Employed suppositionally, mathematical concepts were his key to the removal of the accidental and extraneous *impedimenti* that physically prevent one from arriving at the essences of natural phenomena.

This intuition he owed partially to his discussions with Jacopo Mazzoni, and admittedly it has Platonist overtones, but as it functioned in Galileo's mind it took an anti-Platonist turn also. This turn is found in his second area of innovation, that of experimentation, for he found that the simple observation of natural phenomena was not sufficient to reveal whether or not one had penetrated to their essential characteristics. The investigator of nature would have to experiment, and do this in a quantitative way, that is, make precise measurements, if he were to know with sufficient accuracy when perturbing *impedimenti* had been successfully eliminated. Experiments based on or involving measurements therefore took on the force of *a posteriori* proofs that would, in his mind, establish the truth of the quantitative or mathematical suppositions he had used to define the natural phenomena he was investigating. Here, then, was a new type of *regressus* not mentioned by any methodologist before him, one validated specifically by experimentation and measurement. Through its use Galileo could formulate demonstrative proofs of the properties that these phenomena manifest in sense experience. And because he was convinced of the validity of this procedure he would consistently speak of having true *scienza* and *demostrazioni* of them, a claim pervasive throughout his scientific writings.

Galileo's understandings of suppositional reasoning was therefore much in the background of his methodological innovations. The general techniques of its employment, as has been seen, were outlined in the *Posterior Analytics*, and these had straightforward application in the pure sciences that were clearly recognized as such, namely, physics, mathematics, and metaphysics. The problems became more acute, however, when it was extended to the *scientiae mediae*, those that would employ mathematical premises for the study of nature and artifacts. The main source of the difficulty, as the literature of the day attests, was Ptolemaic astronomy. As mathematical, such astronomy could propose valid geometrical demonstrations, but this did not preclude its using *suppositiones* that were merely fictive in nature, since even so it was capable of "saving the appearances" of the heavens. Yet, to the extent that such mathematical astronomy made use of arbitrary and unverified hypotheses, the demonstrations it formulated *ex suppositione* would generate only an imperfect type of science, a *scientia secundum quid*. On the other hand, *scientiae mediae* such as the medieval science of weights, though likewise making use of suppositional reasoning, were not universally seen as laboring under the same limitations. They could employ *suppositiones* that were true in nature, because they were certified to be so within a specified degree of approximation, with the result that they could generate an applied science such as Archimedean statics that had the rigor of a demonstrative *scientia*. In the climate of opinion that then prevailed, mathematical astronomy could not enjoy the same prerogatives as the science of weights, since there was no way of certifying the suppositions on which the first was based, whereas such a way apparently existed for the second.

It is at this juncture that Galileo's major contribution was made. Although many prestigious mathematicians of his time—among whom can be numbered Commandino, Benedetti, and Guidobaldo del Monte—

were unwilling to accept even Archimedean statics as apodictic because of its approximative character, Galileo broke with them unequivocally, and did so early in his career, in the *De motu antiquiora* and *Le meccaniche*. In this it seems clear that he was anticipating the position of Blancanus and others who came under Clavius's influence, who regarded such *scientiae mediae* not merely as demonstrative but as generating *demonstrationes potissimae*, the most perfect type of causal argument attainable by the human mind. But Galileo went far beyond any of Clavius's disciples in certifying the validity of approximative techniques. He did so mainly through his use of limit concepts, which he saw as applicable not only for the solution of mathematical problems but also for the solution of physical problems that might be tractable by mathematical means.

Early in his career, and certainly by the draft version of his *De motu accelerato* that was later to be incorporated into the *Two New Sciences*, Galileo maintained that, just as a circle may be regarded as a polygon with an infinite number of sides, so rest may be regarded as infinite slowness. Provided variations from the limiting case could be shown to be *insensibile* in observation and experiment, he saw no reason why principles employing such limits could not be used in the study of nature. This opened up for him vast possibilities unthought of by classical mathematicians, particularly when coupled with the use of approximate principles of the type that had earlier been employed to good effect by Archimedes. Through their use Galileo could speak of "a force smaller than any given force" and give this concept meaning within physical contexts that stood on the borderline between statics and dynamics. A force less than any given force, he would maintain, is all that is required to initiate motion on a horizontal surface. Again, a force less than any given force over and above that necessary to support a weight is all that is required to move it. The latter principle, it will be recognized, enabled him to move beyond the medieval concept of positional gravity and solve problems relating to motion along inclined planes that had hitherto escaped quantitative solution. Time and again Galileo made use of both principles to derive important results that could be verified in experiment with an accuracy sufficient to claim for them the rigor of physical science. His own logic was not impeccable, to be sure, and his claims were occasionally extravagant, as in the discourse on floating bodies when he argued that water offers no resistance to motion itself but only to the velocity of motion, apparently unaware that one cannot have motion without some velocity. But the basis on which he reasoned was valid nonetheless, for experiment shows that at velocities that approach rest in slowness the medium through which a body moves offers negligible resistance to its motion.

The importation of such principles into physics, according to the practice of the day, would be acceptable if they were introduced under the guise of *suppositiones*. It is in the domain of suppositional reasoning, therefore, that a good part of Galileo's contribution was made—as has been illustrated throughout this study with repeated citation of his uses of *suppono* and its various inflected forms in Latin and Italian translation. Certainly by 1615, prompted by the renewal of the Ptolemaic-Copernican controversy, he was unequivocally distinguishing between *suppositiones* that are fictive and merely calculational and others that are true in nature and applicable to the physical world. More important, however, was his use of suppositions relating to the removal of the *impedimenti* he regarded as so many accidental causes preventing one from arriving at the true and primary causes of natural phenomena. At first, and this occurs in his physical treatises dating from 1590 or shortly thereafter, he proposed such suppositions simply in the context of thought experiments. Then, as his work progressed through the first decade of the sixteenth century, he was able to introduce them into actual experiments and to show that they were holding true within specified limits of accuracy. Usually he would do this through the use of limiting cases and the certification that his experiments and measurements showed deviations from calculation to be *insensibile* and therefore negligible in a science dealing with sensible matter. On this ground he was disinclined to see his suppositions as arbitrary and unverified hypotheses, such as those that underlay Ptolemaic astronomy. Rather they took on for him the meaning allowed in his logical questions and in the Jesuit notes on which they drew, namely, that of principles or truths established by *a posteriori* demonstration—established independently of the context in which they were being used—and so able to serve as a solid foundation for scientific reasoning.

With regard to Galileo's use of the expression *ex suppositione* the situation is more complex, even though it is closely connected with the teaching on supposition just reviewed. In his early writings up to his reflections on Bellarmine's letter of 1615, Galileo seems to have employed it, following an earlier usage at the Collegio Romano, to designate an argument based on a fictive hypothesis that could generate, at best, a *scientia secundum quid*. Thereafter, he seems to have realized that its continued use in this weak sense would involve him in the fallacy of *affirmatio consequentis*, long known to be associated with hypothetico-deductive reasoning, and so deprive him of any demonstrative claims. Yet some of his uses after 1615 continue to suggest the weak sense. Usually Galileo puts the expression in the mouth of an adversary who is rejecting the apodictic character of one of his own proofs, as Bellarmine effectively had done. This is clearly the case in the *Two Chief World Systems*, where Simplicio moves against Galileo's tidal argument precisely on the basis of Bellarmine's usage. Galileo, in this instance, is not able to meet Simplicio's objection head on, and so diverts the discussion to explain why prevailing westerly winds, which would seem to follow on the supposition of the earth's motion, need not necessarily do so. But in the *Two New Sciences*, when Sagredo and Simplicio both object against his demonstrations of the properties of falling and projectile motion on the grounds that these are made *ex suppositione*, Galileo does not let the imputation of arbitrariness to his proofs go unchallenged. Rather he weaves into his later discussion, and also into his letters explaining the method of demonstrating *ex suppositione* in the *nuova scienza*, a realist defense of his *suppositiones*. This invokes a technique employed by Archimedes, but considerably refined and improved upon by Galileo. The latter's advance over Archimedes is difficult to characterize but would seem to be this. Archimedes was able to justify his *suppositiones* only for the solution of statical problems where simple geometrical magnitudes were being analyzed, whereas Galileo was able to use causal reasoning and experimental techniques to extend Archimedes' methods to the solution of dynamical problems where instantaneous velocities and infinitesimal magnitudes would necessarily be involved.

Exactly how Galileo made this extension is open to a variety of interpretations. A formal justification would, of course, require the use of the calculus and so be open to the charge of anachronism. It seems better, on this account, to restrict oneself to Galileo's own terminology and see his method as invoking a new type of *regressus* already alluded to. His understanding of the Aristotelian *regressus* and its way of avoiding circularity in argument was set forth in his logical questions of 1589, and there it was analyzed as involving a twofold *progressus*, one concluding a process of effect-of-cause reasoning and formulated as a demonstration *quia*, the other concluding a process of cause-to-effect reasoning and formulated as a demonstration *propter quid*. Adapting this methodology to the more sophisticated problems of mathematical physics, one may see Galileo's innovation as essentially located in the first *progressus*. This would serve to certify the truth of his *suppositiones* and approximative principles, and so provide a foundation on which his second *progressus* could be erected in the form of the theorems and propositions constituting his *nuova scienza*.

The first *progressus*, on this understanding, would be a more refined method of resolution than had hitherto been used for the detection of physical causes. Here two of Galileo's innovations seem particularly important. The first may be associated with his experimental interests and employs what might be called a modeling technique to isolate the primary and intrinsic cause of a particular phenomenon. The second can more readily be aligned with his mathematical interests and supplies a quantitative surrogate for extrinsic causes when these prove refractory to experimental detection. Each of these contributions will here be considered in turn.

Galileo's use of experimental models to identify the primary and proper cause of an effect, and thus to certify the validity of an *a posteriori* demonstration, has been analyzed in the recent literature and there correctly identified as a method of causal proportionality.[87] Two features of Galileo's work form the basis of this ascription. The first is his pervasive search for causes and his confident assurance that there can be only one essential cause for any given effect, and the second is his use of proportionality theory deriving from Euclid, which he applied to both extensive and intensive magnitudes, to detect causal influences at work. Galileo's causal maxims, as seen repeatedly throughout this and the preceding chapter, stress not only the doctrine of the *vera causa*, namely, that every effect has one, true, primary, and necessary cause. They emphasize also that there must be a fixed and constant connection between cause and effect, with the result that any alteration in the one will be traceable to a fixed and constant alteration in the other. Coupled with this was his recognition that similar effects have similar causes, and so, in cases where many accidental causes are contributing to an effect being produced, the structure of the complex effect will be isomorphic with, or proportional to, the structure of its complex cause.

Such principles lend themselves to experimental application, particularly in situations over which an investigator has control and so is able to certify the connection between a particular cause and a particular effect. On the basis of tests with a model or an artifact, for example, one may be able to identify a similar effect in nature and reason to the existence of a natural cause alone capable of producing that effect. The most striking application of this technique is Galileo's use of a barge filled with water, and its back-and-forth motion, to simulate the tidal effects of the sea. Another example, also quite striking, is the way he analyzes the tower experiment, that is, the case of an object dropped from a tower, which was being used by his adversaries to prove that the earth must be at rest because the object always falls at the foot of the tower. If the earth were moving, they argued, it should fall some distance from the tower and in a direction opposite to that of the earth's motion. Galileo's reply was to duplicate the phenomenon on a smaller scale by considering the case of an object dropped from the mast of a ship, and noting where the

[87] See Mertz, "On Galileo's Method of Causal Proportionality," especially pp. 236-242. Mertz's insights have proved helpful for drawing together the various strands of Galileo's methodology that continue to reappear throughout this volume.

object falls for the two cases of the ship's being at rest and its being in uniform motion through the water. In the *Two Chief World Systems* Galileo does not state the result as experimentally established, but rather evokes from Simplicio his admission of the *suppositiones* on whose basis one can see that in either case the object must fall at the foot of the mast. It is known from the testimony of the Jesuit Father Andrea, discussed in the previous chapter, that Galileo had performed the experiment and so was doubly certain of his result.[88] Unfortunately, the causal analysis in this case is not sufficient to prove the earth's motion, since all that it permits one to conclude is that the result of the tower experiment will be the same whether the earth is at motion or at rest.

The tidal argument, on the other hand, was in principle capable of manifesting the earth's motion on the basis of the causal analysis it invoked. Yet it, too, proved defective, not because the *progressus* employed was an improper method, but because the effects Galileo alleged for the tides were not completely similar to those he had observed in the waves generated in the barge, nor was the motion imparted to the barge completely similar to the earth's motion he regarded as the cause of tidal phenomena. Had he better analyzed the quantitative modalities of both cause and effect, Galileo might have recognized the disproportion between the model he had constructed and the natural phenomenon he was attempting to explain. Not discerning this, he concluded to a false cause, and it was left to Newton to discover the *vera causa* of the motion of the tides.

The second innovation Galileo made in *regressus* methodology concentrates on the defect just noted, the various quantitative modalities of cause-and-effect phenomena, and uses these to reason mathematically to the existence of a physical cause from observed variations in its effect. For this, as for any physico-mathematical demonstration, certain *suppositiones* are required. In the famous demonstration discussed on the Third Day of the *Two New Sciences*—also sketched in the draft of the *De motu accelerato*—Galileo treats his definition of naturally accelerated motion as such a *suppositio*, not arbitrarily made, as Archimedes' definition of spiral motion had been, but as one acceptable on the basis of experimental evidence. From this supposition he proceeds to deduce, *ex suppositione*, the times-squared law, and then offers his confirmation of the law in terms of experiments with objects moving down inclined planes. Some modern methodologists see in this procedure the very circularity the Aristotelians at Padua attempted to avoid through the use of the *regressus* explained by Galileo in LQ7.3. They do so, apparently,

[88] Sec. 5.4.c.

because they regard the inclined-plane experiments cited in the *Two New Sciences* as the empirical base for both the definition of naturally accelerated motion and the times-squared law deduced from it. As detailed earlier in this and the previous chapter, however, it is known that Galileo performed many experiments with bodies in free fall that were not reported in the *Two New Sciences*, and when these additional experiments are taken into account the circularity disappears. The *naturalia experimenta* Galileo mentions as certifying his definition of natural acceleration need not be seen as referring to his inclined-plane experiments; the very use of *naturalia* to describe them makes it more likely that they refer to his measurements of free fall made from a table top and recorded in 1608 or 1609 on folios of MS 72. These measurements did not involve times of descent directly, as did the inclined-plane experiments, but rather focused on distances—heights of fall and horizontal distances of projection—that gave excellent indirect confirmation of the fact that in free fall, velocity increases directly with elapsed time. Reasoning *a posteriori* from such measurements, Galileo would have his first demonstrative *progressus* conclude to his proposed definition as that which actually occurs in nature, and not as a simply imagined *suppositio*. The accuracy of his measurements was off by a few percent, but this could be accounted for by *impedimenti* and accidental causes and would come within the range of *suppositiones* to which he, as a mathematical physicist, had by then become accustomed.

The reasoning behind the first *progressus*, based as it was on the quantitative characteristics of the initiating motion and its measured effects, must be mathematical. It concludes, however, to the velocity-proportional-to-time definition as the formal expression of how nature causes bodies to fall. Herein would seem to lie Galileo's greatest advance over other investigators in this field. No longer did he have to worry about extrinsic (and accidental) causes that might affect a body's speed of fall, nor did he have to concern himself, as he did in the *De motu antiquiora*, with various mechanisms that would produce the observed velocity increases. In effect he had hit upon a mathematical surrogate for manifesting functional relationships between distances, velocities, and times of fall without having to enter into the details of any causal mechanism. And, once equipped with his definition of naturally accelerated motion, he was in an excellent position to begin the second *progressus* of his new science and devise the many demonstrations it would entail. These would all be *a priori*, indeed *propter quid*, and in his view would constitute a science of dynamics just as rigorous as Archimedes' science of statics or the medieval *scientia de ponderibus*. Such was Galileo's announced objective, a fitting culmination of all the experimentation, calculation, and reasoning that occupied him during the almost fifty years that elapsed between the *De motu antiquiora* and the *Two New Sciences* of 1638.

None of these refinements, it need hardly be emphasized, is to be found in the writings of Valla or Blancanus or Guevara, even though all three worked within a thought context that could easily assimilate them and even appreciate their value. They represent Galileo's unique contribution to the science of mechanics, which indeed transformed that science into a *nuova scienza* on which the Scientific Revolution of the seventeenth century was soon to be erected. Whether or not that "new science" stands in essential continuity with the ideal of *scientia* whose norms are prescribed in Aristotle's *Posterior Analytics* will probably always be a subject of controversy. Obviously Galileo's efforts never did meet with the approval of the peripatetics, the textual Aristotelians who were his perennial adversaries throughout his lifetime. They fared much better with Jesuit scientists, as can be seen in the work of Blancanus's student, Ioannes Baptista Riccioli, who began experiments with falling bodies even before reading Galileo's works, and in his *Almagestum novum* of 1651 supplied detailed experimental confirmation of all of Galileo's results.[89]

b. RESIDUAL PROBLEMS

It goes without saying that the foregoing reconstruction of Galileo's contribution leaves many questions unanswered, and clearly it is not proposed as a definitive resolution of all the historical and philosophical problems that surround the interpretation of his work. The sources of his thought that have been identified in this study are as remarkable as they are unexpected. Now that they have been tagged, as it were, and their influences traced throughout his many writings, the broader question can be raised about other sources on which he might have drawn and other factors that could have influenced his scientific development. The case for a Platonic mind set on Galileo's part, as is well known, has been vigorously argued by Alexandre Koyré.[90] Apart from his early contacts with Mazzoni, what are the evidences that Galileo read Plato or more proximate transmitters of Platonic thought and incorporated their ideas into his notes or other writings?[91] A related but more interesting source, in the author's estimation, would be the tradition in mechanics

[89] These experiments are described by Alexandre Koyré in his *Metaphysics and Measurement: Essays in the Scientific Revolution* (Cambridge, Mass.: Harvard University Press, 1968), pp. 102-108.

[90] Ibid., but especially in his *Galileo Studies*.

[91] Apart from Koyré's writings, see B. T. Vinaty, "La formation du système solaire dans la première journée du *Dialogue*," *Angelicum* 60 (1983): 333-385.

and applied mathematics that flourished in Italy in the late sixteenth and early seventeenth centuries.[92] Only a sampling of such work has been touched on in this study, mainly to connect writers like Blancanus, Guevara, and Valerio with ideas being propagated at that time by Clavius within the context of studies at the Collegio Romano. Much remains to be done before the scientific status of the emerging discipline of mechanics, and its relationship to the *scientiae mediae* of the scholastics, can be properly understood. Galileo undoubtedly played a large part in that development, but he was far from being a solitary figure. His interactions with earlier and contemporary mechanicians still await serious investigation.

A more puzzling aspect of the thesis presented in the foregoing pages is that Galileo emerges from them as an Aristotelian *malgré lui*. His late avowal of a peripatetic mentality in matters methodological in the letter to Liceti is understandable enough in light of the documentation already presented.[93] But how does this square with the many vitriolic attacks on the Aristotelians throughout his many writings, from the early *De motu antiquiora* to the *Two New Sciences*? It is difficult to understand why Galileo usually presents Aristotelian thought in the worst possible light, as completely reactionary and closed to any future development. His early acquaintance with the progressive Aristotelianism of the Jesuits must have alerted him to other possibilities for interpreting and transcending Aristotle's text. Why, then, is Simplicio made the official spokesman for such a complex tradition, far from monolithic by any standard?[94] Perhaps Galileo's contacts with Cremonini and other professorial types at Pisa and Padua were so impressive that they obliterated all memory of alternative approaches to Aristotle.[95] Or perhaps the answer is to be found in Galileo's psychological makeup, in his personality, in his skill at rhetoric and polemics, rather than in any attempt on his part to take account of a reasoned philosophical position.[96]

Related to this is the even more perplexing question of Galileo's own "philosophy of science," to use the modern expression. One may wonder whether he had a consistent epistemology, for example, or whether a

[92] Galluzzi ties these two traditions together in his "Il 'Platonismo' del tardo Cinquecento e la filosofia di Galileo."

[93] Sec. 6.3.f.

[94] Olivieri gives a graphic illustration of this polemical device in his "La problematica del rapporto senso-discorso, tra Aristotele, aristotelismo e Galilei."

[95] On Galileo's relationships with Cremonini, see Charles B. Schmitt, *Cesare Cremonini: Un aristotelico al tempo di Galilei* (Venice: Centro Tedesco di Studi Veneziani, 1980).

[96] See Moss's study, "Galileo's *Letter to Christina*: Some Rhetorical Considerations"; also her "Galileo's Rhetorical Strategies in Defense of Copernicanism," in *Novità Celesti e Crisi del Sapere*, Paolo Galluzzi, ed. (Florence: G. Barbèra Editore, 1984), pp. 95-103.

uniform methodology underlay his investigations during the half-century from 1588 to 1638. There seems to be little evidence in Galileo's writings to qualify him as a systematic philosopher with an express and reflective commitment to one school or another. Attempts to make a positivist or an empiricist of him have not been notably successful. Yet one need not maintain that he was a complete anarchist in matters methodological, or so anti-philosophical as not to have general views on what constitutes correct scientific reasoning. His formal training, while not sufficient to equip him to fill a chair of philosophy in an Italian university, was not insignificant either, and he was certainly capable of entering into dispute with the best philosophers of his day. On the other hand, the extent to which any working scientist employs a method that is consciously and reflectively elaborated and then pursued throughout his investigations is quite open to dispute. The solution implied in this study is that Galileo started out with a general ideal that was scholastic in inspiration, that he attempted to realize this in practice, but that he allowed it to vary in different ways as difficulties arose that had to be overcome. What seems remarkable is the extent to which he preserved his early terminology and continued to make reflective observations consonant with the views recorded in his Latin notebooks. One might expect that the more removed Galileo would get from his intellectual beginnings, the more he would depart from the ideals he had earlier adopted. Yet one of the characteristics of philosophical study is that it introduces a general mind set that is not so easily changed as are views on specific matters. The possibility must be left open for an evolution in his thought, one deriving from a host of sources and contingent circumstances as yet unexplored. Still, in the absence of documentation to the contrary, the likelihood cannot be ruled out that his thinking about science and methodology remained fairly constant over the years.

Problems such as these, to say nothing of those relating to the conflicting claims of faith and reason on Galileo's religious commitments, will engage historians and philosophers for some time to come. Their solution is quite obviously outside the scope of this study. But it seems clear that any solutions to them that *are* attempted will have to take account of the evidence produced herein for a substantial influence from Galileo's early period—the hitherto unknown heritage of the Collegio Romano—in the shaping of his *nuova scienza.*

BIBLIOGRAPHY

Manuscripts

Bufalo, Stephanus del

Lisbon, Biblioteca Nacional, Fundo Geral:

Cod. 1892. A patre Stephano del Bufalo Societatis Iesu Romae anno 1596. In quatuor libros caelorum . . . In duo libros Aristotelis De generatione et corruptione . . . De rebus meteorologicis in communi. . . .

Cod. 2382 [anon.]. In libros Aristotelis De caelo disputationes . . . In libros Aristotelis De generatione et corruptione . . . [Colophon:] Haec etiam sit finis librorum eiusdem Aristotelis De generatione et corruptione . . . die 13 Ianuarii 1597 . . . In primum librum Meteorologicorum . . . In librum secundum Meteorologicorum. . . .

Eudemon-Ioannis, Andreas

Rome, Università Gregoriana, APUG-FC:

Cod. 511. [On spine:] Logica P. Eudemon. [Colophon:] Finis quaestionum et expositionum acutissimi et perspicacissimi Patris Andreae Eudemon Ioannis Graeci Societatis Iesu in universam Aristotelis logicam.

Cod. 555. R. P. Andreae Eudemon Greci questiones in libros Aristotelis De generatione et corruptione, In libros De anima, In libros Metaphysicos. Romae, in Collegio Romano Societatis Iesu anno Domini 1599.

Cod. 713. P. Andreae Eudemon-Ioannis De caelo et mundo . . . In libros De generatione . . . Tractatus [de actione et passione] . . . Quaestio de motu proiectorum. . . .

Cod. 1006. R. P. Andreae Eudemon Graeci Quaestiones in libros Aristotelis De physico audito, De caelo et mundo. Romae in Collegio Romano Societatis Iesu anno Domini 1598.

Cod. 1586. [On spine:] Physica P. Eudaemon Ioannis.

Galileo Galilei

Florence, Biblioteca Nazionale Centrale, Manoscritti Galileiani:

Codices 27, 45, 46, 70, 71, and 72.

Gregoriis, Hieronymus de

Rome, Università Gregoriana, APUG-FC:

Cod. 638. Hieronimus de Gregoriis, 1568, Quaestiones in octo libros Physicorum, in De caelo, et in De generatione.

Jones, Robertus
Lisbon, Biblioteca Nacional, Fundo Geral:
Cod. 2066. Robertus Jones Societatis Iesu Romae 1593. De obiecto totius physicae . . . De modo procedendi . . . De principiis rerum naturalium . . . De natura . . . De communibus affectionibus rerum naturalium . . . De universo, sive De caelo et mundo . . . De impressionibus meteorologicis. . . .
Rome, Biblioteca Casanatense:
Cod. 3611. [On spine:] Logica scripta. [Title page:] Organum Aristotelis a Reverendo Patre Roperto Jones Societatis Iesu professore explicatum anno 1592.
Lorinus, Ioannes
Rome, Archivum Romanum Societatis Iesu, Fondo Gesuitico:
Cod. 654, fols. 275-281: Reverendo Admodum in Christo Patri Nostri Generali De logica Patris Pauli Vallii. . . .
Cod. 664, fols. 464-467: Reverendo Admodum in Christo Patri Nostri Generali . . . Patris Pauli Vallii in primum et secundum librorum Physicorum. . . .
Rome, Biblioteca Apostolica Vaticana, Urb. Lat.:
Cod. 1471. Ioannis Laurinis [= Lorini] Societatis Iesu Logica. [Colophon:] 1584.
Menu, Antonius Maria
Pistoia, Biblioteca Forteguerriana, Chiapelli:
Cod. 235. In primum librum Meteororum Aristotelis per Reverendum Doctorem Antonium Mariam Mendi [= Menu] Societatis Iesu, 25 Iunii anno 1578 Romae.
Ueberlingen, Leopold-Sophien-Bibliothek:
Cod. 138. Quaestiones in philosophiam naturalem datae a Reverendo Patre Antonio Maria Menu Societatis Iesu Sacerdote, anno 1577, 20 Octobris, Romae. [In libros Aristotelis Physicorum] . . . In libros Aristotelis De caelo . . . [In libros Aristotelis De generatione] . . . [Metaphysica] . . . [Colophon:] 6 Octobris 1579.
Parentucelli, Antonius Maria
Pistoia, Biblioteca Forteguerriana, Chiapelli:
Cod. 235. In quartum librum Meteorologicorum Aristotelis tractatus, una cum quaestionibus et explicatione textus a Reverendo Patre Antonio Maria Parentucelli Societatis Iesu Doctore in Collegio Romano [Marg.:] Anno 1583, 11 Iulii.
Pererius, Benedictus
Milan, Biblioteca Ambrosiana:
Cod. D 144 INF, fols. 168r-256v: Propositiones quedam selectae ex primo et secundo Physicorum.

Vienna, Oesterreichische Nationalbibliothek, Vindobon.:

Cod. 10470. Dictata in libros duos Aristotelis De generatione [1566].

Cod. 10476. Commentarius et quaestiones in libros I-IV Physicorum Aristotelis.

Cod. 10478. In libros V-VIII Physicorum Aristotelis.

Cod. 10491. In librum primum Physicorum, anno 1566.

Cod. 10507. Tractatus de generatione et corruptione.

Cod. 10509. Commentarii seu dictata in octo libros Aristotelis Physicorum et in libros De caelo.

Rugerius, Ludovicus

Bamberg, Staatsbibliothek, Msc. Class.:

Cod. 62-1 and 62-2. Commentariorum et quaestionum in Aristotelis Logicam . . . tomus . . . Reverendi Patris Ludovici Rugerii Florentini e Societate Iesu traditus in Collegio Romano 1589.

Cod. 62-3. . . . In octo libros Physicorum . . . anno Domini 1590.

Cod. 62-4. . . . In quatuor libros De caelo et mundo . . . In duos libros Aristotelis De generatione et corruptione . . . anno Domini 1591.

Cod. 62-5. . . . In quatuor libros Meteorologicos, de mixto inanimato . . . Romae . . . 1591.

Cod. 62-6. . . . In tres libros De anima . . . 1592.

Cod. 62-7. . . . In quatuordecim libros Metaphysicorum . . . Romae . . . 1592.

Toletus, Franciscus

Rome, Università Gregoriana, APUG-FC:

Cod. 37. In libros Posteriorum.

Valla, Paulus

Rome, Università Gregoriana, APUG-FC:

Cod. 1710. Commentaria in libros Meterororum Aristotelis Reverendi Patris Pauli Vallae Societatis Iesu . . . Tractatus quintus. De elementis. . . .

Vitelleschi, Mutius

Bamberg, Staatsbibliothek, Msc. Class.:

Cod. 70. Lectiones Reverendi Patris Mutii Vitelleschi in octo libros Physicorum et quatuor De caelo, Romae, annis 1589 et 1590 in Collegio Romano Societatis Iesu.

Rome, Biblioteca Apostolica Vaticana, Lat. Borghese:

Cod. 197. Explicationes in Aristotelis Logicam lectae anno 1588 in Collegio Romano per Professorem Mutium Vitelleschum Societatis Iesu, hodie Societatis eiusdem Praepositum Generalem. A Torquato Riccio eius discipulo conscriptae.

Rome, Biblioteca Nazionale Centrale Vittorio Emmanuele, Fondo Gesuitico:

Cod. 747. R. P. Mutii Vitelleschi Societatis Iesu In libros Meteorologicorum Aristotelis disputationes, Romae 1590.

Rome, Biblioteca Vallicelliana:

Cod. P 144: Reverendi Patris Doctoris Mutii Vitelleschi Societatis Iesu In libros Meteorologicorum Aristotelis commentaria.

Rome, Università Gregoriana, APUG-FC:

Cod. 392. Disputationes in libros De caelo [et De generatione].

Printed Sources

Benedictis, Ioannes Baptista de. *Diversarum speculationum mathematicarum et physicarum liber*. Taurini: Haeres Bevilaquae, 1585.

Blancanus, Iosephus. *Aristotelis loca mathematica ex universis ipsius operibus collecta et explicata . . . Accessere: De natura mathematicarum scientiarum tractatio, atque Clarorum mathematicorum chronologia. . . .* Bononiae: Apud Bartholomeum Cochium, 1615.

———. *Sphaera mundi seu Cosmographia, demonstrativa ac facili methodo tradita, in qua totius mundi fabrica, una cum novis Tychonis, Kepleri, Galilaei, aliorumque astronomorum adinventis continetur. Accessere: I. Brevis introductio ad Geographiam. II. Apparatus ad mathematicarum studium. III. Echometria, idest Geometrica traditio de Echo.* Authore Iosepho Blancano Bononiensi e Societate Iesu, Mathematicarum in Gymnasio Parmensi professore. . . . Bononiae: Typis Sebastiani Bonomii, 1620.

Carbone, Ludovicus. *Additamenta ad commentaria D. Francisci Toleti in Logicam Aristotelis. Praeludia in libros Priores Analyticos; Tractatio de Syllogismo; de Instrumentis sciendi; et de Praecognitionibus, atque Praecognitis.* Auctore Ludovico Carbone a Costacciaro, Academico Parthenio, et in Almo Gymnasio Perusino olim publico Magistro. Venetiis: Apud Georgium Angellerium, 1597.

———. *Introductionis in logicam sive totius logicae compendii absolutissimi libri sex*. Venetiis: Apud Ioannem Baptistam et Ioannem Bernardum Sessam, 1597.

———. *Introductio in universam philosophiam libri quatuor*. Venetiis: Apud Marcum Antonium Zalterium, 1599.

Clavius, Christopher, *In sphaeram Ioannis de Sacro Bosco commentarius, nunc iterum ab ipso auctore recognitus et multis ac variis locis locupletatus*. Romae: Ex officina Dominici Basae, 1581.

———. *Euclidis Elementorum libri XV, accessit XVI De solidorum regularium cuiuslibet intra quodlibet comparatione, omnes perspicuis demonstrationibus, accuratisque scholiis illustrati.* . . . Romae: Apud Bartholomaeum Grassium, 1589.

Commandino, Federico. *Liber de centro gravitatis solidorum.* . . . Bononiae: Ex officina Alexandri Benacii, 1565.

Galileo Galilei. *Le Opere di Galileo Galilei.* Ed. Antonio Favaro. 20 vols. in 21. Florence: G. Barbèra Editrice, 1890–1909, reprinted 1968.

Grassius, Horatius [pseudo: Sarsius, Lotharius]. *Libra astronomica ac philosophica qua Galilaei Galilaei opiniones de cometis a Mario Guiducci in Florentina Academia expositae, atque in lucem nuper editae, examinantur.* Perusiae: Ex typographia Marci Naccarini, 1619.

———. *Ratio ponderum librae et simbellae.* . . . Lutetiae Parisiorum: Sumptibus Sebastiani Cramoisy, 1626.

Guevara, Ioannes de. *In Aristotelis mechanicas commentarii, una cum additionibus quibusdam ad eandem materiam pertinentibus.* Romae: Apud Iacobum Mascardum, 1627.

Guidobaldus e Marchionibus Montis. *Mechanicorum libri.* Pisauri: Apud Hieronymum Concordiam, 1577.

Iordanus [Nemorarius]. *Iordani opusculum de ponderositate, Nicolai Tartaleae studio correctum, novisque figuris auctum.* Venetiis: Apud Curtium Troianum, 1565.

Lorinus, Ioannes. *In universam Aristotelis logicam. Commentarii cum annexis disputationibus Romae ab eodem olim praelecti.* . . . Coloniae: Sumptibus Petri Cholini, 1620.

Pererius, Benedictus. *De communibus omnium rerum naturalium principiis et affectionibus libri quindecim, qui plurimum conferunt ad eos octo libros Aristotelis qui de physico auditu inscribuntur intelligendas.* Romae: Apud Franciscum Zanettum et Bartholomaeum Tosium socios, 1576.

Scaliger, Julius Caesar. *Exotericarum Exercitationum Libri XV. De subtilitate ad Hieronymum Cardanum.* Lyons: Sumptibus viduae Antonii de Harsy, 1615.

Toletus, Franciscus. *Commentaria in librum de generatione et corruptione Aristotelis.* Venetiis: Apud Iuntas, 1602.

———. *Commentaria, una cum quaestionibus, in octo libros Aristotelis de physica auscultatione. Nunc secundo in lucem edita.* . . . Venetiis: Apud Iuntas, 1580.

———. *Commentaria, una cum quaestionibus, in universam Aristotelis logicam.* . . . Venetiis: Apud Georgium Angelerium, 1597.

———. *Introductio in dialecticam Aristotelis.* Venetiis: Apud Iuntas, 1596.

Vallius [= Valla], Paulus. *Logica Pauli Vallii Romani ex Societate Iesu, duobus tomis distincta.* . . . Lugduni: Sumptibus Ludovici Prost haeredibus Rouille, 1622.

Literature and Translations

Badaloni, N. "Il periodo pisano nella formazione del pensiero di Galileo." Preprint of an article to appear in *Saggi* on Galileo and furnished the author by Dr. Carlo Maccagni of the Domus Galilaeana, Pisa, in October 1972.

Berti, Enrico. "Differenza tra il metodo risolutivo degli aristotelici e la 'resolutio' dei matematici." In *Aristotelismo Veneto e Scienza Moderna.* 2 vols. Luigi Oliviero, ed., vol. 1, pp. 435-457. Padua: Editrice Antenore, 1983.

Brown, J. E. "The Science of Weights." In *Science in the Middle Ages,* D. C. Lindberg, ed., pp. 179-205. Chicago: The University of Chicago Press, 1978.

Butts, R. E. and Pitt, J. C., eds., *New Perspectives on Galileo.* The University of Western Ontario Series in Philosophy of Science, vol. 14. Dordrecht-Boston: D. Reidel Publishing Company, 1978.

Carugo, Adriano. "Giuseppe Moleto: Mathematics and the Aristotelian Theory of Science at Padua in the Second Half of the 16th-Century Italy." In *Aristotelismo Veneto e Scienza Moderna.* 2 vols. Luigi Olivieri, ed., vol. 1, pp. 509-517. Padua: Editrice Antenore, 1983.

Caverni, Raffaello. *Storia del metodo sperimentale in Italia,* 6 vols. Florence: Stabilimento G. Civelli, 1891-1900; reprinted New York: Johnson Reprint Corporation, 1972.

Cosentino, Giuseppe. "Le matematiche nella 'Ratio Studiorum' della Compagnia di Gesù." *Miscellanea Storica Ligure* II.2 (1970): 171-213.

———. "L'Insegnamento delle matematiche nei collegi Gesuitici nell'Italia settentrionale: Nota introduttiva." *Physis* 13 (1971): 205-217.

Crombie, A. C. "Mathematics and Platonism in the Sixteenth-Century Italian Universities and in Jesuit Educational Policy." In *Prismata: Naturwissenschaftsgeschichtliche Studien* (Festschrift für Willy Hartner). Y. Maeyama and W. G. Saltzer, eds., pp. 63-94. Wiesbaden: Franz Steiner, 1977.

———. "Philosophical Presuppositions and Shifting Interpretations of Galileo." In *Theory Change, Ancient Axiomatics, and Galileo's Methodology.* J. Hintikka et al., eds., pp. 271-286. Proceedings of the 1978 Pisa Conference on the History and Philosophy of Science, vol. 1. Dordrecht-Boston: D. Reidel Publishing Company, 1981.

———. "Sources of Galileo's Early Natural Philosophy." In *Reason, Experiment, and Mysticism in the Scientific Revolution.* M. L. Righini Bonelli and W. R. Shea, eds., pp. 157-175 and 303-305. New York: Science History Publications, 1975.

Crombie, A. C., and Carugo, Adriano. "The Jesuits and Galileo's Idea of Science and Nature." *Sommari degli Interventi* (Summary of a paper presented at the International Galileo Conference, *Novità Celesti e Crisi del Sapere*, Pisa-Padua-Venice-Florence, March 18-26, 1983), pp. 7-9. Florence: Banca Toscana, 1983.

Davi Daniele, Maria R. "Bernardino Tomitano e la 'Quaestio de certitudine mathematicarum'." In *Aristotelismo Veneto e Scienza Moderna*, 2 vols. Luigi Olivieri, ed., vol. 2, pp. 607-621. Padua: Editrice Antenore, 1983.

D'Elia, P. M. *Galileo in China*. Translated by R. Suter and M. Sciascia. Cambridge, Mass.: Harvard University Press, 1960.

Drake, Stillman. *Cause, Experiment and Science*. A Galilean dialogue incorporating a new English translation of Galileo's "Bodies That Stay atop Water, or Move in It." Chicago-London: The University of Chicago Press, 1981.

———. "The Evolution of *De motu* (Galileo Gleanings XXIV)." *Isis* 67 (1976): 239-250.

———. *Galileo Against the Philosophers*, in his *Dialogue* of *Cecco di Ronchitti* (1605) and *Considerations* of Alimberto Mauri (1606). In English Translations with Introductions and Notes by Stillman Drake. Los Angeles: Zeitlin and Ver Brugge, 1976.

———. *Galileo at Work: His Scientific Biography*. Chicago-London: The University of Chicago Press, 1978.

———. "Galileo Gleanings V: The Earliest Version of Galileo's *Mechanics*." *Osiris* 13 (1958): 262-290.

———. "Galileo Gleanings VII: An Unrecorded Manuscript Copy of Galileo's *Cosmography*." *Physis* 1 (1959): 294-306.

———. "Galileo's Experimental Confirmation of Horizontal Inertia: Unpublished Manuscripts (Galileo Gleanings XXII)." *Isis* 64 (1973): 291-305.

———. "On the Probable Order of Galileo's Notes on Motion." *Physis* 14 (1972): 55-68.

Drake, Stillman, and Drabkin, I. E., eds. *Mechanics in Sixteenth-Century Italy*. Selections from Tartaglia, Benedetti, Guido Ubaldo, and Galileo. Translated and Annotated by Stillman Drake and I. E. Drabkin. Madison: The University of Wisconsin Press, 1969.

Drake, Stillman, and MacLachlan, James. "Galileo's Discovery of the Parabolic Trajectory." *Scientific American*, vol. 232, no. 3 (March 1975): 102-110.

Drake, Stillman, and O'Malley, C. D., eds. *The Controversy on the Comets of 1618: Galileo Galilei, Horatio Grassi, Mario Guiducci, Johann Kepler*. Translated by Stillman Drake and C. D. O'Malley. Philadelphia: University of Pennsylvania Press, 1960.

Eszer, Ambrosius K. "Niccolò Riccardi, O.P., il 'Padre Mostro' (1585-1639)." *Angelicum* 60 (1983): 428-457.

Favaro, Antonio. "La libreria di Galileo Galilei descritta ed illustrata." *Bulletino di Bibliografia e di Storia* 19 (1886): 219-293.

Finocchiaro, M. A. *Galileo and the Art of Reasoning*. Rhetorical Foundations of Logic and Scientific Method. Boston Studies in the Philosophy of Science, vol. 61. Dordrecht-Boston: D. Reidel Publishing Company, 1980.

Fredette, Raymond. "Bringing to Light the Order of Composition of Galileo Galilei's *De motu antiquiora*." Paper delivered at the Workshop on Galileo, Virginia Polytechnic Institute and State University, Blacksburg, Virginia, October 1975.

———. "Galileo's *De motu antiquiora*." *Physis* 14 (1972): 321-348.

———. "Les *De motu* 'plus anciens' de Galileo Galilei: Prologomènes." Ph.D. dissertation, University of Montreal, 1969.

Galileo Galilei. "Bodies That Stay atop Water, or Move in It." In *Cause, Experiment and Science*. Translated by Stillman Drake. Chicago-London: The University of Chicago Press, 1981.

———. *Dialogue on the Great World Systems in the Salusbury Translation*. Revised, annotated, and with an introduction by Giorgio de Santillana. Chicago: The University of Chicago Press, 1953.

———. *Galileo's Early Notebooks: The Physical Questions*. A Translation from the Latin, with Historical and Paleographical Commentary, by W. A. Wallace. Notre Dame: University of Notre Dame Press, 1977.

———. *On Motion* and *On Mechanics*: Comprising *De Motu* (ca. 1590), translated, with Introduction and Notes, by I. E. Drabkin; and *Le Meccaniche* (ca. 1600), translated, with Introduction and Notes, by Stillman Drake. Madison: The University of Wisconsin Press, 1960.

———. *Two New Sciences*, Including *Centers of Gravity* and *Force of Percussion*. Translated, with Introduction and Notes, by Stillman Drake. Madison: The University of Wisconsin Press, 1974.

Galli, Mario G. "L'argomentazione di Galileo in favore del sistema copernicano dedotta dal fenomeno delle maree." *Angelicum* 60 (1983): 386-427.

Galluzzi, Paolo. "Evangelista Torricelli. Concezione della matematica e segreto degli occhiali." *Annali dell'Istituto e Museo di Storia della Scienza* 1 (1976): 71-95.

———. "Il 'Patonismo' del tardo Cinquecento e la filosofia di Galileo." In *Ricerche sulla cultura dell'Italia moderna*. Paola Zambelli, ed., pp. 37-79. Bari: Laterza, 1973.

———. *Momento: Studi galileiani*. Lessico Intellettuale Europeo XIX. Rome: Edizioni dell'Ateneo e Bizzarri, 1979.

Giacobbe, G. C. "Il *Commentarium de certitudine mathematicarum disciplinarum* di Alessandro Piccolomini." *Physis* 14 (1972): 162-193.

———. "Epigoni nel Seicento della 'Quaestio de certitudine mathematicarum': Giuseppe Biancani." *Physis* 18 (1976): 5-40.

———. "Francesco Barozzi e la *Questio de certitudine mathematicarum.*" *Physis* 14 (1972): 357-374.

———. "Un gesuita progressista nella 'Questio de certitudine mathematicarum' rinascimentale: Benito Pereyra." *Physis* 19 (1977): 51-86.

———. "La riflessione metamatematica di Pietro Catena." *Physis* 15 (1973): 178-196.

Hall, Bert S. "The Scholastic Pendulum." *Annals of Science* 35 (1978): 441-462.

Jardine, Nicholas. "Galileo's Road to Truth and the Demonstrative Regress." *Studies in History and Philosophy of Science* 7 (1976): 277-318.

Koertge, Noretta. "Galileo and the Problem of Accidents." *Journal of the History of Ideas* 38 (1977): 389-408.

Koyre, Alexandre. *Galileo Studies.* Translated by John Mepham. Atlantic Highlands, N.J.: Humanities Press, 1978.

———. *Metaphysics and Measurement: Essays in the Scientific Revolution.* Cambridge, Mass.: Harvard University Press, 1968.

Lewis, Christopher. *The Merton Tradition and Kinematics in Late Sixteenth and Early Seventeenth Century Italy*. Padua: Editrice Antenore, 1980.

Lohr, Charles H. "Renaissance Latin Aristotle Commentaries: Authors A-B." *Studies in the Renaissance* 21 (1974): 228-289; continued in the *Renaissance Quarterly*, "Authors C," 28 (1975): 689-741; "Authors D-F," 29 (1976): 714-745; "Authors G-K," 30 (1977): 681-741; "Authors L-M," 31 (1978): 532-603; "Authors N-Ph," 32 (1979): 529-580; "Authors Pi-Sm," 33 (1980): 623-734; and "Authors So-Z," 35 (1982): 164-256.

Maccagni, Carlo, " 'Contra Aristotelem et omnes philosophos'." In *Aristotelismo Veneto e Scienza Moderna*. 2 vols. Luigi Olivieri, ed., vol. 2, pp. 717-727. Padua: Editrice Antenore, 1983.

———. *Le speculazioni giovanili "de motu" di Giovanni Battista Benedetti.* Testimonianze di Storia della Scienza, 5. Pisa: Domus Galilaeana, 1967.

Machamer, Peter. "Galileo and the Causes." In *New Perspectives on Galileo*. R. E. Butts and J. C. Pitt, eds., pp. 161-180. Dordrecht-Boston: D. Reidel Publishing Company, 1978.

Marrone, Steven P. *William of Auvergne and Robert Grosseteste: New Ideas of Truth in the Early Thirteenth Century*. Princeton: Princeton University Press, 1983.

Mertz, Donald W. "On Galileo's Method of Causal Proportionality." *Studies in History and Philosophy of Science* 11 (1980): 229-242.

Moody, E. A., and Clagett, Marshall, eds. *The Medieval Science of Weights.* Madison: The University of Wisconsin Press, 1952.

Moss, Jean Dietz. "Galileo's *Letter to Christina*: Some Rhetorical Considerations." *Renaisssance Quarterly* 36 (1983): 547-576.

———. "Galileo's Rhetorical Strategies . . ." In *Novità Celesti e Crisi del Sapere.* Paolo Galluzzi, ed., pp. 95-103. Florence: G. Barbèra, 1984.

Naylor, R. H. "The Evolution of an Experiment: Guidobaldo del Monte and Galileo's *Discorsi* Demonstration of the Parabolic Trajectory." *Physis* 16 (1974): 323-346.

———. "Galileo and the Problem of Free Fall." *The British Journal for the History of Science* 7 (1974): 105-134.

———. "Galileo: Real Experiment and Didactic Demonstration." *Isis* 67 (1976): 398-419.

———. "Galileo: The Search for the Parabolic Trajectory." *Annals of Science* 33 (1976): 153-172.

———. "Galileo's Simple Pendulum." *Physis* 16 (1974): 23-46.

———. "Galileo's Theory of Projectile Motion." *Isis* 71 (1980): 550-570.

———. "Mathematics and Experiment in Galileo's New Sciences." *Annali dell'Istituto e Museo di Storia della Scienza* 4 (1979): 55-63.

———. "The Role of Experiment in Galileo's Early Work on the Law of Fall." *Annals of Science* 37 (1980): 363-378; this article acknowledges the collaboration of T. B. Settle in duplicating experiments ascribed to Galileo, p. 373.

Olivieri, Luigi, ed. *Aristotelismo Veneto e Scienza Moderna.* 2 vols. Padua: Editrice Antenore, 1983.

———. "La problematica del rapporto senso-discorso, tra Aristotele, aristotelismo e Galilei." In *Aristotelismo Veneto e Scienza Moderna.* 2 vols. Luigi Olivieri, ed., vol. 2, pp. 769-785. Padua: Editrice Antenore, 1983.

Papuli, Giovanni. "La teoria del 'regressus' come metodo scientifico negli autori della Scuola di Padova."In *Aristotelismo Veneto e Scienza Moderna.* 2 vols. Luigi Olivieri, ed., vol. 1, pp. 221-277. Padua: Editrice Antenore, 1983.

Pitt, Joseph C. "Galileo: Causation and the Use of Geometry." In *New Perspectives on Galileo.* R. E. Butts and J. C. Pitt, eds., pp. 181-195. Dordrecht-Boston: D. Reidel Publishing Company, 1978.

Purnell, Frederick. "Jacopo Mazzoni and Galileo." *Physis* 3 (1972): 273-294.

Redondi, Pietro. *Galileo eretico.* Microstorie 7. Turin: Giulio Einaudi Editore, 1983.

Righini Bonelli, Maria Luisa. "Le Posizioni Relative di Galileo e dello Scheiner nelle Scoperte delle Macchie Solari nelle Pubblicazioni edite entro il 1612." *Physis* 12 (1970): 405-410.

Righini Bonelli, M. L., and Shea, W. R., eds. *Reason, Experiment, and Mysticism in the Scientific Revolution.* New York: Science History Publications, 1975.

Rose, Paul Lawrence. *The Italian Renaissance of Mathematics.* Geneva: Librairie Droz, 1976.

———. "Professors of Mathematics at Padua University, 1521-1588." *Physis* 17 (1975): 300-304.

Schmitt, Charles B. *Aristotle and the Renaissance.* Cambridge, Mass.: Harvard University Press, 1983.

———. *Cesare Cremonini: Un aristotelico al tempo di Galileo.* Venice: Centro Tedesco di Studi Venetiani, 1980.

———. *Critical Survey and Bibliography of Studies on Renaissance Aristotelianism, 1958-1969.* Padua: Editrice Antenore, 1971.

———. *Studies in Renaissance Philosophy and Science.* London: Variorum Reprints, 1981.

Settle, T. B. "Ostilio Ricci, a Bridge between Alberti and Galileo." *XIIe Congrès internationale d'histoire des sciences.* Tome IIIB, pp. 121-126. Paris, 1971.

Shea, William R. *Galileo's Intellectual Revolution: Middle Period, 1610-1632.* New York: Science History Publications, 1972.

Sommervogel, Carlos, et al. *Bibliothèque de la Compagnie de Jésus.* 11 vols. Brussels-Paris: Alphonse Picard, 1890-1932.

Villoslada, R. G. *Storia del Collegio Romano dal suo inizio (1551) alla soppressione della Compagnia di Gesù (1773).* Analecta Gregoriana, vol. 66. Rome: Gregorian University Press, 1954.

Wallace, W. A. "Aristotelian Influences on Galileo's Thought." In *Aristotelismo Veneto e Scienza Moderna.* 2 vols. Luigi Olivieri, ed., vol. 1, pp. 349-378. Padua: Editrice Antenore, 1983.

———. "Aristotle and Galileo: The Uses of *Hupothesis* (*Suppositio*) in Scientific Reasoning." In *Studies in Aristotle.* D. J. O'Meara, ed., pp. 47-77. Washington: The Catholic University of America Press, 1981.

———. *Causality and Scientific Explanation.* 2 vols. Vol. 1: *Medieval and Early Classical Science.* Ann Arbor: The University of Michigan Press, 1972. Vol. 2: *Classical and Contemporary Science.* Ann Arbor: The University of Michigan Press, 1974. Both volumes reprinted: Washington: University Press of America, 1981.

———. "Galilée et les professeurs jésuites du collège romain à la fin du XVIe siècle." In *Galileo Galilei: 350 ans d'histoire, 1633-1983.* Paul Poupard, ed., pp. 75-97. Tournai: Desclée International, 1983.

Wallace, W. A. "Galileo and Aristotle in the *Dialogo*." *Angelicum* 60 (1983):311-332.

———. "Galileo's Early Arguments for Geocentrism and His Later Rejection of Them." In *Novità Celesti e Crisi del Sapere*. Paolo Galluzzi, ed., pp. 31-40. Florence: G. Barbèra Editore, 1984.

———. *Galileo's Early Notebooks: The Physical Questions*. A Translation from the Latin, with Historical and Paleographical Commentary. Notre Dame: University of Notre Dame Press, 1977.

———. "Galileo's Science and the Trial of 1633." *The Wilson Quarterly* 7 (1983): 154-164.

———. "Galileo's Sources: Manuscripts or Printed Works?" In *Print and Culture in the Renaissance*. Sylvia Wagonheim and Gerald Tyson, eds. Newark, Delaware: The University of Delaware Press, forthcoming.

———. *Prelude to Galileo: Essays on Medieval and Sixteenth-Century Sources of Galileo's Thought*. Boston Studies in the Philosophy of Science, vol. 62. Dordrecht-Boston: D. Reidel Publishing Company, 1981.

———. "The Problem of Causality in Galileo's Science." *The Review of Metaphysics* 36 (1983): 607-632.

Wisan, W. L. "Galileo and the Emergence of a New Scientific Style." In *Theory Change, Ancient Axiomatics, and Galileo's Methodology*. J. Hintikka et al., eds. Proceedings of the 1978 Pisa Conference on the History and Philosophy of Science, vol. 1, pp. 311-339. Dordrecht-Boston: D. Reidel Publishing Company, 1981.

———. "Galileo's Scientific Method: a Reexamination." In *New Perspectives on Galileo*. R. E. Butts and J. C. Pitt, eds., pp. 1-57. Dordrecht-Boston: D. Reidel Publishing Company, 1978.

———. "Mathematics and Experiment in Galileo's Science of Motion." *Annali dell'Istituto e Museo di Storia della Scienza* 2 (1977): 149-160.

———. "The New Science of Motion: A Study of Galileo's *De motu locali*." *Archive for History of Exact Sciences* 13 (1974): 103-306.

INDEX

LIBRARY OF CONGRESS CATALOGING IN PUBLICATION DATA

Wallace, William A.
Galileo and his sources.

Bibliography: p.
Includes index.
1. Galilei, Galileo, 1564-1642. 2. Science—History—Sources. 3. Science—Philosophy—History—Sources. 4. Collegium Romanum—History. I. Title.
QB36.G2W35 1984 520'.92'4 84-42556
ISBN 0-691-08355-X (alk. paper)

Lightning Source UK Ltd.
Milton Keynes UK
UKHW022000060123
414950UK00009B/655